Chemistry and Technology
of Lime and Limestone

Chemistry and Technology of Lime and Limestone

ROBERT S. BOYNTON

Executive Director
National Lime Association
Washington, D. C.

Interscience Publishers

a division of John Wiley & Sons,
New York • London • Sydney

Copyright © 1966 by John Wiley & Sons, Inc.

Library of Congress Catalog Card Number: 65-24287
Printed in the United States of America

PREFACE

For an industry of such prodigious tonnage and literally countless uses and applications, there has been a surprising dearth of literature on the limestone and lime industry—far less than other materials of comparable size and importance. About 500 million tons of lime and limestone are consumed in the United States alone. World tonnage is incalculable, probably at least 2 billion tons. Only water, air, and sand and gravel exceed it in volume. As evidence of the infinite use pattern there is virtually no manufactured item that does not require lime or limestone directly or indirectly—through intermediates . . . even all food contains some calcium. Actually there have been only three reference texts, one by an American and two of English origin.

E. C. Eckels, *Cements, Limes, and Plasters,* 2nd Ed., John Wiley (New York, 1922).
A. B. Searle, *Limestone and its Products,* E. Benn (London, 1935).
N. V. S. Knibbs, *Lime and Magnesia,* E. Benn (London, 1924).

All three of these books rendered valuable contributions to the limestone and lime-producing and consuming industries and, in effect, elevated this industry from the "Dark Ages." They say a lifetime in the chemical industry is now five years (some contend even less). Evolution has transformed lime into primarily a basic chemical, and limestone is also moving in this direction. Consequently, it is obvious that the time lapse between the dates of the above books and the present is much too long. Tremendous changes in nearly all phases of this industry have occurred during the past thirty-five to forty years, as the reader will discover, and doubtless further changes will continue to occur, even more rapidly, in this dynamic, scientific age. (In fact, significant new developments just while writing this book necessitated numerous revisions). Suffice it to say that the technology of these products desperately needs updating in one convenient volume instead of requiring arduous literature searches that resemble a "will-o'-the-wisp." These materials are too strategic and vital to remain obscure and mysterious. My prime objective in writing this volume is to provide reasonably succinct, reliable, and complete information on these products, primarily for the industrial world and secondarily for academic use. In business this would mean the lime and limestone producers and the consumers, which outnumber the former by at least 10,000 to 1; then, for laboratory and market researchers who have contact with or desire to learn about these products; prospective lime producers who should know the *real facts* before they risk a sizeable capital investment; as an aid to consulting engineers, who specify these products and design equipment and plants for and "around" them; and various governmental agencies that are frequently involved with this industry. I am familiar with many costly miscalculations involving these products and industry due to ignorance (lack of reliable information—not stupidity). Finally, education is, of course, important. Young engineers and chemists emerging from college frequently know very little or nothing about these materials with which they often later, through fate, become enmeshed. Possibly, and hopefully, this book will help to fill this void. Unlike many reference texts, an earnest attempt has been made here to enable the reader to understand the industry (to have a "feel" for it), not just the products that it makes.

Of course, all chapters of this book could have been written in greater detail, but this would be impractical, since this volume would necessarily expand into many volumes. For those desiring further elaboration, bibliographies follow each chapter. I have generally endeavored to distill the facts and present a composite picture rather than all of the conflicting and controversial facts and theories, with emphasis on *present* know-how. Disagreement with some of my conclusions is inevitable, but all opinions are based on sincere, unbiased conviction. Although the book is primarily written for North American perusal, frequent allusions are made to foreign technology where pertinent.

I have been intimately associated with this industry for nineteen years, including studying it in Europe, and am indebted to hundreds of knowledgeable technical and executive personnel in this industry, both in the United States and abroad, for much of the information and background that made this book possible. However, private and proprietary information, given to me in confidence by many industry friends, was fully respected and not divulged.

July 1965   ROBERT S. BOYNTON

CONTENTS

Chapter One   Introduction   1

  Basic definitions   2
  History   3

Chapter Two   Formation and Properties of Limestone   5

  Origin   5
  Mineralogy   7
  Classification of limestones   8
  Geologic nomenclature   12
  Occurrence   13
  Impurities   17
  Physical properties   20
  Chemical properties and fundamental reactions   29

Chapter Three   Limestone Exploration and Extraction   34

  Exploration criteria   34
  Extraction of Limestone   38
    Stripping, 38 • Disposal of overburden, 41 • Quarry
    layout, 41 • Mining layout, 43 • Drilling, 46 •
    Blasting, 47

Chapter Four   Limestone Processing   55

  Primary crushing   58
  Secondary crushing   59
  Micron-sized limestone   60
  Conveying   61
  Screening   62
  Washing   64
  Heavy-media separation   65
  Storage   65
  Portable plants   68

Costs   70
Stone safety record   73

Chapter Five   Limestone Uses   75

Chemical and metallurgical   78
  Ferrous metals, 79 · Nonferrous metals, 85 · Chemical-process industries, 86 · Industry fillers (whiting),
  92 · Portland cement manufacture, 97 · Lime manufacture, 100 · Miscellaneous industrial uses, 101
Agriculture 103
  Direct liming, 103 · Fertilizer fillers, 113 · Mineral
  feed, 114 · Poultry grits, 115 · Miscellaneous uses,
  115
Construction (aggregates) 116
  Competitive aggregates, 116 · Gradation, 118 · Specifications, 120 · Requirement, 121 · Portland cement
  concrete, 121 · Road stone, 122 · Asphalt filler,
  125 · Limestone sand, 126 · Railroad ballast, 126 ·
  Rip rap, 127 · Fill, 127 · Roofing granules, 127 ·
  Stucco, 128 · Concrete products, 128 · Precast concrete, 128

Chapter Six   Theory of Calcination 132

Dissociation temperature 133
Recarbonation 135
Heat consumed 136
Decrepitation and preheating 137
Rate of heating 138
Rate of dissociation 139
Loss of weight 140
Core 141
Maintenance of temperature 142
High versus low temperature 142
Shrinkage 143
Porosity and density 146
Effect of stone size 150
Effect of calcination on ionic spacing 153
Effect of salts 153
Influence of stone impurities 154
Effect of steam 157
Dead-burned lime 157
Recapitulation 158

Chapter Seven   Definitions and Properties of Limes 165

Nomenclature 165
Physical properties of quicklimes 167
Physical properties of hydrated limes 172
Chemical properties of quick and hydrated lime 182

Chapter Eight   Lime Manufacture 201

Kilns 204
Vertical kilns 206
Rotary kiln 226
Miscellaneous kilns 237
Chemical analyses of limes 243
Refractory linings 244
Flexibility 246
Fuels and combustion 246
Empirical survey on combustion 252
Heat balance 253
Instrumentation 260
Classification of quicklime 263
Dust collection 267
Dead-burned dolomite production 268
Oystershell lime 270
Precipitated calcium carbonate 272
Hydraulic lime 273
Selective calcination 277
Manufacturing costs 279
Lime plant safety 281

Chapter Nine   Theory of Lime Hydration 287

Chemical reaction 288
Water content 289
Hydrated forms 290
Rates of hydration 294
Rate of Carbonation 298
Particle size 298
Surface area 303
Plasticity 306
Water retentivity 311
Putty volume 313
Sedimentation 314

Reactivity 317
Dehydration of hydroxides 318
Preparation of CP-grade Ca(OH)$_2$ 319
Conclusion 320

Chapter Ten   Methods of Hydration 322

Slaking methods 322
Commercial (dry) hydration 326
   Plant layout, 327 • Equipment, 327 • Pressure hydra-
   tion, 333 • Purity—chemical analysis, 337

Chapter Eleven   Uses of Lime 340

Chemical and Industrial 344
   Metallurgy, 344 • Pulp and paper, 354 • Chemicals
   manufacture, 355 • Water and sanitation, 360 • Neu-
   tralization considerations, 367 • Ceramics and building
   materials, 374 • Food and food by-products, 377 •
   Miscellaneous industrial uses, 379
Agricultural uses 381
   Soil liming (farms), 381 • Miscellaneous uses, 385
Construction (structural) 388
   History, 388 • Masonry mortar, 390 • Plaster and
   stucco, 403 • Road construction, 410

Chapter Twelve   Economic Factors of Lime and Limestone . . . 430

Lime Industry 430
   Competition, 430 • Prices, 432 • Transportation, 434
   • Bulk versus packages, 434 • World Trade, 435 •
   Number, size, location of plants, 437 • Profits, 442 •
   World production, 443 • Captive lime, 445 • Re-
   search, 447
Limestone industry 451
   Competitive factor, 451 • Prices, 453 • Transporta-
   tion, 454 • Packaging, 454 • World trade, 454 •
   Number, size, location of plants, 455 • Taxes, 455 •
   Profits, 457 • Captive production, 458

Chapter Thirteen   Analytical Testing of Limestone and Lime . . 459

Physical testing of limestone 460
Limestone: use specifications 468
Limestone: chemical analyses 471

Physical tests of lime 472
Lime materials specifications 486
Lime (and limestone): chemical analysis 487

Appendix A 500

Appendix B 504

Appendix C 506

Appendix D 508

Index 509

CHAPTER ONE

Introduction

There are a few absolutely indispensable forms of matter, like air and water, that are generally taken for granted by mankind but without which life simply would cease. Analogizing on this there are other almost equally basic and essential materials without which life would be, at the very least, completely barren (and possibly not sustainable) and modern commerce and industry could not exist. Among these ancient, prosaic, unglamourous, and low-cost materials are lime and limestone. Literally any object that exists in man's home, his office (or virtually any manufactured product) has required lime or limestone in some phase of its manufacture, directly or indirectly, either as a prime or incidental processing material, as Chapters 5 and 11 on the interminable uses of these materials testify. In fact, lime—limestone's basic—essentiality has been likened to one leg of a six-legged stool on which industry revolves, the other essential legs being iron ore, salt, sulfur, petroleum, and coal. Stated another way, these are the six essential building blocks of commerce and industry. Of these, limestone is the greatest in physical volume, surpassing coal for the first time in 1962. The strategic importance of these materials in peace or war is frequently overlooked or erroneously minimized in favor of other scarcer, more costly, and more dramatic products that are essential for a few important purposes but not for industry as a whole.

Limestone's abundance is evidenced by the fact that an estimated 3.5 to 4% of the elements in the earth's crust contain calcium and 2% contains magnesium. Of all elements calcium ranks fifth in abundance, only exceeded by oxygen, silicon, aluminum, and iron. Of course, as with most elements, there are strata of limestone laminated between layers of shale and sandstone that are so deep in the earth's crust as to be inaccessible. The life-sustaining aspect of calcium can be appreciated when one realizes that in varying amounts

1

this element is also present in all soils, most water, and all plant and animal life. Human and animal assimilation of it through food is vital to support life; inadequate amounts consumed causes poor health, bone deformities, like rickets in babies. In fact, 99% of all bones and teeth are composed of calcium and phosphorus, with the calcium percentage predominant. There are a great many different forms and types of limestone, varying in color, chemical composition, mineralogy, crystallinity, texture, and hardness. There are also other important calcium- and magnesium-bearing rocks and minerals, like gypsum, phosphate rock, serpentine, and fluorspar, that should not be confused with limestone since they are not carbonate rocks, an essential characteristic of limestone, but none of these natural resources are as basic. Next to sand and gravel, limestone is the greatest tonnage material produced in the United States and had an annual production of 488 million tons in 1963.

Basic Definitions

Lime and limestone are much abused terms. Often users, and even some sellers, will refer to some form of limestone as lime and, less often, vice versa.

Limestone is a general term embracing carbonate rocks or fossils; it is composed primarily of calcium carbonate or combinations of calcium and magnesium carbonate with varying amounts of impurities, the most common of which are silica and alumina. In contrast, *lime,* which is invariably derived from limestone, is always *a calcined or burned form of limestone,* popularly known as "quicklime" or "hydrated lime." The calcination process expels the carbon dioxide from the stone, forming calcium oxide (quicklime), and when water is added, calcium hydroxide (hydrated or slaked lime) is created. While all three of these products have some similarities, the correct connotation of these terms is important since there are many pronounced differences. Webster recognizes the above distinction in defining both of these materials.

An analogy to the close relationship of limestone and lime would be sulfur and sulfuric acid. The former are the native ores, which in spite of processing are still used in their original chemical composition; the latter are their primary manufactured products, basic chemicals, of different chemical composition. The limestone group consists of the most basic alkalies; sulfur is at the other end of the pH scale—the most basic acid. The uses of these two sets of materials are at the same time totally diverse, complementary, and often interchangeable. There are a sufficient number of overlapping uses for lime to regard limestone as its principal competition. However, from a tonnage standpoint the main competition for limestone would be other aggregates, like sand and gravel, granite, basalt, etc.

History

Unquestionably lime and limestone are prehistoric and among the oldest materials used by mankind. Limestone uses are an outgrowth of the Stone Age, in which primitive man utilized limestone and other forms of rock to confine fires, construct stone shelters, and make crude tools and weapons. It is conjectured that lime was discovered later in the same age. One intriguing theory on its discovery is that primitive man probably found his slabs of limestone, used for fireplaces, disintegrating into a white paste-like putty following a hard rain. In summation the stone had been calcined by the heat from his fires, and the resulting quicklime was simply hydrated by the rain water. The pyramids of the Egyptians, however, are the first recorded use for limestone and employ huge nummulitic limestone blocks and lime along with alabaster (gypsum) for mortar and plaster between 4000 and 2000 B.C. Marble statuary and luxurious wall construction was initiated shortly after this period. Although lime's use originated for construction, it was employed by the Greek and early Roman empires also as a chemical reagent. Xenophon in 350 B.C. records how a ship carrying a cargo of linen and lime "for its bleaching" was wrecked near Marseilles. Cato mentioned the burning of lime in kilns in 184 B.C. The medical use of saturated solutions of lime water was cited by the Roman, Dioscorides, in 75 A.D. The first great highway builders, the Romans, used both limestone and lime extensively. There was even some application and recognition of the virtues of agricultural liming at the time of Christ.

One of the early forms of chemical warfare was the reported gruesome practice of the English in throwing quicklime in the faces of the French, their enemy in a war in A.D. 1217. Shakespeare and earlier English writers, like Layamon, frequently mentioned lime in their writings, prose and plays. European alchemists in the Middle Ages are known to have causticized the potassium carbonate present in wood ashes with lime to produce a crude form of lye as a base for soapmaking.

The first sound technical explanation of lime calcination, including its expulsion of CO_2 gas, was by the British chemist, Joseph Black, in the eighteenth century. Lavoisier, the French scientist, a few years later confirmed Black's theory and elucidated further on it. Other renowned French scientists—Vicat, Debray, and Le Châtelier—in the nineteenth century are credited with measuring most of the earliest fundamental data, such as the dissociation temperature and pressure incident to calcination as well as evaluating structural limes and mortars for construction.

Strangely, many of the ancient civilizations appear *independently* to have discovered lime and perfected their own uses from it. Some civilizations undoubtedly learned from each other, like the Greeks from the Egyptians and the Romans from the Greeks. But who educated (at that time) such isolated civilizations as the Incas, Mayas, Chinese, and the Mogul Indians? Their ancient applications of lime have been repeatedly confirmed by archaeologists through the artifacts they unearthed. Somehow they all "stumbled" onto it by primitive ingenuity or happenstance. Fragmentary decipherings of their scarce records reveal some barbarous methods of making lime mortar, such as slaking quicklime in barley water, mixing blood from animals into the lime and sand, etc. Yet some of their old edifices stood for two to three thousand years, disheveled from the erosion of time but still intact and often structurally sound. Restoration societies have rescued many of these old structures from desuetude and have remortared their eroded joints with modern cementing materials. Ironically often these modern scientifically prepared mortars disintegrate in five to twenty years, requiring repeated expensive maintenance, in contrast to the thousands of years that the ancients' bizarre mortars have endured. Possibly modern civilization has not advanced in some respects as much as we think.

CHAPTER TWO

Formation and Properties
of Limestone

Origin

In defining and describing limestone and its properties more emphasis will be placed on its physical and chemical than on its geological characteristics.

This most important and abundant of all *sedimentary* rocks that is employed commercially is usually of organic origin.[1,2] Fossiliferous, marine sedimentation in oceans and fresh bodies of water, consisting of shells and skeletons of plants and animals, were gradually accumulated through deposition, layer upon layer to form in some instances massive beds of limestone. Some of this sediment was deposited by natural chemical reactions, namely, the infinitesimal dissolution of these calcium carbonate fossils through the solvent action of carbon dioxide, forming calcium bicarbonate, which was subsequently reprecipitated in carbonate form. Sometimes this precipitation is indirect through the intermediary of plant and animal organisms. Direct precipitation of the carbonate through a saturated solution is caused by either an increase in temperature, which reduces the solubility of the calcium carbonate, or through evaporation. Originally the fossils were formed by gradual assimilation of the carbonate as a bicarbonate from the seawater, inducing a very slow but inexorible build-up of the shell or skeletal structure. On the demise of the organism its shell of almost pure calcium carbonate remains as a nucleus to perpetuate the above phenomena. It is a true reversible reaction with dissolution and precipitation in approximate equilibrium.

Huge coral reefs are gradually accumulated in this manner over

5

thousands of years, and in the millions of years of geologic ages that have reshaped the geography of the world, they have formed mountains in the interior of continents. Many mountains in Europe and North America are coralline in origin. Pressure and heat have supplemented chemical precipitation in consolidating the minute carbonate particles into these imposing compact masses.

In a similar manner inland streams and rivers are carriers of soluble calcium bicarbonate through the leaching effect of rainwater percolation through soils in watersheds and the gradual dissolution of carbonate rocks as the streams cascade over them. The co-reactant, carbon dioxide, must be present to effect this phenomena. Most soils contain at least traces of calcium and magnesium; some are highly calcareous. Such soluble calcium is the most common type of mineral hardness that exists in varying amounts in fresh-water bodies.

The texture and crystalline form of the limestone depends on the size and shape of the calcareous particles or grains that are deposited. Usually this deposition is contaminated with varying amounts of impurities, like silica, that through time becomes an intimate part of the stone, seemingly cemented to the carbonate particles. Generally the purest limestone formations occur more often in the thickest masses or beds of up to several hundred feet thick. However, frequently these thick beds will contain strata of relatively impure stone. Contamination of the stone with soil usually occured at the commencement of deposition, but in some instances impurities were absorbed through pores and interstices during deposition. These impurities occur both vertically and laterally in the bed, but usually a change in purity is much more gradual laterally than vertically.

The calcareous sediment may remain relatively soft, like chalk or marl, or it may be metamorphosed through high temperature and fused into a highly crystalline form, such as marble. Examples of limestone that are not of organic origin would be stalactites and stalagmites in caves and some oolitic limestone, travertine, and calcareous tufa. However, the presence of fossils in varying degrees is apparent in most commercial limestone, and this enables the geologist to determine from which geologic age it was derived. Often the prehistoric fossils are found intact in the stone in an almost perfect state of preservation (Fig. 2-1).

The other common constituent of limestone, magnesium carbonate, is also formed in a related manner, but it is also theorized that much of it was created by chemical displacement of calcium with magnesium from magnesium salts that abound in seawater.[3,4] De-

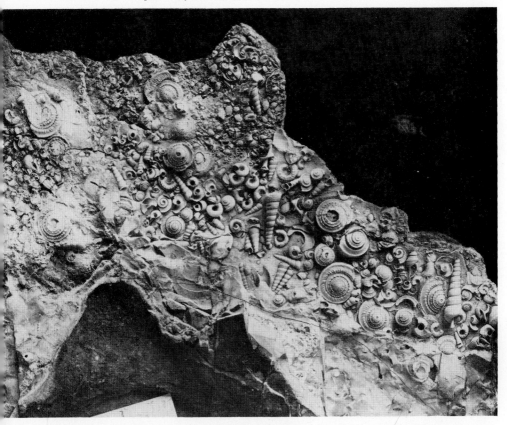

Fig. 2-1. Lower Ordovician limestone from Missouri displaying wide assortment of fossils in unusual clarity: gastropods, crinoids, brachipods, pelecypods, and other prehistoric shell and vertebrae of marine organisms. (Courtesy, U. S. National Museum Collections)

composition of serpentine (hydrous magnesium silicate) would be another possible origin.

Mineralogy

All geological authorities are in agreement that limestone may be composed of four minerals,[5,6] exclusive of impurities, having the following physical characteristics:

Calcite. $CaCO_3$; rhombohedral; molecular weight—100.1; specific gravity—2.72; molecular volume—36.8; hardness—3; may be colorless, but often variously tinted by impurities.

Aragonite. $CaCO_3$; orthorhombic; specific gravity—2.94; molecular volume—34; hardness—3.5 to 4; usually white but often tinted by impurities.

Dolomite. $CaMg(CO_3)_2$; rhombohedral; molecular weight—184.4; specific gravity—2.83; molecular volume—65.2; hardness—3.5 to 4; usually colorless, but often tinted pink or tan.

Magnesite. $MgCO_3$; rhombohedral; molecular weight—84.3; specific gravity—3; molecular volume—28.1; hardness—3.5 to 4.5; white, tan, or brown.

Of the above minerals the only point of disagreement is the rival contentions that dolomite is (1) a double carbonate of calcium and magnesium and (2) a mechanical mixture of calcite and magnesite.

There is an academic curiosity, the mineral *vaterite* (or μ $CaCO_3$), that occurs in nature. It is of no commercial significance since it only occurs in small, scattered amounts and warrants only brief mention. It usually appears amorphous but it crystallizes in hexagonal plates. It is quite unstable, tending to convert rapidly to calcite when subject to increases in temperature or saturated moisture conditions. It can be produced synthetically by slow precipitation at 60°C.

All of these minerals can be identified with X-ray diffraction instruments. Aragonite can be distinguished from calcite since it alters the color of hot cobalt nitrate solution to violet whereas calcite yields a blue color. Calcite is easily the most abundant of these minerals.

As evidence of the many divergent crystal structures of limestone, see micro-photographs of thin sections (Figs. 6-10 to 6-17) on p. 159.

Classification of Limestone

Obtaining the proper perspective of limestone is desirable at the outset since there are confusingly numerous varieties of forms, types, and purities. Complicating the situation further are the names used to delineate various types, many of which are repetitious or even seemingly misnomers. Even though its color, texture, and crystallinity is radically different, *marble* is a form of limestone since it primarily contains calcitic and dolomitic crystals, the same chemical composition. So limestone is a *broad* term. However, in this book calcareous sandstones of only 35 to 50% calcium carbonate or other highly impure limestone, chalk, or marl will not be considered. Even magnesite, pure magnesium carbonate, which might conceivably be considered limestone, will be omitted since its value, utilization, and tonnage are so different.

The two most fundamental types of limestone are *high calcium* and *dolomitic*. Pure high calcium (calcite or aragonite) is 100% calcium carbonate. Pure dolomite is 54.3% $CaCO_3$ and 45.7% $MgCO_3$, or stated another way, it is composed of 30.4% lime (CaO), 21.8% magnesia (MgO), and 47.8% CO_2. No limestones of these purities are commercially available. For practical purposes high quality high calcium is 97 to 99% $CaCO_3$ and dolomitic is 40 to 43% $MgCO_3$, with the $CaCO_3$ component slightly higher than theoretical, and 1 to 3% impurities.

Following is a list of varieties of limestone in common use, some of which have different nomenclatures, together with definitions:

Argillaceous limestone is an impure type of limestone, containing considerable clay or shale, and, as a result, has a relatively high silica and alumina content.

Bastard limestone is a commercial rather than geological term, synonymous with magnesian limestone; it is neither high calcium nor high magnesium (dolomitic), but intermediate in calcium and magnesium carbonate content.

Bituminous limestone is synonymous with carbonaceous limestone and contains various types of organic material, such as peat, natural asphalt, and even petroleum, as impurities. Such stone is often black and may exude a fetid odor due to the presence of organic matter.

Brecciated limestone is composed of fragments of hard limestone cemented together by calcium carbonate.

Calcitic limestone is a term generally used by agronomists to convey a high calcium stone, but it could be misleading since its connotation might infer *pure* calcite, which it usually is not.

Carbonaceous limestone is similar to bituminous.

Cementstone is an impure limestone, usually argillaceous, possessing the ideal balance of silica, alumina, and calcium carbonate needed for Portland cement manufacture. When calcined it produces a hydraulic cementing material.

Chalk is a soft, fine-grained, fossiliferous form of calcium carbonate, varying widely in color, hardness, and purity. White chalk may be extremely pure, 98 to 99% calcium carbonate, whereas grey chalks may contain as much as 20% impurities. The grain size is so minute that it was regarded as amorphous but is now considered to be cryptocrystalline. The microparticles tend to be more rounded than octagonal, like calcite, and as a result, have more surface area.

Chemical-grade limestone is a pure type of high calcium or dolomitic limestone that is used by many of the chemical process industries or where exacting chemical requirements are necessary. It contains a minimum of 95% total carbonate content. (In a few areas of the United States where an abundance of high-quality stone is available, the minimum may be 97 or 98% total carbonate.)

Compact limestone is a general term depicting a dense, fine-grained, homogeneous, usually hard type of limestone.

Coral limestone is a general term designating stone in which the dominant fossil is coral.

Cherty limestone is an impure stone of unstable physical qualities, containing considerable silica.

Clam shells are usually a very pure form of calcium carbonate, although the purity of such shells may be impaired by incrustations of silica sand bonded to the shell. It is strictly of marine origin.

Coquina is a type of fossiliferous calcium carbonate shell found in tropical areas of high purity, but it is relatively soft.

Dolomitic limestone contains considerable magnesium carbonate. True, pure dolomitic stone contains a ratio of 40 to 44% $MgCO_3$ to 54 to 58% $CaCO_3$. However, the term is more loosely used so that any carbonate rock containing 20% or more $MgCO_3$ is regarded as "dolomite." It varies in purity, hardness, and color.

Ferruginous limestone contains iron as an impurity and is yellowish or reddish in color.

Fluxstone is a pure form of limestone used as a flux or purifier in metallurgical furnaces. It can be a high calcium, magnesian, or dolomitic type, providing the stone is low in impurities and contains at least 95% total carbonate content. It is synonymous with metallurgical grade limestone.

Fossiliferous limestone is a very general term connoting any carbonate stone in which fossil structure is visually evident.

Glass stone is a pure form of high calcium or dolomitic stone, low in iron, used expressly for glass manufacture. It qualifies as chemical grade stone.

Glauconic limestone contains crystals of glauconite (hydrated silicate of iron and potassium).

High calcium limestone is a general term for stone containing largely calcium carbonate and not much magnesium carbonate (only 2 to 5%, depending on point of view). It occurs in varying degrees of purity.

Hydraulic limestone is an impure, argillaceous carbonate, somewhat similar to cement stone except that it may be high calcium, magnesian, dolomitic, and usually it produces cement-like materials of lower hydraulicity.

Iceland spar is the purest form of limestone available, comprising virtually pure calcite of about 99.9% $CaCO_3$. It is also known as optical calcite. Its occurence is rare; it is pearly white, quite transparent, and has large crystals.

Indiana limestone refers solely to a dense type of smooth-textured high calcium, oolitic limestone that occurs largely in Bedford County in Indiana. It is utilized almost exclusively as a large sized building or dimension stone for facing large buildings.

Lias limestone is a term used in Great Britain to describe a hydraulic limestone. It received its name from the Lias formation of the Jurassic system.

Lithographic limestone is a drab or yellowish compact stone, usually of magnesian type, that is employed in slabs for lithographic printing. Its smooth surface may be etched with weak acid.

Magnesian limestone contains more magnesium carbonate than high calcium stone but less than dolomite. Authorities are not in full agreement as to its range of magnesium carbonate, but the consensus favors 5 to 20%. It is also referred to as bastard limestone and occurs in varying purity.

Marble is a metamorphic, highly crystalline carbonate rock which may be high calcium or dolomitic. It may be extremely pure or rather impure. It occurs in virtually every color in varying mottled effects and is the most beautiful form of limestone. Because of its unique texture it can be cut more precisely and polished to a smoother surface than any stone. It is usually very hard; some types approach even granite in hardness.

Marl is an impure, soft, earthy rock of marine origin in which nature has intermixed varying amounts of clay and sand into the loosely knit crystalline structure. Some marls contain more silica and alumina (soil) than calcium carbonate and are not regarded as limestone.

Metallurgical grade limestone is the same as fluxstone.

Oolitic limestone is composed of small rounded grains of calcium carbonate, precipitated in concentric laminates around a nucleus piece of calcium carbonate or even silica. It is frequently quite pure, with a $CaCO_3$ content ranging up to 99%+.

Oyster shell is another of the many fossiliferous limestones similar to clam shell in purity, chemical composition, and hardness.

Phosphatic limestone is a type of rock, usually of high calcium variety, which contains a relatively high percentage (up to 5%) of phosphorus. It is derived from invertebrate marine organisms which contain phosphorus to a lesser degree but are analogous to animal bone.

Pisolitic limestone is similar to oolitic except that the grains of $CaCO_3$ are much larger—the size of a pea.

Shell limestone is a more general term for oyster, clam shells, etc. It is 100% fossiliferous.

Siliceous limestone is synonymous with cherty limestone.

Stalactite and stalagmite are conical or icicle-like shapes of $CaCO_3$ formed on the roofs and floors, respectively, of caverns. They are precipitated from cold groundwater that drips from limestone crevices.

Travertine is a calcium carbonate that occurs through chemical precipitation from natural hot-water springs. Since it is deposited in strata, a banded appearance results. Of all limestones it is the most closely akin to marble in appearance and use and is usually quite hard and compact.

Tufa (calcareous) has the same derivation as travertine, but it is much softer and more porous.

Whiting at one time connoted only a very finely pulverized, purified form of chalk, dust-like $CaCO_3$ of micron sizes. Now the term is used more broadly to include all finely divided, meticulously milled carbonates that are derived from high calcium or dolomitic limestone, marble, shell, or chemically precipitated $CaCO_3$. Often it is referred to as "limestone whiting," etc. thus revealing its origin. Unlike all of the above *natural* forms of limestone, it is strictly a *manufactured* product.

Geologic Nomenclature

In addition to the above definitions geologists prefer more specific descriptions of deposits based on the geologic age from which they are derived.[7-9] Thus, the large northwestern Ohio dolomitic deposits are described as the Niagara formation from the Upper Silurian age (or system) or "Niagaran dolomite,"; the large high calcium stone industry in the northern part of lower peninsula Michigan is the Dundee formation of the lower Devonian age; the high calcium industry in Shelby County, Alabama has the Longview and Newala formations of the Ordovician age; etc. Most of the formations were named by the state geological surveys from established geographical names near the formation as a permanent means of identification.

Much of the limestone in the United States is derived from the Paleozoic era that is 220 to 470 million years old as well as some dolomite from north and southeast New York state from the earliest of all eras, the pre-Cambrian, that is, at least 550 million years old. Other early ages in this Paleozoic era are the Cambrian and Ordovician in central and southeast Pennsylvania and eastern Tennessee, the Berkshire Mountains of Massachusetts, and the Shenandoah area of Virginia. The Ordovician age is also well entrenched in southeast Missouri. Then there is the Mississippian age in western Illinois, the Ozark Mountains of Missouri, and in Indiana. A continuation of the same Niagaran dolomite formation that exists so abundantly in northwest Ohio extends throughout northern Illinois and Wisconsin. The Cretaceous period in the later Mesazoic era usually denotes a chalky limestone and occurs in central Texas (Austin chalk formation), Alabama, Kansas, Nebraska, and in small amounts in a few other states.

Occurrence

In virtually every state of the United States and every country in the world, there is some occurrence of limestone, although in some states and countries its existence is largely negated by economic inaccessibility. Either it is lodged too deeply or intricately in the earth or is located in areas too distant from transportation and/or markets. Other factors obviating its utility are the occurrence of small, scattered lodes of stone that individually are too small to justify the necessary capital investment for exploitation and collectively are too widely disseminated; also the stone may be deficient in certain desired chemical or physical qualities.

Many of the state geological surveys have plotted most of their limestone deposits, along with other mineral resources, on highly detailed state, county, or sectional maps, but there is a dearth of national and regional maps that delineate limestone formations, probably since such maps could not be produced in sufficient detail to be of material assistance in geological studies. In fact, the only complete map on limestone for the United States was compiled in 1913 by Burchard and Emley. This map is now of questionable value since it does not include some significant limestone formations that were subsequently discovered. Moreover, such a map does not differentiate between the exploitable and the valueless deposits. Consequently, from a national viewpoint a map that depicts the *active* deposits that are being "worked" is of greatest significance and is the logical prelude to a study of large-scale state and sectional maps. To this end the author has prepared a map depicting most of the *active* high calcium limestone quarries and mines in the United States (Fig. 2-2). A companion map[4] had previously been drawn locating the active dolomitic and magnesian limestone deposits (Fig. 2-3). Contrasting with these two maps is a geological map of the high calcium limestone and chalk formations, including those that are unexploitable (Fig. 2-4). A comparison of these maps quickly reveals that the largest limestone-producing states are not necessarily those states that nature has endowed with the most limestone, or vice versa. In any event these maps pinpoint the so-called "limestone country" in both a practical and academic manner.

Note that some states are predominantly either high calcium or dolomitic—Wisconsin and Minnesota are almost entirely dolomitic, and Kentucky and Arizona are high calcium—or possess abundant exploitable amounts of both types—Ohio and Illinois. Yet the occurrence of these two types tends to be erratic and without a predictable pattern. For example, in some concentrated dolomitic and high calcium areas, only one of these types of stone exists over a broad expanse of many square miles; in other concentrated limestone areas both types may occur with the deposit of one type indescriminately contiguous to the other; and in still others both types may occur jumbled together heterogeneously or even with one side of a quarry being high calcium and the other dolomitic. Purity of the stone similarly varies over a wide range. A few states, notably New Hampshire and North Dakota, are almost devoid of limestone per se and what little that exists is unusable. Supplementing these maps are statistics on limestone production by states[10] contained on p. 455.

United States reserves of commercially usable stone have never been systematically estimated and would probably be so vast as to be incalculable. However, such is not the case with the pure, chemical grade deposits that comprise only about 2% of the total limestone revealed on these maps. Although reserves for this type have never been estimated either, one authority[12] conjectures that they are definitely limited and intimates that the United States should not be complacent with its existing proven reserves. Certainly quite a number of strategic deposits have been exhausted in the past, others are approaching depletion, and continual exploration is needed to replace these losses. At the present tremendous annual rate of consumption

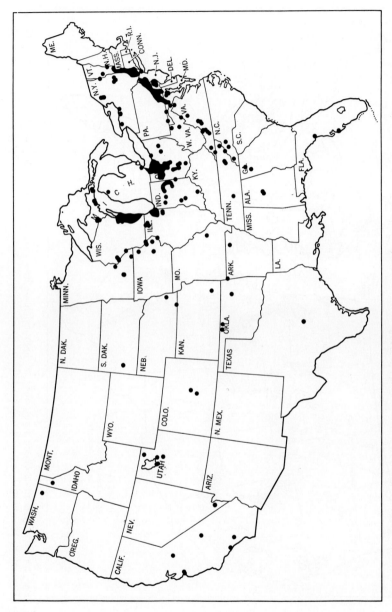

Fig. 2 3. Location and concentration of dolomitic and magnesian limestone deposits in the United States, most of which are in operation.

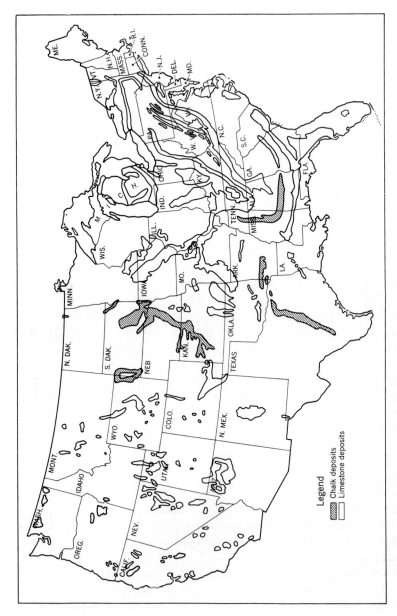

Fig. 2-4. Location of geologic formations of high calcium limestone and chalk in the United States. (From Herbert P. Woodward).

it is possible that some new economical methods of stone beneficiation for marginal deposits may be urgently needed by the year 2000 unless some new extensive deposits are discovered. It is believed that a somewhat parallel situation prevails in most of the other major nations of the world.

Impurities

There are two classifications of impurities: (1) *homogeneous,* in which the impurities in the form of clay, silt, and sand (or other forms of silica, like quartz), contaminate the stone when it is first deposited and in which the impurities are well dispersed throughout the formation and (2) *heterogeneous,* in which the impurities collect only in the crevices or between the strata or as siliceous pieces or nodules of sand, chert, or flint loosely embedded in the limestone. This is the source of *silica* and *alumina,* the major impurities. A minor source of silica is derived from feldspar, mica, talc, and serpentine.

Iron is the third major impurity and can be homogeneously disseminated after the limestone has started to form by chemical displacement of the calcium, making iron carbonates. This frequently occurs in oolitic limestone. Another source of iron would be iron oxide, distributed heterogeneously from such minerals as pyrite, limonite, magnetite, and hematite, but only rarely in the latter two.

Phosphorus and *sulfur* usually occur but generally in small quantities. Chemical displacement of the carbonate radical by oxy-acid radicals of these two elements is a homogeneous source of impurity. Additionally, phosphorus is also derived from certain fossiliferous skeletons (invertebrates) forming the nucleus of calcareous deposits. Pyrite is another source of sulfur.

The remaining impurities are so minute that they could be considered as *trace elements* in the relatively pure stones. These might be manganese, copper, titanium, sodium and potassium (as oxides), fluorine, arsenic, strontium, and others. Table 2-1 shows the extent of these trace elements in a typical commercial limestone.[11] (More on trace elements on p. 156.) Of the above the one exception would be a higher content of the soluble alkali salts present in the clay or shale seams of the more impure types, such as argillaceous stone. With the impure stones these salts are usually carbonates and chlorides of sodium and potassium. In studying the trace amounts of soluble salts quantitatively in pure limestone one investigator concluded that calcium sulfate and sodium chloride usually occur in

18 Chemistry and Technology of Lime and Limestone

Table 2-1. Trace elements found in a Missouri limestone as a result of spectrographic analysis of the rock layers.

Layer or Specimen	Mn	Cu	Co	Zn	B	Ni	V	Mo	Cr	Ag	Sr	Ba	Ti	Sn	Pb
A	5^{-3}	3^{-5}	—	—	2^{-3}	3^{-4}	—	—	.01	5^{-5}	.2	.03	.1	—	—
	0.1	.0006	—	—	.04	.006	—	—	0.2	.001	4.	0.6	2	—	—
B	2^{-3}	5^{-5}	—	5^{-3}	3^{-3}	3^{-4}	2^{-3}	—	.01	5^{-5}	.3	.03	.3	—	—
	.04	.001	—	0.1	.06	.06	.04	—	0.2	.001	6	0.6	6	—	—
C	5^{-3}	1^{-4}	—	2^{-3}	1^{-3}	—	—	—	5^{-3}	5^{-5}	.1	.02	.2	—	—
	0.1	.002	—	.04	.02	—	—	—	.01	.001	2	0.4	4	—	—
D	.02	2^{-4}	—	3^{-3}	5^{-3}	—	5^{-3}	—	.02	2^{-5}	.2	.2	.3	—	—
	0.4	.004	—	.06	0.1	—	0.1	—	.4	.0004	4	4	6	—	—
E	.03	5^{-4}	—	—	3^{-3}	3^{-4}	5^{-3}	—	.03	2^{-5}	.05	.03	.2	—	2^{-4}
	.6	.01	—	—	.06	.006	0.1	—	.6	.0004	1.	0.6	4	—	.004
F	3^{-3}	5^{-4}	—	.02	5^{-3}	5^{-4}	3^{-3}	—	.03	2^{-5}	.02	.03	1.0	—	—
	.06	.01	—	0.4	0.1	.01	.06	—	.6	.0004	0.4	0.6	20	—	—
G	.01	5^{-4}	—	1^{-3}	2^{-3}	2^{-4}	—	—	.02	2^{-5}	.02	.03	.2	—	2^{-3}
	0.2	.01	—	.02	.04	.004	—	—	.4	.0004	0.4	0.6	4	—	.04
H	.01	3^{-4}	—	1^{-3}	1^{-3}	2^{-4}	1^{-3}	—	.02	5^{-5}	.03	.02	.2	—	—
	0.2	.006	—	.02	.02	.004	.02	—	.4	.001	0.6	0.4	4	—	—
I	3^{-3}	5^{-4}	—	—	2^{-3}	2^{-4}	—	2^{-3}	5^{-3}	—	.05	.03	.1	—	1^{-4}
	.06	.01	—	—	.04	.004	—	.04	.01	—	1	0.6	2	—	.002
J	.02	1^{-3}	—	—	3^{-3}	5^{-4}	1^{-3}	3^{-3}	.01	2^{-5}	.3	.2	.1	2^{-4}	3^{-4}
	.4	.02	—	—	.06	.01	.02	.06	0.2	.0004	6	4	2	.004	.006
Bin	1^{-3}	2^{-4}	—	—	1^{-4}	1^{-4}	—	—	.02	2^{-5}	.05	.03	.2	—	2^{-4}
	.02	.004	—	—	.002	.002	—	—	.4	.0004	1	0.6	4	—	.004
Chats and road rock	3^{-3}	1^{-3}	—	1^{-3}	1^{-3}	5^{-4}	1^{-3}	—	.01	—	.2	.2	.2	—	—
	.06	.02	—	.02	.02	.01	.02	—	.02	—	4	4	4	—	—
Gr. qy.	.02	2^{-4}	—	—	—	—	—	—	2^{-3}	—	.1	.02	.01	—	—
	0.4	.004	—	—	—	—	—	—	.04	—	2	0.4	0.2	—	—

Upper figure: amount of each element reported in percent.
Lower figure: element reported in fractional pounds per ton of limestone.
Elements not detected in any of the above samples: Cd, Li, Bi, As, Sb, Be, and Ga.

high calcium stone and that magnesium chloride, magnesium sulfate, and potassium chloride are more likely to occur in dolomitic stone.

Some organic matter, usually well under 1%, often exists in limestone. This alleged impurity is *carbonaceous*, but its presence is usually inconsequential since the amount is usually so minute, and when the stone is calcined, this organic matter is quickly volatilized. However, even traces of it can color limestone a light gray, and with higher amounts a blackish stone results on up to bituminous or carbonaceous limestones, previously defined. It is derived from bituminous matter or the residue of marine organisms that are not completely oxidized or decomposed.

Usually *magnesium* carbonate is *not* regarded as an impurity except for certain special chemical purposes that demand a very high calcium stone. It is sometimes regarded as the essential constituent, in others of equal value to its calcium counterpart; and still in others as valueless but not deleterious.

Quantitatively a so-called chemical or metallurgical grade of stone would require a minimum combined carbonate (calcium and magnesium) content of 95% in the United States.[12] In some areas that are favored with numerous high-grade deposits, 97 or 98% would be the minimum. In fact, the requirements of the chemical-process industries on impurities have become increasingly stringent. A 95% carbonate stone, widely used in the 1930's, may be completely unacceptable for certain uses in the 1960's . . . or it could only be used with substantial price concessions. Invariably the principal impurity is silica and secondarily alumina. The combined total of these two substances comprises usually 85 to 95% of the total impurities. Typical chemical analyses of some representative, conventional commercial limestones are contained in Table 2-2. Further, more specific information on impurity tolerances is contained in Chapter 5.

Table 2-2. Representative chemical analyses of different types of U. S. limestones

				Limestone				
	1	2	3	4	5	6	7	8
CaO	54.54	38.90	41.84	31.20	29.45	45.65	55.28	52.48
MgO	0.59	2.72	1.94	20.45	21.12	7.07	0.46	0.59
CO_2	42.90	33.10	32.94	47.87	46.15	43.60	43.73	41.85
SiO_2	0.70	19.82	13.44	0.11	0.14	2.55	0.42	2.38
Al_2O_3	0.68	5.40	4.55	0.30	0.04	0.23	0.13	1.57
Fe_2O_3*	0.08	1.60	0.56	0.19	0.10	0.20	0.05	0.56
SO_3†	0.31	—	0.33	—	—	0.33	0.01	—
P_2O_5	—	—	0.22	—	0.05	0.04	—	—
Na_2O	0.16	—	0.31	0.06	0.01	0.01	—	—
K_2O	—	—	0.72	—	0.01	0.03	—	—
H_2O	—	—	1.55	—	0.16	0.23	—	n.d.
Other	—	—	0.29	—	0.01	0.06	0.08	0.20

1 = Indiana high calcium stone.
2 = Lehigh Valley, Pa. "cement rock."
3 = Pennsylvania "cement rock."
4 = Illinois Niagaran dolomitic stone.
5 = Northwestern Ohio Niagaran dolomitic stone.
6 = New York Magnesian stone.
7 = Virginia high calcium stone.
8 = Kansas cretaceous high calcium (chalk).

* Includes some Fe as FeO.
† Includes some elemental S.

Specific Gravity. Pure calcite is considered to have a specific gravity of 2.7102 at 20°C and infinitesimally higher, 2.7112, at 0°C. The rarer mineral, aragonite, is heavier—2.929. However, most commercial limestone has a range in values—2.65 to 2.75 for high calcium and 2.75 to 2.90 for dolomitic. Values for magnesian are intermediate but closer to high calcium in value. Chalk is much less—between 1.4 and 2.

Hardness. Pure calcite is standardized on Moh's scale at 3, whereas aragonite is harder, ranging between 3.5 and 4. Commercial limestone varies considerably—2 to 4. Dolomite is generally harder than high-calcium stone, but its crystal is relatively brittle and tends to break with an uneven to conchoidal fracture. Most limestone is soft enough to be readily scratched with a knife.

Strength. There is also a wide variance in the strength of limestones. Generally marble and travertine are among the highest values; chalk and marl are among the lowest. Some of the cherty, impure stones have very high strengths. *Compressive or crushing strengths*[17] range between 1200 and 28,400 psi. *Shear strength* values typically range between 600 and 3000 psi; however, some marble has been measured as high as 6500 psi. *Tensile strength* has been reported between 350 to 900 psi.

Modulus of Elasticity. One authority[17] estimates this value between 3 to 6 million psi. Another investigator[18] calculates 1900 to 3000 kg./mm.^{2}; 2600 kg./mm.^{2} has been given as a fair average value for marble.

Modulus of Rigidity. As the reciprocal to the modulus of elasticity, one authority[15] reported values for this characteristic extending from 1.5 to 5.1 million psi for eleven different American high calcium and dolomitic limestones and 0.23 to 0.57 million psi for two South Dakota chalks.

Coefficient of Expansion. From 0 to 300°C the expansion of limestone is minute, but the coefficient increases with rises in temperature. Between 20 and 50°C the coefficient of expansion of oolitic limestone[17] was measured at 0.000005/°C. Another researcher reports slightly higher values for aragonite than calcite—0.00001 to 0.000035.

Young's Modulus. Values range between 3.2 and 12.3 million psi.[14]

Poisson's Ratio. Values range between 0.07 and 0.35.[14]

Refractive Index. Calcitic or aragonitic crystals possess much greater optical or refractive properties than dolomite. Increasing iron content causes the refractive index of the latter to heighten. Calcitic crystals exhibit double refraction; in contrast, dolomite produces negative double refraction.

Refractive indices[19] for the sodium D line at 18°C are:

Calcite	*Aragonite*
$\mu\omega 1.6584$	$\mu\alpha 1.5300$
$\mu\epsilon 1.4864$	$\mu\beta 1.6812$
	$\mu\gamma 1.6857$

Luminescence. Both calcite and dolomite have limited luminescent qualities. Fragments of dolomite transmit light poorly or not at all, depending on impurities present. Both pure forms of dolomite and calcite are luminescent when exposed to cobalt-60γ-radiation; calcite when exposed to short-wave radiation. Some varieties of calcite will phosphoresce after exposure to ultraviolet light and can be made to fluoresce and appear purple. In contrast, pure dolomite appears colorless under ultraviolet light except for small bands of purple caused by an excess of $CaCO_3$ in the dolomitic lattice.

Diffuse Reflecting Power. This is ratio of total luminous flux reflected to that received, measured for the various regions of the spectrum. The wavelengths are those of maximum energy.

Material	Reflecting Power %					
	0.54	0.60	0.95	4.4	8.8	24.0
$MgCO_3$		85.2	89.4	10.8	4.1	8.8
Limestone*		42.9	—	20.3	5.0	—
$CaCO_3$ Marble†		53.5	—	6.4	5.1	—

* Indiana, dense, fine-grained.
† White color, ground, unpolished.

Velocity of Sound. The velocity of sound through marble has been calculated to be 3810 m./sec., or 12,500 ft./sec.

Thermal Conductivity. Values for white marble in cal./cm.³/sec./°C. are given as follows: 50–100 = 0.00614; 100–150 = 0.00524; 150–200 = 0.00415. No values are found for limestone or dolomite, but they are conjectured to be slightly lower.

Another investigator[13] reports comparable calculations as follows:

	Temperature °C	Cal./cm.^{3}/sec./°C
High calcium limestone	130	0.0039
Dolomitic limestone	123	0.0034
Chalk	—	0.0022
Seventeen varieties of marble	30	0.0050–0.0077

Thermal Expansion. Values[13] for the average linear expansion coefficient $\dfrac{I\,\Delta L}{L\,\Delta T}$ are given for: limestone $8 \pm 4 \times 10^{-6}$; marble $7 \pm 2 \times 10^{-6}$.

Heat Capacity. Values[13] in Joules/gram are cited for: limestone 1.00 at 58°C, marble 0.79 at 0°C, marble 1.00 at 200°C.

Specific Heat. At a temperature of 100°F and stated in Btu./lb., the specific heats for high calcium and dolomitic limestones are 0.205 and 0.215, respectively.[21, 22] The specific heats increase as the temperature rises, as depicted in Fig. 2-7, but a specific heat value of 0.01 has been measured for calcite at —250.8°C.

Specific heats of pure calcite have been measured and are similar to Fig. 2-7. More limited values given for aragonite appear to be very similar to calcite. The values for dolomite average about the same as the above, but one researcher obtained 0.222 between 20 and 98°C.

Melting Point. Since all limestone is transformed into oxides before melting or fusion can actuate, no values are given here. For the melting point of quicklime, see p. 170.

Heat of Formation. In the formation of calcite by $CaO + CO_2 = CaCO_3$, the heat of this reaction is 42,600 cal./mole at 25°C or 1368 Btu/lb.[21]

The only value for magnesite that appears plausible is 28,900 cal./mole. Dolomite is intermediate between these two values.

Electrical Resistivity. In ohms/cm. values[13] are reported as follows: limestone (Mammoth Cave, Kentucky) 10^5; Cambrian limestone (Missouri) 10^4 to 10^5; marble (Carrara, Italy) 10^{10}; marble (Arudy, France) 4×10^8.

Dielectric Constant. Values[13] are given: limestone 8 to 12; dolomite 7.3; marble 8.3.

Solubility. In pure distilled water that is devoid of dissolved CO_2 and in a CO_2-free atmosphere, limestone is virtually insoluble, but with increasing partial pressures of CO_2 solubility gains steadily to the point where it might be characterized as faintly soluble.

There have been many measurements by many researchers on solubilities under varied conditions, such as different temperatures, widely varying partial pressures of CO_2, and in aqueous solutions containing diverse soluble salts in varied concentrations. Many of the results are somewhat discordant, but this is believed to be due to divergent laboratory procedures, conditions of tests, and the quality of the limestone. The following values represent largely the concensus among the investigators or those who have established the greatest credence in their findings. [23-26]

CO_2-FREE. The solubility of calcite at temperatures of 17 to 25°C, free of CO_2, is 0.014-5 g./l. As the temperature is increased, the solubility also increases gradually until at 100°C the value is about 0.03-4 g./l. One authority, who reported 0.0375 g./l. at 100°C, discovered that on boiling this saturated solution from $1\frac{1}{2}$ to 48 hours that the solubility was further increased about five times. This, of course, explains how scale forms in kettles and boiler tanks. Magnesium carbonate is slightly more soluble, but measurements are more

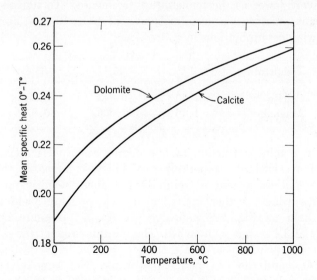

Fig. 2-7. Mean specific heats of calcite and dolomite at different temperature.

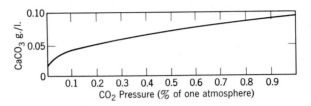

Fig. 2-8. Solubility of CaCO₃ increases as fractional gains occur in CO₂ pressure at constant temperature of 25°C.

sparse and they range from about 0.06 to 0.1 g./l., depending on temperature. At 100° C a solubility of 0.063 g./l. was reported and on boiling for sustained periods the solubility was increased up to tenfold—twice as much as with calcite. There is a dearth of dolomitic values, and the few reported are contradictory. They are believed to be intermediate between calcite and magnesite, but closer to the former. Aragonite has been confirmed as about 7% more soluble than calcite at all temperatures. Solubility can be increased for *all* limestones by about 13% with fine pulverization.

EFFECT OF CO₂. The solubility of all limestone is markedly augmented by increasing increments of CO_2 pressure, but contrary to the above measurements that are CO_2-free, the extent of solubility becomes progressively lower as the temperature is elevated. To illustrate this, at constant pressure of one atmosphere of CO_2, Bäckström[25] made the following determinations in g./l.:

Temperature	9°C	25°C	35°C
Calcite	1.30	0.943	0.765
Aragonite	1.46	1.066	0.876

At only a minute fraction of one degree atmospheric pressure of CO_2 and at constant temperature, solubility is increased three- to four-fold, and the extent of solubility advances steadily up to one full degree where further increases in CO_2 pressure yield a more gradual gain as Fig. 2-8 indicates.[28,29] At 35° pressure optimum solubility of 3.4 g./l. is approached. The solution product (or saturated solution) is calcium bicarbonate that is derived by hydrolysis from the $CaCO_3$- and the CO_2-permeated water. The same investigator reports as high as 3.93 g./l. of $CaCO_3$ solubility at 18°C and CO_2 pressure of 56°. Thus, the solubility of $CaCO_3$ may be increased up

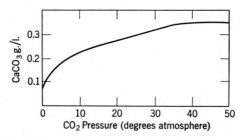

Fig. 2-9. Effect of CO_2 pressure on solubility
of $CaCO_3$ at constant temperature of 25°C.

to about three hundredfold by CO_2 saturation and temperature
correlation.

Kline[27] calculated calcium carbonate's *ratio of solubility* at various
temperatures, independent of a pressure of one atmosphere (with
25°C = r or 1). He proves that solubility decreases steadily as temperature is raised above 0°C as follows:

Temperature	0°	10°	20°	25°	30°	50°
Ratio	1.8	1.4	1.1	1.0	0.9	0.6

Increasing increments of partial CO_2 pressures have the same effect
on magnesite except that the degree of dissolution is even more pronounced. Increasing temperatures also reduce solubility. At 19.5°C
and at 9°CO_2 atmospheric pressure, 56.59 g/l of solubility have been
reported. Figure 2-9 and 2-10 show the effect of CO_2 pressure on
increasing solubility and Fig. 2-11 displays the influence of tempera-

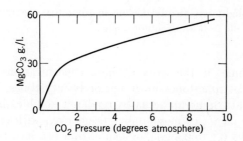

Fig. 2-10. Rise in solubility of $MgCO_3$ with
increasing CO_2 pressure at constant temperature of 19°C.

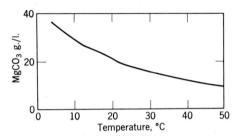

Fig. 2-11. Decrease in solubility of $MgCO_3$
as temperature rises and at constant CO_2
pressure of 1 atmosphere.

ture on dissolution of $MgCO_3$.[28, 29] Thus, $MgCO_3$ is about fifteen to
twenty times more soluble than $CaCO_3$ on an equivalent basis, but
dolomite is regarded as being only two to three times more soluble.[30]

EFFECT OF SALTS. Aqueous salt solutions of various concentrations
affect the solubility of calcium and magnesium carbonate in different
ways, with or without CO_2 pressure. Results of many investigators
are consolidated as follows:

		Solubility of		
	CaCO₃			MgCO₃
Increase	Slight Increase*	Decrease	Increase	Slight Increase*
Na_2SO_4	$MgCl_2$	NaOH	Na_2SO_4	NaCl
NH_4NO_3	NaCl	Na_2CO_3	Na_2CO_3	$K(HCO_3)_2$
NH_4Cl		K_2SO_4	NaOH	
KCl		$CaSO_4$		

* Usually a slight but steady decline in solubility occurs with increasing
salt concentrations past an optimum point.

Although some of the salts at high concentrations will increase
the solubility of limestones up to ten or twenty times, in no instance
is the carbonate mineral actually rendered soluble, in the popular
sense. Modest increases in solubility occur even with very dilute concentrations, like 0.1 to 0.5 g./l. of solute. In a few instances dissolution
of the carbonates progressed until very high salt concentrations of
200+ g./l. were reached. Only a few salts appear to vitiate the carbonate's scant solubility, the most pronounced being the negative

effect exerted by sodium carbonate and hydroxide on calcium carbonate.

Gains in solubility with calcium carbonate, have been measured in aqueous solutions of such *organics* as albumin, globulin, and tri-ammonium citrate, which should be of medical or pharmaceutical significance.

CHEMICAL PROPERTIES AND FUNDAMENTAL REACTIONS

Since most of the chemical properties of limestone relate to its primary end product, lime, which is described in detail in Chapter 7, the following discourse will be necessarily brief and will only include its few *direct* chemical properties. Carbonates are not nearly as chemically reactive as limes (oxides and hydroxides), so in a sense, except for its role as the sole source of lime, its physical properties may be of greater significance than its chemical characteristics. Its *predominant chemical property* is that it can be thermally decomposed (calcined) into lime. This property will be fully described in Chapters 7 and 8. Appendix A, which tabulates the important calcium and magnesium compounds, directly or indirectly derived from limestone, is designed to supplement this section and lime's corresponding Chapter 7 on chemical properties.

Chemical Stability. High calcium and dolomitic limestone are among the most chemically stable substances. Decomposition never occurs at ordinary temperatures (in fact, a wide range in temperatures from absolute zero to about 600°C). These minerals are also un-affected by CO_2-free water, except for the infinitesimal dissolution described on p. 25. Decomposition can only occur at very high temperatures or through reaction with strong acids. Weak carbonic acid or CO_2 produces a mild destabilizing effect, as previously cited on p. 26. The dissociation at elevated temperatures is fully discussed in Chapter 7 under Theory of Calcination.

Aragonite, however, is not as stable as the other two minerals. If sustained moisture is present, it tends to transform into calcite—a recrystallization phenomenon—but with no change in chemical composition. On heating dry aragonite to 400°C, well below the dissociation temperature of these carbonate minerals, it changes irreversibly to calcite. This mineral is likewise vulnerable to acid attack.

Hydrates of Calcium Carbonate. It is possible to synthesize in the laboratory some unique and highly unstable hydrates of calcium carbonate through precipitation that are of no practical value,[31] just academic curiosities.

A hexahydrate ($CaCO_3 \cdot 6H_2O$) may be prepared in a few ways, such as by the slow, carefully controlled absorption of CO_2 in milk of lime, containing sucrose, at 5 to 10°C or by about 300 hours of retention at 10°C. Large monoclinic crystals of 10 mm. and longer have been formed in cold seawater that has been treated with lime.

A pentahydrate ($CaCO_3 \cdot 5H_2O$) has been reported in experimental laboratory work but is very fugitive and impossible to isolate for any length of time.

A monohydrate ($CaCO_3 \cdot H_2O$) has been detected as the hexahydrate decomposes in a transitionary manner to anhydrous calcite.

Bicarbonates. This highly unstable compound is formed through the slight solvent action of carbonic acid on limestone [$CaCO_3 + H_2CO_3 \rightleftarrows Ca(HCO_3)_2$]. The same reaction occurs with the magnesium carbonate component of dolomite. It only occurs in solution form and attempts to isolate it as a solid have invariably failed or proven inconclusive. It is really reversible in reaction, see p. 5.

In contrast, magnesium bicarbonate has been isolated experimentally as a very unstable solid, with $MgCO_3$ occupying the role of a base in forming salts.

pH Value. All limestones are invariably alkaline with pH values ranging between 8 and 9, depending upon the various solubilities of their saturated solutions. Dolomite has higher values than calcite because of its more soluble $MgCO_3$ constituent, about 9 to 9.2: $MgCO_3$ has been reported at 9.5 to 10.

Reaction with Acids. High calcium limestone dissolves in most strong acids readily, accompanied by the liberation of CO_2 gas from the stone. It even effervesces freely in cold, very dilute hydrochloric acid, but dolomitic stone will not effervesce unless the weak HCl is heated, since it is not nearly as reactive as high calcium stone. As evidence of this, if a polished surface of dolomite is etched by cold, weak acid and then examined under a microscope, the true dolomitic areas are unaffected, whereas the calcite crystals exhibit erosion. However, there is also a slight difference in the rate of reactivity with high calcium stones, depending on the impurities present and to a lesser extent on the porosity and crystal size. Copious impurities and large crystals retard reactivity. (See Appendix A for the reaction of limestone with a wide range of mineral and organic acids.)

The reaction with *acid gases* is somewhat different. There is no reaction of dry sulfur dioxide with $CaCO_3$ at ordinary temperature, but at 95°C the sulfur is readily absorbed as a sulfate. With wet sulfur dioxide the reaction is slow, but the reactant is instead a sulfite as demonstrated in the following equation:

$$2CaCO_3 + 3SO_2 \rightarrow 2CaSO_4 + S + 2CO_2 + 74 \text{ cal.}$$

$$\rightarrow CaH(SO_3)_2$$

High calcium limestone reacts as rapidly as lime with gaseous or liquid nitrogen peroxide as follows:

$$CaCO_3 + 3NO_2 \rightarrow Ca(NO_3)_2 + CO_2 + NO + 27 \text{ cal.}$$

Nitrous oxide (NO) is practically unreactive with $CaCO_3$, even at high temperatures.

Dolomitic stone reacts to some extent with most strong liquid and gaseous acids too, forming similar magnesium salts; its rate of reactivity is simply much slower. However, its rate of reactivity accelerates with greater acid concentrations, agitation, and elevated temperatures.

Heat of Solution in Acid. At 25°C and stated in cal./mole, the heats of solution for calcite and aragonite in HCl have been measured[21] as follows:

	in Pure HCl	HCl Saturated with CO_2
Calcite	4495	3245
Aragonite	4472	3204

Data for dolomite is scanty and inconsistent, but the values are less than high calcium.

Solubility Products. This property is also described as the ion product constant. It is the product of the concentrations of the Ca^{++} and Mg^{++} ions in the saturated solution of salts of low solubility, like $CaCO_3$ and $MgCO_3$. In terms of activity as moles per liter of solution, values are given for: $CaCO_3$—0.99×10^{-8} (at 15°C), 0.87×10^{-8} (at 25°C), $MgCO_3$—2.6×10^{-8} (at 12°C). Slightly different values are reported by European researchers.

Electromotive Force Potentials. Both calcium and magnesium as ions or elements rank high in chemical activity in the standard electromotive series table. Among all elements calcium ranks sixth, just behind barium, and has an electrode potential estimated at +2.87. Lithium, which is first, is not much higher, +2.9595. Magnesium ranks eighth among all elements, just behind sodium, and possesses an electrode potential estimated at $+2.40$. All values are at $25°C$.

Such activity applies to *metallic* calcium and magnesium—not compounds of these metals, like the carbonates. The oxides and hydroxides of these elements, however, have a chemical activity more like the metals, see Chapter 7.

Chlorine. Finely pulverized high calcium limestone or whiting will react with chlorine very slowly in a limited manner in cold water and more rapidly in warm water but not nearly as readily as hydrated lime. The reactant is calcium chlorate, an unstable chemical of questionable value.

Phosphorus. With phosphorus compounds it has been concluded that under no circumstances will dolomite react with superphosphate (fertilizer grade) when intimately mixed to form either tricalcium or trimagnesium phosphate.

There is no reaction between $CaCO_3$ and metasodium pyrophosphate, but a slow reaction will occur with $CaCO_3$ and Na_2HPO_4 with vigorous agitation at room temperature. $CaCO_3$ will slowly decompose $(NH_4)_2HPO_4$ by liberating the ammonia due to its alkalinity.

Experiments in mixing colloidal suspensions of finely micronized $CaCO_3$ with tricalcium phosphate indicate no chemical reaction, but the mixture of these two salts yields a more stable colloidal solution than $Ca_3(PO_4)_2$ alone.

Dilute sodium hexametaphosphate solutions exert a pronounced dispersion effect on a $CaCO_3$ precipitate or solid.

Ammonia. $CaCO_3$ is peculiarly sensitive to ammonia compounds for such a material of relatively low activity. It hydrolyzes ammonium sulfate or ammonium chloride readily to form ammonium carbonate with either calcium sulfate or calcium chloride as a co-product.

Thermal Dissociation. *This is the most important chemical property of limestone.* All carbonate rocks dissociate at high temperature, forming oxides of calcium and magnesium through the expulsion of CO_2 gas. To avoid being repetitious this will be described fully in Chapters 6 and 8.

SELECTED BIBLIOGRAPHY

1 F. Pettijohn, *Sedimentary Rocks,* Harper & Row, 660 pp., 2nd Ed. (1957).

2. W. Krumbein & L. Sloss, *Stratigraphy and Sedimentation,* Freeman, 479 pp. (1951).

3. V. Clee, *Bibliography on Dolomite,* Nat'l Res. Council, Washington (1950).
4. S. Colby, "Occurrences and Uses of Dolomite in U. S.," *I. C. 7192,* 21 pp. (1941).
5. O. Bowles, *Limestone and Dolomite,* U. S. Bu. mines I. C. 7738, 79 pp. (1956).
6. D. Graf and J. Lamar, "Properties of Ca and Mg Carbonates and Their Bearing on Some Uses of Carbonate Rocks," Econ. Geol., **50,** pp. 639–713 (1955).
7. J. Gilson et al., *Industrial Minerals & Rocks,* A.I.M.E. publ., pp. 123–201 (1960).
8. G. Gazdik and K. Tagg, *Annotated Bibliography of High Calcium Limestones in the U. S.,* U. S. Geological Surv. *Bull. 1019-1* (1957).
9. M. Goudge, "Limestones in Geology of Canadian Industrial Mineral Deposits," *Sixth Comm. Mining & Metall. Cong. Proc.,* pp. 144–158 (1957).
10. P. Cotter, "Crushed Stone," *1963 Minerals Yearbook,* U. S. Bu. Mines.
11. W. Keller, A. Klemme, and E. Pickett, "Survey of the Chemical Composition of Rock Layers in an Agricultural Limestone," Econ. Geol. **45,** pp. 461–469 (1950).
12. K. Landes, *Metallurgical Limestone Reserves in the U. S.,* Nat'l Lime Assn. Bull., 30 pp. (1963).
13. F. Birch, J. Schairer, and H. Spicer, *Handbook of Physical Constants, Geol. Soc. Am.,* Special Paper 36, 2nd Ed. (1950).
14. S. Windes, "Physical Properties of Mine Rock," U. S. Bu. Mines *R. I. 4459 & 4727* (1949 and 1950, resp.).
15. B. Blair, "Physical Properties of Mine Rock," U. S. Bu. Mines *R. I. 5130 & 5244* (1955 and 1956 resp.).
16. R. Mayer and R. Stowe, *Physical Characterization of Limestone & Lime,* Azbe Award #4A, Nat'l Lime Assn. (1964).
17. D. Kessler and W. Sligh, U. S. Bu. of Stand., *Tech. Paper 349* (Pt. of Vol. 21) pp. 497–590 (1927).
18. Nagaoka, Phil. Mag. (5) **50,** p. 53 (1900).
19. Martens, Ann. Phys. (4) **6,** p. 603 (1901).
20. Nelson, Phys. Rev. (2) **18,** p. 113 (1921).
21. H. Bäckström, J. Am. Chem. Soc., **47,** pp. 2432, 2443 (1925).
22. J. Murray, "Specific Heat Data for Evaluation of Lime Kiln Performance," *Rock Prods,* p. 148 (Aug. 1947).
23. J. Johnston, J. Am. Chem. Soc., **37,** p. 2001 (1915).
24. G. Frear and J. Johnston, J. Am. Chem. Soc., **51,** p. 2080 (1929).
25. H. Bäckström, Z. Phys. Chem., **97,** p. 179 (1921).
26. A. Seidell, *Solubilities of Inorganic & Metal Organic Compounds,* Van Nostrand, pp. 264–276, 951–956, Vol. 1, 3rd Ed. (1940).
27. Kline, Discussion (1923 research) by Frear & Johnston (ref. #24 above).
28. O. Haehnel, J. Prakt. Chem. (2) **107,** p. 165–176 (1924).
29. O. Haehnel, J. Prakt. Chem. (2) **108,** p. 61–74 and 187–193 (1924).
30. A. Mitchell, J. Chem. Soc., **123,** pp. 1887–1904 (1923).
31. J. Johnston, H. Merwin, and E. Williamson, Am. J. Sci. (4) **41,** p. 473 (1916).
32. H. Woodward, Map of U. S. limestone deposits. *U. S. Bu. Mines Bull. 395* (1937).

CHAPTER THREE

Limestone Exploration

and Extraction

Exploration Criteria

Some limestone operations have been doomed to failure at the outset because of haphazard or incomplete prospecting methods. It should be axiomatic that before money is invested in limestone property and processing equipment that the owner should be certain of both the *quality* and *quantity* of the stone that he plans to extract and process. This means *"proving"* the deposit. Attempts to economize on exploration costs, of $20,000 to $25,000, have jeopardized and eroded ultimate multimillion dollar investments. Thus, careful, methodical exploration methods can be likened to insurance; they remove much of the speculation inherent in this mining venture. There are three basic decisions that should be concluded positively after prospecting and before any investment is made.

1. *Is there a sufficient reserve of stone to justify further investment?* Five to fifteen years supply of stone is insufficient. There should be proven reserves of at least twenty-five years—and ideally forty or more years, except in very unusual circumstances. A generous allowance of time to amortize the capital investment for this low-priced, low-profit-margin commodity is judicious.

2. *Does the desired or required quality of stone exist in the deposit?* Is the quality of sufficient abundance and uniformity? Oftentimes deposits have strata of highly desirable stone but adjacent to or even interspersed in the good vein is poor-quality stone or shale that is costly to remove and must be wasted. Selective quarrying and/or stone beneficiation methods are costly.

3. *Is the deposit economical to exploit?* There are excellent deposits of stone in the far western United States that may never be developed.

They may be located forty miles from the nearest railroad and are located too distant from existing markets. The cost of constructing forty miles of rail track through rugged, mountainous terrain is prohibitive. Even if this obstacle is overcome the high shipping costs to distant markets is insurmountable. Then there are favorably located deposits of good-quality stone that may never be developed since the extraction of the stone is too arduous and costly. Nature, in forming these deposits, has simply made them economically inaccessible. Such deposits might have too thick an overburden or have beds of stone that dip at weird angles, frequent cavities and faults, inclusions of shale that resist economical beneficiation, etc. In short, the high cost of extraction renders such deposits hopeless to exploit.

OBJECT OF SEARCH

Cost of exploration can be minimized if the prospector knows exactly the type of limestone he desires. With this decision made, the prospecting can proceed objectively with a minimum of wasted effort and time. If the objective is to obtain chemical or metallurgical grades of stone or stone for lime manufacture, then the chemical characteristics of the stone are paramount. Low silica, alumina, sulfur, and iron in particular are desired. If a high magnesia content is necessary, then the search can be narrowed to dolomitic and magnesian stones. If stone is desired for concrete or road metal, then the physical characteristics (hardness, soundness, low porosity) are predominant. If dimension stone is the object, uniform texture, attractive color, strata of stone free of cracks, and susceptibility to polishing are essential.

HOW TO PROCEED

Since exploration of mineral deposits is a highly specialized field, an experienced geologist is indicated in locating deposits. However, before making field trips it is advisable first to contact the state geological survey in the state or states of particular interest. Most states have such departments, which have usually minutely detailed maps of all of the mineral resources of their state and have at least roughly classified the limestone by origin, type, and occasionally even chemical and physical properties.[1] In addition, the U. S. Geological Survey possesses a vast amount of data on limestone formations and deposits, and in effect they have coordinated much of the information developed by states.[2-4] An initial study of this readily available information can prevent wild goose chases and can pinpoint the prospecting targets.

Fig. 3-1. Small abandoned limestone quarry, partially flooded, in rugged terrain.

Abandoned quarries should not be overlooked (Fig. 3-1). Some often contain substantial reserves of good stone. Possibly the former operator was forced to cease production due to some extenuating circumstances or he was one of those who was too "penny-wise and pound foolish" in proving his deposit. Encountering a large submarginal bed of stone might easily discourage a quarryman to the point of abandoning his enterprise yet the next adjacent bed might be of excellent quality.

When stone is exposed in bare outcrop or in an abandoned quarry, sampling for quality is, of course, simple. Even a few random samples will often reveal with reasonable accuracy the quality of the whole bed. Most sedimentary rocks, as a rule, are fairly constant in composition through the *same* bed or zone; the greatest variations occur in passing from one bed to another. The important thing is to sample *all* beds. Facilitating such sampling would be a cliff or an escarpment along a river or ravine in which the exposed cross section permits sampling of various levels and numerous beds. When the stone is unexposed, systematic surface core drilling is necessary to provide the needed quality data.

CORING

When a high-quality chemical grade is desired, core drilling[4, 5] is performed at regular intervals of approximately every 3 to 20 feet. The space between drillings is usually determined by the uniformity of the stone. In flat lying beds of uniform thickness and composition and where chemical quality is not so important, spacings of 100 to 500 feet may be warranted. Lack of uniformity justifies closer spacing. Where the rock formation is folded or dips at steep angles, closer spacing is necessary. Detailed maps of the deposit under study should be made as sampling is begun. In addition to plotting the position, thickness, and slopes of the bed on the map, each core drilling should be plotted and numbered. Such maps are of great permanent value to a company. Usually too the cores, even after testing, are permanently stored, carefully numbered to conform with the maps. A *core drill* is preferred over a *churn drill* since the former cuts out precise cylindrical samples that are suitable for both chemical and physical tests and it may be employed at any angle. Churn drills can only be used vertically. The diamond-tipped core drills, which are the most efficient, consist of a rotating steel drum with black diamonds set in the lower edge. Usually the cylinder produced by a core drill is 2 to 3 inches in diameter and 6 to 12 inches in length. Shot drills and drills with special alloy steels can be used in softer stones.

In flat-bedded deposits drill holes are projected perpendicularly into the rock. With steeply dipping beds inclined drilling at right angles to the bed is usually preferred. The cost of core drilling, of course, depends on the hardness of the rock, but will range around $2 to $5 per foot.

Coring is also valuable in ascertaining the thickness and character of the overburden, a vital consideration on whether the deposit can be stripped economically.

The size of the deposit in tons of limestone reserves can be calculated after coring and testing. The density of stone can be calculated from the samples or estimated at 160 lb./cu. ft. From the map the net length, width, and depth of the usable deposit, deducting for the inclusions of poor stone, shale, or chert, are multiplied, and this total is again multiplied by its weight/cu. ft. This result is then divided by 2000 to obtain the tonnage reserve. $\left(\dfrac{L \times W \times D \times \text{Wt./cu. ft.}}{2000}\right)$

Then the final question is answered. Does the deposit have at least *twenty-five times as much stone* as the projected average annual requirement over the next twenty-five years? If it does not, then in all probability the search should continue.

Extraction of Limestone

From a detailed map of the deposit it is now possible to plan a long-range systematic program of stone extraction. (1) Selection of the most strategic section of the deposit to open the quarry. (2) Depending on the thickness and uniformity of the various beds of stone, determination of the depth and length of the quarry face. (3) Whether quarrying should be pursued on two or more levels instead of from one large face. (4) Contingent plans for possible future underground mining. (5) Selection of an efficient area to haul overburden and waste stone which will not interfere with future quarrying or mining or the stone-processing and shipping operations.

If possible, such detailed plans should provide some latitude for possible subsequent revisions that might be necessitated due to a miscalculation, act of God, change in markets or use of stone, etc. All mining ventures have varying inherent risks, and limestone is no exception; hence, the importance of mobility in quarrying or mine layouts.

STRIPPING

The removal of overburden in the form of topsoil, clay, silt, sand, gravel, or loose rock from a deposit is called "stripping," the first basic step in stone extraction. Only rarely are deposits found in bare outcrop that require little or no stripping. In most cases stripping is regarded as a necessary phase of quarrying. It often presents a difficult and costly operation where the overburden is thick or irregular. Frequently, limestone and marble deposits are difficult to strip because of the eroding effect of slow-percolating groundwater over thousands of years, which seeps into interstices and joints in and between beds and causes considerable irregularities in the rock surface. Tortuous cavities that are filled with siliceous material and even steep pinnacles of rock may result. Such irregular surfaces are particularly difficult to strip when chemical grades of stone are required and clean stripping is necessary to prevent the entrapped seams or cavities of clay from degradating the quality of the stone. In contrast, these clay pockets offer little or no problem in quarrying stone for cement manufacture since siliceous material is a requisite in the portland cement kiln feed. In fact, most quarries for cement utilize all or some of their overburden in preparing their cement raw-material conglomeration. The depths of overburden may range from a few inches to ten to twenty inches and in extreme cases up to forty or fifty feet. In some areas the overburden may be composed of "rotten" rock and hardpanned clay that is so hard that blasting is required for removal. If stripping becomes too arduous and expensive, it may be more economical to mine the stone underground to avoid this operation.

Stripping Methods. Probably no quarry process is as variable as stripping; this is because of the widely divergent conditions confronting the operator. Most operations employ a combination of the following methods:

1. The *hydraulic method* consists of washing the overburden away with a stream of water under pressure. Although this is theoretically the simplest method, the conditions surrounding its use can be exacting. An abundant supply of water is paramount, since about ten tons of water are required to remove one ton of overburden. By building settling basins, however, the water can be repeatedly recirculated for reuse. To avoid stream contamination this is almost essential. It is also necessary to apply the water away from the quarry face, preferably down an incline, to avoid flooding the quarry floor. Generally it is only efficient in washing away friable types of soils, sands, and silts. Hardpanned clays, rock rubble, and boulders usually resist the force of the water. Hydraulicking is particularly effective in washing out cavities of clay and sand that may honeycomb the surface of irregular deposits. Hydraulic equipment needed is a high-capacity pump, a pipeline, a mounted nozzle, an auxiliary dredging pump, and power.

2. The *dragline scraper* is employed as an excavator and is effective in scooping out clay pockets from cavities. It lacks flexibility in lateral movement, and there should be a convenient dumping ground nearby for the overburden.

Chemistry and Technology of Lime and Limestone

3. The *power shovel* is commonly employed, both steam, diesel and electric types. Revolving types are most productive. This equipment will handle all material with equal facility. However, large boulders and rock projections are usually shattered by blasting in advance of the shovel. Such shovels are of various capacities of 0.5 to 2.5 cu. yd. Shovels with smaller buckets of 0.5 to 0.75 cu. yd. may be used in clay pockets. Where the overburden is only three feet or less, the shovel is not too efficient.

4. The *tractor-propelled bulldozer* is almost invariably employed to some extent because of its great mobility in lateral movement, usually in conjunction with the shovel. It facilitates the shovel operation by blading off rock-free soils and other loose material into piles. It is ineffective in working close to irregularly surfaced beds and in removing clay from pockets and seams. It is particularly effective in initial stripping and in stripping flat lying beds of fine-grained soils.

5. *Other equipment* employed are tractor excavators (loaders), rubber-wheeled loaders, clamshell buckets with derrick arms, and motor graders.

Labor with picks and shovels is too costly now to be employed. Costs of stripping vary tremendously, depending upon thickness and character of the overburden, type of equipment employed, and skill of the equipment operators. Costs would range from $.15 to $1/yd.,[3] including removal of the overburden to a nearby dump. Table 3-1

Table 3-1. Power shovel productivity in yardage per hour

Class of material	Shovel dipper capacity in cu. yd.								
	.375	.5	75	1	1.25	1.5	1	2	2.5
Moist loam and light sandy clay	85	115	165	205	250	285	320	355	405
Sand and gravel	80	110	155	200	230	270	300	330	390
Good, common earth	70	95	135	175	210	240	270	300	350
Clay: hard & tough	50	75	110	145	180	210	235	265	310
Rock, well blasted	40	60	95	125	155	180	205	230	275
Common soil, with rocks and roots	30	50	80	105	130	155	180	200	245
Clay, wet and sticky	25	40	70	95	120	145	165	185	230
Rock, poorly blasted	15	25	50	75	95	115	140	160	195

Above values assume: (1) suitable depth of shovel cut for maximum effect; (2) no delays in operation; (3) 90° shovel swing; (4) all material loaded into hauling units. (Courtesy of Power Crane and Shovel Association.)

shows productivity of revolving shovels by size for different materials.

DISPOSAL OF OVERBURDEN

A common failing is to accumulate an overburden pile too close to the quarry operation as a temporary expedient. Ultimately quarries must be widened. In cases where beds are narrow and steeply inclined, they must necessarily be expanded laterally at possibly a rapid rate. Thus, if the waste dump is in the path of the logical direction of the quarry development, it must again be transported to another more distant site. Consequently, proper initial quarry planning should anticipate this contingency and plan a dump site a sufficient distance away that is not located on good quarriable stone and will not necessitate costly rehandling of the wastes. Furthermore, if the quarry excavation is located too close to the spoil banks, a hazard exists in the possibility of rock slides that might be disturbed by blasting.

In extreme cases it might be necessary to haul these wastes a considerable distance, in which case an efficient truck-haulage program will have to be devised. In some cases land developers and contractors will haul the material away free in order to fill in nearby swamps or ravines in rehabilitating land for agricultural or other useful purposes. If the waste contains considerable broken rock and not too much fine-grained soil, it might find usage in nearby construction projects, such as dams, road bases or fills, and railroad beds. It may prove useful as fill in leveling the grades on the plant's own roads. A network of serviceable roads is essential in modern quarry operations, both in transporting stone within the plant and quarry confine and from plant access roads to main highway arteries.

QUARRY LAYOUT

The shape and depth of the strata of stone and the general topographic terrain have a decided influence on the quarry layout.[1, 4, 7] There are two fundamental types of open quarries with many modifications. These are the open-shelf and open-pit quarries.

Open Shelf. A ledge of desirable stone may exist bare or unexposed on the side of a hill above the level of the surrounding terrain. By literally quarrying into the hillside, the quarry floor may be little, if any, lower than the adjacent land surface. This open-shelf quarry is usually the simplest, lowest-cost type, providing ready access into and transport of stone from the quarry. Drainage is no problem, and there is no quarry pumping expense. Usually any overburden is very thin. However, such deposits of high-grade stone are relatively rare.

Open Pit. In most cases an open-pit quarry[8, 10] must be developed in which the quarry floor is well below (50 to 200 ft.) the natural grade of the land (Fig. 3-2). The depth of quarrying naturally depends largely on the thickness of the strata of good stone. If the bed is thin and flat or at only a slight angle, which commonly occurs in the Middle West, the pit must be enlarged laterally. An extreme example of this type of deposit is an old, high-producing deposit in Marblehead, Ohio, in which a 22-foot horizontal bed of stone with a thin veneer of overburden has been stripped and quarried over an area of over 1.5 square miles. While this type of operation is low-cost, land acquisition costs are generally much higher because of the extensive area of operation.

In contrast, if the beds are deeper and more narrow, the open pit must be much deeper and is expanded laterally at a much slower rate. More complicated quarrying is necessary when the beds are folded and exist in steep angles, sometimes being practically vertical.[9] This type of quarry frequently occurs in the Appalachian mountain section of the eastern United States. When such beds are thick, deep quarrying is necessarily pursued. If these angled beds are thin, any lateral extension must move in the direction of the outcrop. With tilted beds in deep quarries, loose overhanging rock presents a peril, and removal of the overhang becomes progressively more expensive as the pit is deepened. A narrow working face in steeply dipping beds of limited thickness is cumbersome to work and rather unproductive. This condition can be alleviated by cutting *benches* into the face which permits quarrying at several levels and provides additional active working faces (Fig. 3-3).

Fig. 3-2. Panorama of typical open quarry with stone-processing and lime plant in background.

Fig. 3-3. Multilevel open quarry.

Some massive stone formations have laminated beds of inconsistent quality. This leads to *selective quarrying* in which one bed of high purity may be used for chemical stone and the next bed of lower purity, but adequate strength, sold as road stone. Different benches are established to quarry the stone separately from different levels.

MINING LAYOUT

In cases of thick overburden or complex, narrow, and inclined beds of considerable depth, underground mining[11, 12] may be necessary for greater efficiency and even safety, although it is employed in only about 9% of the limestone operations in the United States. Generally, mining stone is much more expensive than open-pit quarrying. Yet there are some mines that are most efficient and relatively low-cost, depending upon circumstances. In fact, there is a slightly increasing trend to go underground in the United States, usually motivated by the demand for purer limestone.

The advantages in mining are the following:

1. Stripping is eliminated.
2. Generally better chemical quality is obtained, since the stone is not contaminated with overburden.

3. Year-round operation is possible, regardless of rain and snow. As a result, a smaller inventory of stone can be maintained.

The principal disadvantages of mining are enumerated as follows:

1. Drilling and blasting are more costly because of the greater precision required in maintaining the proper roof height and correct spacing of pillars. Big blasts are unsafe.

2. Because of different blasting practices, the proportion of stone fines is usually increased.

3. At least 20 to 25% of the stone is unusable because it must be retained to form pillars and ceilings for roof support.

There are two basic types of mining with various modifications:

The conventional *room and pillar* method[13] consists of a horizontal or slightly inclined tunnel from the outcrop or the side of an open pit into the bed. The tunnel is widened by precision blasting in fanlike fashion systematically in several directions. Thick supporting pillars of the stone are preserved regularly at 25 or 35 ft. centers in "checkerboard" fashion. The pillars usually average about 8 ft. in diameter, and the rooms thus formed usually have roofs at 25 ft. Drilling and blasting can proceed simultaneously on faces in rooms that are widely separated. In large operations a network of roads between the pillars results in the mined-out areas offering fast truck transit of the stone to the primary crusher located at or outside the main tunnel exit.

This method is most successfully employed at low cost by a number of companies that have mined into a steep hillside at grade level. The majority of such mining operations, however, are below grade level. Usually such mines are rather shallow—only 25 to 150 ft. in depth.

Stimulating increased interest in limestone mining has been the success of several companies in renting their mined-out areas as strategic bombproof warehouses for storing valuable government and industry records, Office of Civilian Defense medical supplies, industrial materials, etc. Further advantages over conventional warehouses are extraordinary year-round uniformity in temperature and humidity. The pioneer of this lucrative quarry by-product business is the Southwest Lime Company of Neosho, Missouri; it has converted about one million square feet of space into a much-sought-after warehouse.[14]

The other method is called *"stope mining,"*[16, 17] which consists of a vertical or steeply inclined shaft down into the bed (Fig. 3-4). The simplest stope method is the single-breast stope. The tunnel is worked out in all directions on a steadily enlarged circumference, with pillars for roof support retained at irregular intervals. In a more complicated stope-mining operation in Bellefonte, Pennsylvania, the central vertical shaft is cut quite deep—as low as 750 ft. below ground level. However, at different levels horizontal stopes have been tunneled at different directions from the central shaft. It is from these narrow stopes that stone is obtained by precision blasting. Mining proceeds simultaneously at different levels, and stone is transported by small rail cars, reminiscent of coal mining, into the central shaft, where the primary crusher is located. The deepest limestone mine is located at Barberton, Ohio and is over 2300 ft. deep.[15] Only a small percentage of the total stone in the bed can be extracted with this method.

Naturally this scheme of mining is considerably more expensive than most room-and-pillar mines. The high cost can only be justified by the need to obtain high purity, and such stone is usually not competitive at the market place. Therefore, most stope miners will

Fig. 3-4. Stope miner; chemical-grade limestone deposit.

consume the stone themselves for making higher-priced products, like chemical lime.

DRILLING

Since World War II equipment for primary drilling has undergone quite a revolutionary change.[7] Before the war, churn (well) drills, steam-driven piston, and hand-operated air hammer drills were the vogue. However, in most large modern quarries these drills have now been frequently replaced by much larger and higher-speed *rotary*[18] and *percussion*[19] drilling equipment that cuts through rock with large-diameter holes of 6 to 12 in. and depths of 150 ft. at a production rate of four to six times the former equipment. Popularizing this equipment has been the advent of new lightweight, portable rotary air compressors, driven by diesel engines that supply the energy for these drills. The engines are attached to or pulled along by the drill carriage or they are installed on the same mounting as the drill. Even air-motor drives for tractor mounts are being employed for drilling rigs. Stone cuttings are rapidly removed by a high volume of air, thus enabling high drilling speeds to be attained. Improved detachable drill bits of tungsten carbide or special case-hardened steel permit the same diameter of hole to be maintained in deep holes even in hard granite deposits, a failing of some of the former equipment. Generally larger diameter holes are obtained with the rotary than the percussion drill.

However, with some rock better fragmentation from blasting is achieved with small diameter holes (2 to 4 in.) spaced closer together than with the larger diameter holes. Limestone *frequently* varies greatly in the way it fractures, and this characteristic plus the size of stone that is desired requires a certain amount of trial and error in drilling and blasting in order to achieve optimum results at the lowest cost. In any event, coordination of the size of drill holes with the amount and type of explosive is paramount. One fact to remember is that the volume of the drill hole varies as the square of its diameter. This means that a 6 in. diameter hole of a given depth has nine times the volume of a 2 in. hole of the same depth and can accommodate nine times more explosive. There is a general preference for small diameter holes with the narrow ledges (or benches) and for large diameter holes with the thick benches. The small diameter holes must be spaced much closer together.

The greatly improved productivity of these new large drilling rigs is a major factor in counteracting other inflationary costs in quarrying in spite of their much higher capital cost of $100,000 to $200,000 each. There are still many miscellaneous uses for the former smaller drills but they are used now mostly for *secondary drilling* purposes, such as removing rock pinnacles from the quarry surface, removing rock overhanging the face, assisting in the stripping, and generally for drilling shallow holes. Doubtless there will always be a need for pneumatic hammers and well drills for such purposes. The latter are employed for larger holes than the former. Generally, drilling costs increase as the hardness and abrasiveness of the stone increases.

BLASTING

The initial shattering of a burden of stone from a quarry face is called *primary blasting*. Any further blasting that is required to obtain the desired degree of fragmentation in the rock is *secondary blasting*. If the primary blasting is accomplished with optimum efficiency, usually as a result of considerable experimentation in a particular quarry or mine, there is usually no need for secondary blasting. However, if the initial charge is miscalculated or misfires, secondary blasting is always required to topple down to the quarry floor the remaining rock face that is intact in probably a badly fractured, dangerous condition. So it is prudent to plan primary blasts carefully even though drilling and explosive costs as a consequence are seemingly increased.

Blasting has undergone sweeping changes in the United States since 1955, both in materials and methods. Because of much lower explosive materials costs, *fertilizer grades of ammonium nitrate*[20,21] with fuel oil have largely replaced dynamite, which previously was the most commonly used explosive.

AMMONIUM NITRATE. This chemical, which is employed as pellets, prills, or granules, is not an explosive by itself. It can be handled and stored almost as safely as any fertilizer material. However, when fuel oil (preferably No. 2 diesel grade) is added, or any other sensitizer, it must be handled with the same caution as dynamite. There are several methods of using ammonium nitrate:

1. After nitrate is poured into a hole by hand, fuel oil is added at an approximate proportion of 3 to 6% of the nitrate by weight.

2. The nitrate is added dry by hand from a bag after the fuel oil is added in about the same proportion indicated in 1 above (Fig. 3-5).

3. When nitrate prills are added manually to the hole, the mixture of nitrate and fuel oil is permitted to age for two days before igniting.

Fig. 3-5. Pouring fertilizer grade ammonium nitrate into 6-inch drill hole preparatory to AN/FO blasting. (Courtesy, Spencer Chemical Division, Gulf Oil Corp.)

4. Specialized pneumatic explosive loaders have been perfected to feed the nitrate-fuel oil mixture into drill holes through hoses inserted into the blast holes by air pressure. Advocates of this practice contend that loading holes with the explosive mixture is faster (two materials are added at one time) and that due to air pressure the hole is filled more compactly than by hand (no hand tamping of the mixture is necessary).

Generally it is observed that the smaller-sized nitrate prills have more blasting power than larger sizes. This blasting technique is still so relatively new that doubtless improved applications will be developed as more experience is gained. However, among the many postwar developments for improved processes in stone extraction and processing, this one development has been, at least percentage-wise, the greatest production cost-saver of all process improvements. In some cases quarry blasting costs have been cut by half the cost of the traditional dynamite method. This method was first introduced in 1955, and seemingly in only one or two years the limestone industry was converted to it. In the past carbon black was used as a sensitizer with ammonium nitrate, but it has proven to be not nearly as effective or economical as plain fuel oil.

DELAYED ACTION. There is steadily increasing use of delayed-action blasting,[5] millisecond delays of 0.015 to 0.040 seconds between drill holes, since seismic vibrations in ground movement due to the blast are greatly minimized. With this method seismographs have recorded no more vibration with eight holes than with one hole without delayed action. With habitation encroaching closer to quarries, reducing seismic forces to a minimum is essential to avoid damage suits from nearby home and property owners. In fact, seismograph records of blasts have proven invaluable to quarries in court cases on alleged damage claims and have disproven spurious claims. This method permits use of larger holes and heavier burden blasts than would otherwise be feasible and safe. In addition, with many companies improved fragmentation with less secondary blasting has resulted from this method. Again experimentation is required to determine the optimum millisecond delay interval to be employed along with the hole spacing and type and strength of explosive. The delays are obtained by using electric caps or primacord connectors, which are purchased with various time intervals.

The traditional stone detonation methods without delayed action are described briefly as follows:[23]

"Upper" or "blanket" shooting involves a second blast at the same section of a face, but before the broken stone on the quarry floor has been loaded. The theory is that the bank of loose stone prevents the succeeding blast from scattering on the floor, thereby facilitating loading of the stone with the shovel and with less clean up. Some contend that better fragmentation also results.

Snake-hole blasting is opposite from the above in that the shooting is done after the preceding shot has been loaded and removed. It is more applicable with long quarry faces where the shovel for loading is alternated at different sections of the quarry floor. In this way, drilling, blasting, and loading proceed simultaneously.

Tunnel ("gopher-hole") blasting is practiced with certain types of hard rock and where high faces exist. Small tunnels are driven into the face parallel to the quarry floor; holes are then drilled parallel to the face in both directions. The resulting blast produces a wedge-shaped cut into the face with the upper stone lifted slightly from the detonation before it cascades on the lower stone. To prevent possible overhangs in deep quarries with this method, holes are also drilled on top into the quarry surface and are shot simultaneously with the tunnel blast.

Next to ammonium nitrate *ammonium dynamite* probably is most commonly used. It is safer, easier to handle, less costly, and has a lower freezing point than nitroglycerine dynamite. However, it is soluble in water and cannot be used in wet holes. In such cases gelatin dynamite or ammonia gelatin are employed, since they are insoluble. Black powder, nitrostarch, and liquid oxygen explosive (LOX) have limited, intermittent use in stone quarries. Generally with limestone and softer rocks lower-strength (velocity) explosives are employed than with harder stone, like granite—30 to 60% strength depending on extent of blast, hardness and fracturing qualities of stone, and height of face. Lower velocity detonations are also achieved by employing explosives with coarser particle sizes. Often explosives of greater strength are injected into the bottom of deep holes to achieve better fragmentation at the toe. The magnitude of the consumption of explosives is about 2 to 6 tons of stone per pound of explosive, depending on the several variables discussed. Explosive manufacturers provide excellent technical service on the most efficient use of their products.

Secondary Breaking. Even in highly successful large primary blasts, there are often some large massive blocks or boulders that resist fragmentation. These must be broken since they clutter up the quarry floor and impede the efficiency of the shovel in the loading operation. Before World War II these boulders were broken by pneumatic hammer drilling and secondary blasting. This practice has now been largely replaced by the *drop-ball* method.[22]

This consists of dropping a 3.5- to 4-ton forged or cast iron ball from the 60 ft. boom of a crawler-type crane directly onto the boulder. One or two such blows, when dropped accurately, will generally shatter the large block into sizes suitable for the shovel and primary crusher to accommodate. This technique has significantly lowered quarry costs. Alloyed cast iron balls have a very long life, much more than ordinary forged balls.

Loading. *Power shovels* are almost universally employed in loading the broken stone onto trucks or inclined rail-type cars for transport from the quarry or mine floor to the primary crusher. The main difference is the capacity of the shovels. One vital consideration on the size is that it should be coordinated with the size of the primary crusher.

Sizes of shovels vary widely from about .5 to 5 cu. yd. However, the trend is away from the smaller construction shovels (less than 1 cu. yd.) because of the lower payload and greater maintenance required. Diesel, gasoline, and electric shovels are employed, but the electric is generally favored for sizes of 3 cu. ft. and larger. Even when constructed with special alloy steels, attrition on the shovel buckets, particularly the digging teeth, is most severe from abrasive action of the stone, necessitating considerable maintenance. Repointing worn teeth by welding on new manganese steel teeth reduces maintenance.

In mining, ceilings are often too low to permit use of shovels, at least the large-bucket types, so rubber-wheeled front loaders are utilized.

Haulage. Rugged rubber-wheeled *off-highway haulage trucks* have largely replaced small-gauge *rail cars* that are hauled up inclines by cable; this is because they are more productive. In fact, none except small quarries can afford to use the cable car. Capacities of trucks have steadily increased up to 50 to 65-ton payloads, although the average is around 25 to 30 tons. There are several frequently used types. When trucks are also employed for hauling overburden and stone spalls, rear dump trucks are usually preferred. However, side dump bodies predominate on regular haulage to a fixed crusher location in order to avoid time-consuming turning and backing. In some cases trucks are of the semi-trailer or full-trailer mounting type. The era of donkey haulage ended several decades ago in the United States, but it is still employed in a number of foreign countries that are lacking in automotive mechanization.

The majority of large modern stone plants have their primary crusher located above the quarry, but usually in close proximity to it. This means that roads must be built and maintained for the truck haulage (Fig. 3-6). Usually this is accomplished in deep quarries by ramps that hug the quarry face and wind around it until grade level is reached. Usually these ramplike roads are built at a 6 to 10° grade. Steeper inclines will slow the rate of haulage and increase truck maintenance. In large deep multi-level quarries these roads are often rather intricate and require considerable planning for safe, efficient haulage. Primary crushers are also located on the quarry floor. Usually in such cases the stone from the discharge hopper of

27. "Safety Recommendations for Sensitized Ammonium Nitrate Blasting Agents," *U. S. Bu. Mines I.C. 8179* (1963).

28. P. O'Brien, "Seismic Energy from Explosives," *Geophys. J.* **3** (1960), pp. 29–44.

29. Nat. Safety Council, "Ammonium Nitrate—Fuel Mixtures as Blasting Agents," *Data Sheet 536,* (1963).

30. *Rock Products* magazine, Chicago, Ill.—11/58, p. 132; 2/59, pp. 91, 108; 7/59, p. 131; 6/61, p. 92; 2/62, p. 78; 10/63, p. 87; 2/64, pp. 72, 84.

31. *Pit & Quarry,* magazine, Chicago, Ill.—5/58, p. 185; 9/58, p. 122; 5/59, p. 136; 7/59, p. 126; 9/59, p. 88; 11/59, p. 114; 10/60, p. 102; 5/63, p. 95; 12/63, p. 89.

CHAPTER FOUR

Limestone Processing

The need for individualization that has prevailed in exploration for deposits and the extraction of the stone is still just as compelling in the design of a plant for conversion of the stone into marketable or usable products. The same variables exist—the physical characteristics of the stone to be crushed and screened, the chemical requirement (impurity tolerances) that demand conformance, and lastly, and at this juncture the most important requirement, the sizes or gradations of stone that are desired. Thus, a plant with adequate capacity, tailor-made to the particular terrain and sizes and quality of stone required is essential.

There are no existing foolproof, detailed blueprints for such a plant. In fact, if an existing modern plant of highest efficiency was duplicated exactly, it might easily prove to be totally unsuitable at the second location. So in designing a new plant it is prudent to seek advice from engineering consultants specializing in this field and from manufacturers of equipment for stone processing. In view of this situation, the succeeding pages will treat this subject generally with emphasis on practical plant-design considerations and suggestions on avoidance and correction of operating pitfalls.

For the specifics of plant design, past issues of *Pit and Quarry*[1] and *Rock Products* contain numerous valuable "case history" articles on many varied plants, providing specifications (and brand names) of equipment and performance data, and clearly demonstrate how individualized each plant is. For the countless modifications in the basic types of equipment, the most complete information source is the annual *Pit and Quarry Handbook and Purchasing Guide*, published by *Pit and Quarry* in Chicago. Literally almost every modern type and model of equipment now in commercial use in the limestone, lime, and aggregate industries are illustrated and described in detail.

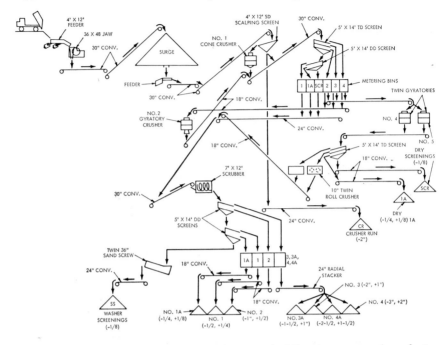

Fig. 4-1. Flow diagram of large, modern, crushed-limestone processing plant, featuring flexibility in sizes and gradations. (Courtesy, *Pit & Quarry.*)

Figs. 4-1 and 4-2 depict flow diagrams of typical, modern stone-processing plants.

Obviously even though numerous ranges of stone gradations may be desired as primary products, necessarily other plus, minus, and possibly intermediate sizes will be accumulated. Ideally these undesirable sizes should be marketed as by-products, the return from which enhances the profits of the primary products. This principle is equally valid whether the stone products are sold or are consumed captively to produce more highly refined products, like whiting, lime, and portland cement. In the event that these submarginal sizes and gradations are unmarketable or unusable, they will have to be stockpiled at some convenient location on the premises, like overburden. Such material is classified by the industry as *spalls* (or spall piles). Again, as in the case of overburden, spall accumulations should be deposited in an area that will not interfere with plant processing and quarrying. A malpractice is to intermingle overburden and spalls together. These piles should be segregated, since subsequently a

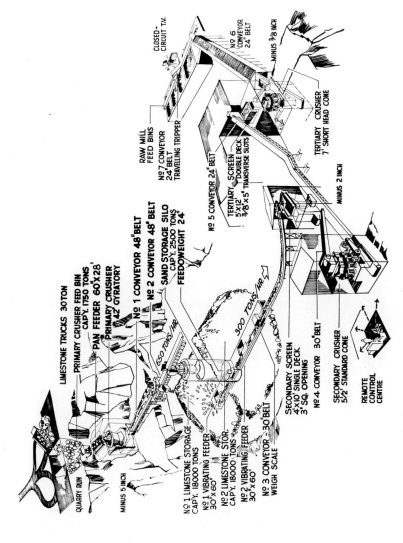

Fig. 4-2. Isometric flow diagram of a limestone crushing operation. (Courtesy, *Pit & Quarry*.)

profitable outlet for this stone may emerge; otherwise this potential value is usually dissipated in a heterogeneous waste pile. The cost of subsequent screening and classifying is often too formidable.

Primary Crushing

The type of primary crusher to employ depends on the hardness and fracturing qualities of the stone and particularly the gradation that is desired in the primary product. The two types of primary crushers that are most frequently employed are the *jaw* and *gyratory,* particularly with moderate to hard rock. These crushers can often be modified for better adaptation to the desired quality and requirements of the stone through adjustments in setting the discharge opening. Primary crushing is much more productive and far less costly than secondary blasting in achieving size reduction. The size of the crusher is usually governed by the maximum size of the stone that flows through the shovel dipper in stone loading, such as for example, stone measuring up to 2 to 3 ft. Proper correlation of the shovel and crusher sizes is important for productivity in order to minimize the number of oversized rocks that may jam up or obstruct the crushing operation. Too many large-sized rocks, even though they may individually pass through the crusher, will tend to interlock, "bridge," and impede the flow of stone through the crusher and necessitate cumbersome hand labor to pry the rocks loose with long metal crowbars. So it is safer to err on the overcapacity side in specifying the crusher's size. In fact, many experienced quarrymen prefer oversized primary crushers to avert jamming delays, in spite of absorbing higher capital equipment and power costs, since less labor and maintenance are necessary.

In plants crushing limestone for portland cement kiln feed flour or other pulverulent purposes, the jaw and gyratory crushers are not preferred since they do not crush as finely as *hammer mills* and other special, related *impact crushers.* But where fines are undesirable and where the primary product is lime kiln feed, coarse aggregate, or fluxstone of a minimum .5 to 1 in. size, jaw or gyratory crushers are virtually economically essential. Hammer mills are usually only practical with soft, nonabrasive stone that does not contain more than 5% quartz. The hammer produces a more cubical-shaped particle than any other crusher.

Jaw Crusher. This crusher is built with various feed openings up to 6.5×8 ft. Although such crushers can be fed with stone directly from side or rear dump trucks, oftentimes they are equipped with a *pan or vibrating feeder* to provide a steady, smooth flow of stone through the crusher and to prevent "choke feeding," which increases the content of fines, and to minimize rock "bridging" delays. Employment of such a feeder permits use of a slightly smaller primary crusher. *Grizzlies* are also optionally employed to remove undersized stone and clay. The jaw crusher tends to produce flatish sizes because of its straight-line crushing surfaces, but this tendency can be reduced with corrugated metal sides, which yield a more cubical-shaped stone. Jaw crushers utilized are the older *Blake* types and newer improvisations with overhead eccentric and single toggle. The latter are preferred since greater capacity is achieved with the same size opening. The discharge end opening of the jaw is adjustable, providing some latitude in accommodating different top-sized stone.

Gyratory Crusher. This primary crusher has nearly the same maximum opening size as the jaw (6×6 ft.), but for the same size it possesses much greater capacity—three to four times as much. However, it is about twice as heavy and requires about twice as much headroom and horsepower to operate. It also represents a greater capital expenditure. Another disadvantage is that when operated at maximum capacity, the gyratory must be geared to close settings. This means that when crushing surfaces become worn, very little adjustment is possible before surfaces require rebuilding or resetting, leading to costly maintenance. But because of its high capacity it is generally preferable to the jaw in very large operations.

When the stone is still too large and difficult to convey by chutes or belt conveyor from the discharge end of the primary crusher directly to a belt conveyor, plants will either feed the material to a secondary crusher or to another feeding hopper in close proximity to the belt. This conserves the life of the conveyor belt from the attrition of heavy falling stone by reducing the distance of drop. This practice is also particularly important if much clay overburden or fines are present, since these fines produce a lubricating effect, causing the stone to slide at high velocity down the chute to the belt. An alternative practice would be use of a grizzly chute that permits the clay and fines to fall on the conveyor belt first, cushioning the fall of the larger stone.

Secondary Crushing

Because of the substantial demand for limestone fines, many plants utilize secondary crushing, if for no other reason than to make their by-product stone salable.

Gyratory crushers are also utilized in secondary crushing, particularly when the top stone size is 6 in. or more.

However, for smaller sized stone of —6 in., a gyratory type with concave crushing surface, *cone crushers,* is preferred because of its higher proportionate capacity, simplicity of adjustment, and low maintenance. It is very effective with hard, abrasive stone. Other secondary and tertiary crushing equipment are the *hammer mill, rod mill,* and specialized *impact breakers.* These latter crushers are not preferred for stone that is very hard and abrasive because of their high maintenance. Double and triple *roll crushers* are not employed nearly as much as in former years because the rolls tend to wear unevenly, creating high maintenance. Hammer mills tend to produce the highest percentage of screening sizes, whereas rod mills are ideal for grinding limestone sand (or sizes of about $\frac{3}{8}$ in. to no. 100 mesh, with about 50% in the no. 15 to 50 mesh range).

Secondary or even *tertiary crushing* (fine grinding or pulverization) is practiced with *ball* or *pebble mills, colloid mill*s that contain revolving plates, and *tube mills.*[3, 4] These mills are particularly useful for finely subdivided *stone* products of no. 100 to 200 mesh, and even finer. For successful fine pulverization with some equipment the stone should be dry. Consequently, porous limestone that may contain 5 to 12% absorbed water may require drying in rotary driers before it can be successfully pulverized. Most of these mills are operated as closed systems in which *air separation* is employed to remove the finest material during the operation. By removing this flour or dust the succeeding pulverization is more effective, since these dust particles tend to cushion the grinding effect on coarser particles.

Other stone-grinding operations thrive best with wet processes, such as the production of limestone sand with rod mills in which the no. 100 mesh sizes are floated off in countercurrents of water at various velocities. So the feasibility of dry or wet processes will vary with the type of milling equipment, the physical and chemical character of the stone, and the desired size distribution. Some preliminary experimentation with different milling equipment in concert with dry or wet processes is highly desirable before a plant is designed.

Micron Sized Limestone

Limestone whiting that competes with chalk derived whiting and precipitated calcium carbonate (described on page 92) requires the most meticulous grinding and milling procedures of all limestone products in order to achieve the exacting micron particle size distributions, shapes, and surface areas that are required. The following techniques are generally applicable to chalk, marble, shell, or limestone whiting, although the grinding of chalk, because of its greater softness, is simpler. Either wet or dry grinding is employed.

Modern plants employ *wet grinding* in continuous ball and pebble mills or conical mills with bowl and cone classifiers. The finely comminuted limestone is floated out of these mills as milky suspensions into a series of settling tanks. After sedimentation the wet sludge is dewatered in vacuum filters and/or thickeners, and the material is dried in rotary or tunnel driers. Any agglomerated material is then easily pulverized. The whole operation resembles a typical chemical operation. Wet grinding is often necessary to achieve high chemical purity and a high degree of whiteness in the stone dust.

Dry grinding is less costly than wet grinding. Ball, pebble, roller, and hammer mills are also utilized. Impact pulverizers with closed-circuit air separation and micronizer mills, which circulate particles at high velocity in a stream of air, are employed in modern plants that strive for maximum fineness. The latter type is claimed to produce material with 99.5 to 100% passing a No. 325 (43-micron) sieve down to 1 micron, with an average particle size of 10 microns.

One possible source of contamination to the stone in these processes would be an iron build-up from the milling equipment (balls, rods, and general abrasion of the iron and steel impacting against the stone). This would be minimized in the wet process, but in the dry process, if rigid limits are placed on iron content, probably only stone of soft to moderate hardness should be attempted.

Conveying

The most prevalent stone-conveying equipment at modern plants is the *rubber belt conveyor* accompanied by *bucket elevators*. Generally the belt conveyor is preferred to receive the discharge from the primary crusher since bucket conveyors tend to wear and deform more readily from the coarse, hard stone than belts.

Most belts are supported on troughing idlers with ball bearing rollers that require very little maintenance. Generally 18° is considered to be the maximum practical incline for crushed limestone; however, improvements in the reinforcement of the belts are permitting slightly higher elevations in some cases. The thickness of the rubber belt varies, depending on the size and abrasiveness of the stone and volume conveyed. Smooth, uniform loading of the belt permits use of narrower belts and/or greater capacity through operating belts at higher speeds. Minimizing spillage is desirable.

Fig. 4-3. Poidometer discharge of limestone to incline belt conveyor for kiln feed. Trough underneath belt collects dripping from wet stone.

For vertical or steep inclines enclosed gravity or centrifugal discharge bucket elevators are usually employed, particularly for medium to fine sized stone. This type has largely replaced the *skip hoist* conveyor in most plants.

To solve individual conveying problems other specialized conveyors are used in part by some plants, such as *oscillating trough* conveyors, *screw* conveyors, *apron* conveyors, *drag* conveyors, *elevating wheels, pointed bucket carriers, air slides,* and *air-blowing hydraulic systems.* The apron and vibrating trough conveyors are used in place of rubber belts for conveying hot limestone from rotary dryers. The air slides and air-blowing mechanisms are employed increasingly in loading or unloading of stone. A conveying system is displayed in Fig. 4-3.

Screening

Vibrating screens of different types have almost completely replaced the *rotary* or *trommel screens* that prevailed in most stone plants in the 1920's and most of the 1930's. The latter screens are still employed in some plants as scalping screens for large stone, but invariably the modern trademark of an efficient stone plant features vibratory screens almost exclusively for *all* sizes of stone. (At first it was conjectured that vibrating screens were only efficient in classifying small fractions.) *Shaking screens* and *grizzlies* have also been largely replaced with vibrating equipment. However, where accurate sizing is not important, bar or live-roll grizzlies are still useful in scalping off fines and retaining large stone of 3 to 10 in.

Vibrating Screens. This type, size, and frequency of vibration is determined by the size gradation of the stone to be screened. The three basic types of vibrating screen are:

1. Mechanical vibration with inclined screen surfaces.
2. Mechanical vibration with horizontal screen surfaces.
3. Electromagnetic vibratory types.

Inclined surface screens are probably the most widely used, but horizontal screens are ideal for areas of low headroom. Vibration may be imparted mechanically to the screen frame or directly to the screen cloth. Hardened steel punched plate is often used in place of woven wire or square mesh screens where large sizes of hard, abrasive stone are classified. It possesses much longer life, but produces a very high decibel count—an *extremely* noisy operation. Wire-cloth screens are generally more satisfactory for small sizes, particularly with moderate to soft stone.

Based on the volume and degree of attrition of the stone, the gauge of wire cloth is adjusted accordingly. Vibrating screens are supplied in single, double, triple, and quadruple connecting decks, which when installed resemble compartments, and with diverse specified sieve openings conforming to a plant's individual requirements.

The only operating problem with these screens of consequence would be in fine screen meshs below $\frac{1}{4}$ in. If the stone is even slightly damp or contains clay, it tends to clog ("blind") the screen openings. Corrective measures used to alleviate this problem would be rotary driers; greater care in quarrying and/or at the primary crusher to reduce clay or soil content; heating the wire cloth, since fines do not cling to heated metal as readily; employing rubber balls in the compartment below which bounce up against the screen and dislodge clogged openings; applying sprays of water against the screen which may also serve to rinse some of the clay and stone dust from the larger stone particles.

The stone and equipment industries generally subscribe to the nominal dimensions, permissible variations, and limits for woven wire-cloth screens contained in ASTM specification E-11-58T.

Washing

Before World War II there was relatively little washing of stone. However, in recent years washing of stone has increased due to more *stringent* quality requirements of both the chemical and construction industries. Contents of clay, shale, and fine sandy silt can be appreciably reduced by various washing processes. Yet this is usually a relatively costly operation. So other methods of upgrading stone quality should be carefully analyzed before installing a washing operation. It is possible that selective quarrying, partial hand selection before or after the primary crusher, scalping off of the clay with grizzlies or other equipment, assuming stone fines can be wasted, may prove to be more economical solutions to this problem than washing. However, often the deposit is so honeycombed with clay that washing must be pursued—or it may simply be more efficient with narrow gradations, like stone sand, where fines can be ejected more easily.

The following are the most common washing practices:

Scrubbing. Stone, containing clay or stone dust that is not desired, can be washed by agitating or rubbing the stone particles against each other in water with blades or paddles in conical, horizontal, revolving scrubbers. Suspended chains are sometimes used to intensify scrubbing action.

Water Jets or Sprays. The force of a jet of water applied to stone will soften even most "hardpanned" clay, washing it off of the stone. Such water jets can be trained on the stone as it is conveyed over screens.

Softening and Disintegration. A more pronounced scrubbing action is obtained with *log washers*. Stone is fed into the lower end of a slightly inclined trough, equipped with paddles. Impelled by the paddles the stone moves up the incline and is discharged at the other end. A countercurrent of water floats off the clay and stone dust which is removed by the churning action of paddles that promote intensive friction of the stone fractions rubbing against each other.

However, the most elaborate washing operations are performed in Europe. Some plants in West Germany and England actually use *flotation processes* in which washing occupies a role as important (and even more expensive) in stone processing as crushing and screening. In some plants all of the stone of more than $\frac{1}{2}$ in., on leaving the primary and secondary crushers, is introduced into the washing system and from there into driers and then into the screen classifiers. Such intricate plants are necessary since large deposits of high-grade limestone are not nearly as abundant as in the United States and Canada. Such stone does not have intermittent clay seams, which might be scalped off by dry methods; the deposits literally contain a labyrinth of clay and to such an extent that unwashed stone might contain only 85% average carbonate content. By washing, the carbonate content is beneficiated to 95 to 97%. Flotation, as such, is rarely practiced in the United States.

Heavy-media Separation

Another wet method for chemical beneficiation of stone is the heavy-media separation process practiced successfully by one company on the West Coast.[5] This is employed in lieu of selective quarrying, and provision is made in the plant design to divert all high-quality stone from the added expense of the heavy-media separation step. This process is based on the sink-float principle. Stone of the desired gradation of $4\frac{1}{2} \times \frac{3}{16}$ in. is fed from a conical scrubber by belt to the heavy-media separation plant into a large cone separator. This vessel is filled with the medium, finely ground ferrosilicon plus a small amount of magnetite in water. The medium is agitated so as to maintain a specific gravity of 2.70 to 2.75. The desired stone concentrate sinks in the medium while the impurities (tailings) rise and float to an overflow and thence to a waste pile. The beneficiated stone is elevated by air lift and rinsed. About 95% of the medium is recovered for recyclical use after reconcentration by thickeners and dewatering equipment. Stone that contained 3 to 15% SiO_2 and up to 2% R_2O_3 is upgraded to 0.5 to 1% SiO_2 and less than 1% R_2O_3. Mineral contaminants removed from the stone are quartz, opalite, serpentine, and granite. Economics preclude use of this process in most areas.

Storage

Provision is made in plant layouts for storage of huge tonnages of processed or semi-processed stone since it is impossible to maintain shipments and/or other plant usage in constant balance with production due to fluctuation in demand. Furthermore, large inventories of stone are the only safeguard against quarry shutdowns due to floods, severe weather, and general plant breakdowns. In cold weather areas stockpiles of stone must necessarily be accumulated to satisfy peak seasonal demands for agriculture and construction in the spring that often exceed daily plant capacities.

Economics virtually dictate that most of this storage be maintained in *stockpiles* or *surge piles,* uncovered on the ground. In most large modern plants, after crushing and screening, the stockpiles are accumulated from elevated conveyors with trippers in which a steady flow of stone is discharged onto stockpiles below. Minimizing the distance that the stone falls reduces subsequent breakage and stone dust. To reduce breakage some plants employ a mobile vertical chute filled with baffles which break the fall of the stone onto the piles.

Fig. 4-4. Stockpile of semiclassified stone that will be screened further.

Fig. 4-5. Storage pile of processed commercial stone underneath elevated conveyor; stone-processing plant in background.

In this manner a radial elevated conveyor can maintain three or four stockpiles of different stone sizes from a central point. Such piles are usually conical (Figs. 4-4 and 4-5).

When the primary crusher is located on the quarry floor, surge piles of crushed stone at ground level are essential. Such stone may then be conveyed to the secondary crusher and screening plant for more processing. This stone inventory enables the plant to operate continuously. Other withdrawals from the piles can be made by shovel, bucket loader, or drag scraper.

Stone can be withdrawn from these stockpiles by gravity to underground tunnels below the piles, and from there belts convey the stone for loading (shipments) or further processing or blending of different sizes within the plant. In the tunnel system it is possible to blend different sizes onto the conveyor to meet intermittent or special gradation requirements. A few of the largest commercial plants will maintain as many as ten different sizes of stone in the stockpile-tunnel system, permitting great flexibility in blending. Table 4-1 lists the ten most common sizes stocked in such plants. A final screening to remove fines may be necessary before the stone is conveyed to the loading bin or hopper. Dust and fines can also be removed by rinsing or washing at this point. In the handling and conveying of classified

Table 4-1. Sizes of commercial stone frequently stockpiled in large modern plants.

1. 5×3 in.	6. $\frac{1}{2} \times \frac{3}{8}$ in.
2. 3×2 in.	7. $\frac{3}{8} \times \frac{1}{4}$ in.
3. 2×1 in.	8. $\frac{1}{4} \times \frac{1}{8}$ in.
4. $1 \times \frac{3}{4}$ in.	9. Limestone sand
5. $\frac{3}{4} \times \frac{1}{2}$ in.	10. Agstone or 90% — no. 100 mesh

stone, the efficient plants strive to minimize subsequent degradation in stone sizes to obviate secondary screening. Segregation of many size fractions tends to occur in stockpiling, particularly in small sizes below one inch. Generally, accumulating stone layer upon layer reduces this tendency.

Small plants operate from fewer stockpiles and are obliged to change their screens frequently in order to be able to meet different gradation requirements. With lower tonnage this means that dump trucks are often used in place of conveyors in stockpiling and reshuffling plus or minus sized material back to the piles.

Enclosed steel or concrete storage silos or bins are utilized primarily for immediate rail or truck loading (shipments) or for subsequent process requirements. In most plants usually less than 5% of the stone inventory is stored by this means.

Table 4-2 provides gradation information and recommendations on sizes of coarse aggregates, contained in the *Simplified Practice Recommendation R163-48,* published by the Department of Commerce. For the uses recommended for these various gradations, see Table 5-14 on page 119.

Portable Plants

Since World War II there has been an increased use of portable plants, mounted on pneumatic-tired wheels.[6] Contractors were the first to employ them as a source of aggregate on large construction projects where high haulage of commercial stone is costly. However, some limestone producers have purchased such plants to offer this service to contractors as well as to add standby capacity for their plant if peak demands or breakdowns eventuate. They have also proven useful on a supplementary basis when selective quarrying is performed.

Generally these plants cannot compete costwise with a large, modern stone operation. Proportionately more hand labor is required, and

Table 4-2. Sizes of coarse aggregates, applicable to all forms of crushed stone, gravel, and slag

Amounts finer than each laboratory sieve (square openings), percentage by weight

Size number	Nominal size square openings [a]	4	3½	3	2½	2	1½	1	¾	½	⅜	No. 4	No. 8	No. 16	No. 50	No. 100
1	3½ to 1½	100	90 to 100		25 to 60		0 to 15		0 to 5							
1-F [b]	3½ to 2	100	90 to 100			0 to 10	0 to 2									
2-F [b]	3 to 1½		100	90 to 100			0 to 15									
2	2½ to 1½			100	90 to 100	35 to 70	0 to 15	0 to 5								
24	2½ to ¾			100	90 to 100		25 to 60		0 to 10		0 to 5					
3	2 to 1				100	90 to 100	35 to 70	0 to 15	0 to 5							
357	2 to No. 4				100	95 to 100		35 to 70		10 to 30		0 to 5				
4	1½ to ¾					100	90 to 100	20 to 55	0 to 15	0 to 5						
467	1½ to No. 4					100	95 to 100		35 to 70		10 to 30	0 to 5				
5	1 to ½						100	90 to 100	20 to 55	0 to 10	0 to 5					
56	1 to ⅜						100	90 to 100	40 to 75	15 to 35	0 to 15	0 to 5				
57	1 to No. 4						100	95 to 100		25 to 60		0 to 10	0 to 5			
6	¾ to ⅜							100	90 to 100	20 to 55	0 to 15	0 to 5				
67	¾ to No. 4							100	90 to 100		20 to 55	0 to 10	0 to 5			
68	¾ to No. 8							100	90 to 100		30 to 65	5 to 25	0 to 10	0 to 5		
7	½ to No. 4								100	90 to 100	40 to 70	0 to 15	0 to 5			
78	½ to No. 8								100	90 to 100	40 to 75	5 to 25	0 to 10	0 to 5		
8	⅜ to No. 8									100	85 to 100	10 to 30	0 to 10	0 to 5		
89	⅜ to No. 16									100	90 to 100	20 to 55	5 to 30	0 to 10	0 to 5	
9	No. 4 to No. 16										100	85 to 100	10 to 40	0 to 10	0 to 5	
10	No. 4 to 0 [c]										100	85 to 100				10 to 30
G1 [d]	1½ to No. 50						100	80 to 100	50 to 85			20 to 40	15 to 35	5 to 25	0 to 10	0 to 2
G2 [d]	1½ to No. 8						100	65 to 100	35 to 75			10 to 35	0 to 10	0 to 5	0 to 5	
G3 [d]	1½ to No. 4						100	60 to 95	25 to 50			0 to 15	0 to 5			

[a] In inches, except where otherwise indicated. Numbered sieves are those of the United States Standard Sieve Series.

[b] Special sizes for sewage trickling filter media.

[c] Screenings.

[d] The requirements for grading depend upon percentage of crushed particles in gravel. Size G1 is for gravel containing 20 percent or less of crushed particles; G2 is for gravel containing more than 20 percent and not more than 40 percent of crushed particles; G3 is for gravel containing crushed particles in excess of 40 percent.

there is greater drilling and blasting expense since smaller crushers must be employed.

Costs

Since World War II tremendous mechanization has occurred in this industry, stimulated largely as a means of combating mounting wages and other inflationary pressures. As a result, the capital investment in most stone plants has also spiraled. However, the net result is that costs and prices of stone products have risen far less than on most products.

The cost saving or productivity advances in quarry and stone plant operations are summarized as follows:

1. Larger earth-moving equipment in stripping operations similar in size and type to large construction equipment.

2. Much higher speed, larger diameter, and deeper drilling with large rotary- and percussion-type drilling rigs.

3. Expanded use of delayed-action blasting.

4. Widespread substitution of ammonium nitrate-fuel oil mixtures in place of dynamite for blasting.

5. Use of the drop ball and less secondary blasting in the quarry.

6. Larger shovels for quarry loading.

7. Truck haulage of stone from quarry to primary crusher replacing cable cars on an incline; increases in truck payload capacity.

8. Larger capacity primary crushers.

9. Marked increase in extent and efficiency of conveying systems —virtual replacement of hand labor.

10. Vibratory screens replacing revolving screens.

11. Increased dieselization for the generation of electric utility power for the whole plant, individual hoisting, earth-moving, and truck equipment.

Information on stone-processing costs has been sparse, and what little that has been reported is of questionable value because of inflationary factors. Sporadic cost information should be viewed with a jaundiced eye, because rarely are the conditions on which the costs are calculated clearly explained. Without complete details such cost information is practically meaningless. Costs naturally vary greatly, depending upon the size and gradation of the stone—the finer the stone and more restricted the size distribution, the higher the costs because of increased grinding, screening, and auxiliary processing. Where high chemical purity is a requisite, costs are further increased because of selective quarrying, hand selection, washing, and other beneficiation methods. However, the above maxim can be misleading because of by-product stone fines, the cost of which may be allocated to the primary products. Consequently, any cost figures should be predicated strictly on *primary stone products* in order to be of significance.

The major factor in stone-production costs is the quarrying, or the cost of extracting the stone up to the point it enters the primary crusher. In virtually every plant this cost will exceed the subsequent cost of crushing, grinding, and classifying to obtain primary marketable products. In some instances the cost of extraction may be three to five times more costly than stone processing. This, of course, does not usually occur with most of the refined and pulverized stone products that are subject to so much intense grinding and classifying and it never occurs with whiting.

It is possible to indicate the magnitude of and the approximate range in total cost, without allowance for percentage depletion, of a typical primary (marketable or usable) product, such as 1.5×3 in. stone, which might be used as fluxstone, coarse aggregate, or lime kiln feed. As of 1965 the following wide variation is prevalent, $\pm 10\%$: $.60 to $1.80/ton (approx. average—$.95/ton).

Those plants with stone costs of $1/ton or more on the above basis are usually captive stone plants, where the higher-cost stone can be more easily absorbed into much higher-priced products and where high chemical purity is essential, or else they are plants located in remote or mountainous areas where the terrain is not conducive to a simple quarry and plant layout. Possibly in such cases the quarry or mine is located some miles distant from the stone-processing plant, adding materially to truck haulage costs. Also this does not necessarily mean that the highest-cost producers are the least efficient. The author is familiar with a few inherently high-cost operations that are actually more efficient and more highly mechanized than other plants whose costs are about 50% less. As an example, compare a quarry that has an average of 25 ft. of irregular overburden, a steeply tilted, weirdly folded deposit of stone with frequent faults, and stone of erratic chemical uniformity or a stope mining enterprise 1000 ft. underground with, say, a quarry that has only a few feet or no overburden; a natural vertical quarry face; and consistent chemical uniformity extending vertically and laterally in a massive deposit. There simply is a *vast* difference in respective costs. The former could not possibly compete with the latter at the same location.

Fig. 4-6. Aerial view of integrated lime operation: quarry, stone-processing, and vertical lime kiln plant.

This wide disparity in costs does not apply to modern stone-processing plants. Differences in costs are relatively minor (an estimated 25% spread) for plants comparably mechanized. The main distinction would be in the fracturing qualities of the stone that may yield excessive spalls or fines in obtaining a given size of stone or may impose heavy attrition on equipment due to abnormal hardness and abrasiveness. Of all stone-processing plants probably on an average plants operated by cement companies would have the lowest cost on a comparable basis. Such plants are only interested in reducing the stone to a flour consistency as rapidly as possible. To facilitate this objective many cement mills utilize, as much as possible, soft forms of limestone, including calcareous marl, chalk, oyster shell, and precipitated calcareous wastes. With such soft materials they can bypass conventional primary crushing equipment and use normal secondary and tertiary grinding equipment to achieve their goal. Figure 4-6 illustrates aerially an integrated quarry, stone, and lime operation.

Stone Safety Record

The only safety statistics on stone operations are compiled by the U. S. Bureau of Mines in cooperation with the National Crushed Stone Association, which sponsors an annual plant and quarry safety contest. Included in this study and contest, along with limestone, are granite, trap rock, and basalt aggregate plants. Statistics include both the plant and quarry as a unit. The safety record indicates that a stone operation is considerably more hazardous than the average United States industrial operation. However, improvement in plant and quarry safety has been achieved gradually since 1950. Generally, the two most frequent causes of accidents are: (1) *Falls,* including rock slides, and (2) *Materials handling.* In fact, 103 of the 300 lost-time accidents of the 225 competing plants in 1962 fell into these two categories. During the same year there were eight fatalities and twelve permanent partial disabilities. Statistically, plants with underground mines had higher frequency rates, but much lower severity rates, than plants with open quarries. A recapitulation of the industry's safety record in 1961, 1962, and the average over the period between 1926 and 1962 is summarized below:

	Basis	*1926–1962*	*1961*	*1962*
Injury—frequency rate	injuries/mill. manhours	25.2	14.2	19.8
Injury—severity rate	days lost/mill. manhours	4653	2412	3782

REFERENCES

1. *Pit & Quarry* (on plant design & general)—10/58, p. 96; 2/59, pp. 74, 112; 5/59, p. 130; 6/59, p. 88; 7/59 p. 122; 9/59, p. 76; 10/59, p. 124; 12/59, p. 76; 6/60, p. 80; 4/62, p. 146; 9/62, p. 118.
   (on screening)—8/58, p. 82; 5/60, p. 112.
   (on crushing)—10/58, p. 96; 9/59, p. 128; 4/62, p. 159.
   (on conveying)—8/61, p. 32.
2. *Rock Prod.* (on plant design & general)—12/55, p. 59; 4/59, pp. 82, 94, 124; 11/59, p. 120; 12/60, p. 90; 1/63, p. 89; 7/63, p. 92.
   (on screening)—5/55, p. 62; 8/59, p. 90; 11/59, p. 97; 12/59, p. 90.
   (on conveying)—9/61, p. 113; 7/62, p. 78.
   (on grinding)—9/61, p. 132.
3. C. Berry, "Modern Machines for Fine Grinding," *Ind. & Eng. Chem.* (July 1946), p. 672.
4. J. Smith, "Size Reduction," *Chem. Eng.* (Aug. 1952).
5. H. Utley, "Heavy Media Separation Plant at Kaiser Operation," *Pit & Quarry* (Nov. 1952), p. 92.
6. B. Herod, "Iowa Producer Keeps on the Move," *Pit & Quarry* (June 1961), p. 109.

General References on Stone Processing

7. O. Bowles, "Stone," *Mineral Facts & Problems, U. S. Bu. Mines Bull. 556* (1954).
8. O. Bowles, *The Stone Industries*, McGraw-Hill (New York), 2nd Ed. (1939), pp. 377–472.
9. G. Brown, "High Capacity Rock Crushing Plants: Selection, Installation, & Operation," *Quarry Managers J., 46* (Jan. 1962), p. 19.
10. N. Eilertsen, "Mining Methods & Costs, Kimballton Limestone Mine," *U. S. Bu. Mines I.C. 8214*, 50 pp. (1964).
11. A. Goldbeck, "Crushed Stone," *Industrial Minerals & Rocks*, AIME publ. (1949), pp. 245–293.
12. K. Gutschick, "Operation Fluxstone," *Rock Prod.* (Nov. 1955), p. 78.
13. W. Key, "Chalk & Whiting," *Industrial Minerals & Rocks*, AIME publ. (1960), pp. 233–241.
14. R. Mann, "Methods & Costs of Quarrying, Crushing, & Grinding Limestone," *U. S. Bu. Mines I.C. 6608* (Feb. 1933).
15. Nat'l Crushed Stone Ass'n, *Proc. of Conventions on Operating Problems & Discussions* (1946–1964).
16. N. Severinghaus, "Crushed Stone," *Industrial Minerals & Rocks*, AIME publ. (1960), pp. 285–302.
17. H. Wilson, "Chalk & Whiting," *Industrial Minerals & Rocks*, AIME publ. (1949), pp. 182–193.
18. H. Wilson & K. Skinner, "Occurrence, Properties & Preparation of Limestone & Chalk for Whiting," *U. S. Bu. Mines Bull. 395*, 155 pp. (1937).
19. W. Bickle, "Crushing & Grinding," (A bibliog.), *J. of Am. Ceram. Soc., 42* (Nov. 1959), p. 294.

CHAPTER FIVE

Limestone Uses

Total tonnage of crushed and broken limestone produced in the United States is so prodigious (461 and 488 million tons in 1962 and 1963 respectively) that it is difficult to comprehend. However, possibly the following practical illustration will dramatize its magnitude. If this vast tonnage in 1962 was loaded in 47 ft. gondola rail cars of 50 net tons each, allowing 6 ft. between cars, the resulting fanciful hypothetical train would circumscribe the earth at the equator by nearly four times, or comprise 95,472 miles of continuous train. If annual world-wide limestone tonnage was similarly equated, this homely analogy would have to be related in distance to the moon. It's world tonnage is so tremendous as to be practically incalculable, and it apparently has never been estimated. The author will hazard a guess that it lies somewhere between *2* and *2½ billion tons.*

In total tonnage *limestone ranks second after sand and gravel of all United States commodities,* having usurped this position from coal for the first time in the early 1960's. Petroleum has also forged ahead of coal and is just behind limestone with 452 million tons of production in 1962 as against 438 million tons for coal. Undoubtedly limestone ranks at least third and possibly second in the world production of *all* commodities.

Unlike oil, coal, and sand and gravel, it possesses a larger, more diversified end-use structure than these other materials, although strangely not as diverse in uses and functions as its first derivative product—lime. In contrast to lime (see pp. 340–341) its functions as categorized below are more physical than chemical, at least in proportion to its total tonnage.

Physical uses

Refractory	
Raw material	Construction aggregate
Filler	Masonry unit
Filter aid	Ornamental construction

Table 5-1. Limestone and dolomite (crushed and broken stone) sold or used by producers in the United States, by uses (Thousand short tons and thousand dollars)

	1962		1963	
Use	Quantity	Value	Quantity	Value
Concrete and roadstone	276,878	$365,098	292,976	$380,893
Flux	26,081	36,821	27,185	39,322
Agriculture	23,029	39,348	25,956	44,195
Railroad ballast	5,065	6,578	4,923	6,410
Riprap	10,016	12,253	10,690	13,229
Alkali manufacture	2,840	3,188	2,955	3,282
Cement—portland and natural	83,318	92,886	86,842	92,646
Coal—mine dusting	400	1,667	539	2,268
Fill material	440	330	383	296
Filler (not whiting substitute):				
Asphalt	3,208	6,955	1,994	5,012
Fertilizer	448	1,132	457	1,133
Other	351	1,567	419	1,921
Filtration	79	141	62	117
Glass manufacture	1,337	4,294	1,492	4,781
Lime and dead-burned dolomite	19,356	32,959	21,450	36,024
Limestone sand	1,706	3,103	1,759	3,234
Limestone whiting[a]	838	9,639	785	9,298
Mineral food	692	3,847	618	3,793
Paper manufacture	271	821	358	1,099
Poultry grit	161	1,333	160	1,342
Refractory (dolomite)	322	563	769	1,297
Sugar refining	623	1,506	646	1,580
Other uses[b]	1,741	4,253	2,125	5,472
Use unspecified	1,753	2,518	2,805	3,289
Total	460,953	632,800	488,348	661,926

[a] Includes stone for filler for abrasives, calcimine, calking compounds, ceramics, chewing gum, fabrics, floor coverings, insecticides, leather goods, paint, paper, phonograph records, plastics, pottery, putty, roofing, rubber, wire coating, and unspecified uses. Excludes limestone whiting made by companies from purchased stone.

[b] Includes stone for acid neutralization, calcium carbide, cast stone, chemicals (unspecified), concrete products, disinfectant and animal sanitation, electrical products, magnesia, magnesite, magnesium, mineral wool, oil-well drilling, patching plaster, roofing granules, stucco, terrazzo, and water treatment.

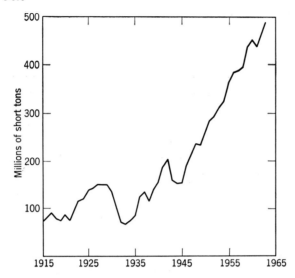

Fig. 5-1. Upward trend of U. S. limestone consumption, including meteoric rise since World War II.

Table 5-2. Use of crushed and broken miscellaneous carbonate aggregates in 1962

| | In units of one thousand | | | | | |
| | Marble | | Shell | | Calcareous marl | |
Use	Tons	$	Tons	$	Tons	$
Concrete and road material			12,792	18,611		
Cement (portland)			5,117	5,531	956	855
Lime			1,441	1,876		
Poultry grit			587	4,635		
Mineral food			4	22		
Terrazzo	397	4,535				
Agriculture (liming)					226	156
Other uses	1,243[a]	9,512	113[b]	566		
Total	1,623	14,378	20,054	31,241	1,182	1,011

[a] Includes agriculture, asphalt filler, flux, poultry grit, roofing granule, stone sand, stucco, and whiting.

[b] Includes agriculture, asphalt filler, and whiting.

Its chemical functions are largely indirect and are primarily predicated on its being the *prime source of lime* (decomposes when calcined). However, in its own right it will also (similar to lime) *neutralize acids*.

Its end-use breakdown in 1962 and 1963 is provided in Table 5-1. Trends in total United States consumption over the years have been generally on the ascendancy, as Fig. 5-1 indicates. Table 5-2 presents the end-use picture of marble, shell, and calcareous marl in 1962. All statistics used in these tables and graphs (and generally throughout the chapter) are obtained from the U. S. Bureau of Mines' annual Minerals' Yearbooks.[1]

CHEMICAL AND METALLURGICAL

About 31%, or 150 million tons, of all limestone produced is consumed by the chemical-process industries, including metallurgical and portland cement plants. Nearly all of this tonnage (over 99%) is employed directly in high-temperature thermal processes—smelting, sintering, calcining, etc.—embracing numerous industries in which the limestone is converted or calcined into lime before the desired chemical reaction can be effected. In other words, limestone is only employed since it is the most basic, economical source of lime (calcium oxide). In these processes in its carbonate form it is inert, but when thermally converted to lime, the resulting calcium and magnesium oxides are highly reactive.

Of course, the same approximate end reaction could be obtained, usually with greater celerity and with less fuel consumption, if commercial lime was substituted for limestone, but for reasons of economy, expediency, or tradition, huge tonnages of limestone are consumed. In fact, some companies will utilize both lime and limestone in these thermal processes, even in the same plant; other plants will employ only one or the other, depending upon performance, economics, or biased preference. However, since chemical-grade limestone averages between three and four times less in delivered cost than chemical lime on an equivalent CaO basis, economics obviously favors limestone. But because of this overlapping and interchangeable usage the succeeding discourses on chemical uses are often necessarily complementary with the corresponding use sections in Chapter 11.

These uses of limestone also should not be confused with captive lime production, which is pursued by some large consumers. Again

in this section it is limestone, not lime, that is charged into furnaces
or kilns along with other raw materials and fuel.

Ferrous Metals

Following portland cement manufacture, the metallurgical uses[15]
of limestone, the prime function of which is for fluxing, is the second
largest chemical use category and had **27.2** million tons in **1963**.
The major amount, about 90%, is for fluxing pig iron and steel,
as depicted in Table 5-3.

Pig Iron. In 1963 on an average .301 ton of limestone flux was
required to produce one short ton of pig iron, so that this material
followed fuel (usually coke) and iron ore as the third major raw
material in pig iron production. The flux stone, after calcining in
the blast furnace, purifies the crude molten iron by removing principally silica and alumina and secondarily manganese and sulfur.
Phosphorus is largely separated in the ensuing steel-making processes.
These primarily acidic oxide impurities, derived largely from the ore
and secondarily from the fuel, unite chemically with the basic calcium
and magnesium oxides to form a molten slag, which is tapped off
gravimetrically from the liquified iron. The purified iron is then cast
into pigs or charged into contiguous furnaces for further refinement
in making steel. With 54 to 55% Mesabi range iron ores about 0.5
ton of slag is formed per ton of pig iron.[48] However, since blast

Table 5-3. Shipments of metallurgical (fluxing) limestone by use—in
thousands of tons

Year	Blast furnace	Open hearth	Other smelters[a]	Other metal- lurgical[b]	Total
1953–1957 average	29,814	7112	1055	276	38,257
1958	19,427	4777	866	546	25,616
1959	19,752	6439	965	1050	28,206
1960	21,627	7409	997	1382	31,415
1961	18,129	6412	896	1761	27,198
1962	16,996	6411	646	2028	26,081
1963	—	—	—	—	27,185

[a] Includes: flux for copper, gold, lead, zinc and unspecified smelters.

[b] Includes: flux for foundries, cupolas, and electric furnaces.

furnace slag is only about one-third the weight of iron, its volume is about one and one-half times that of the iron. With such traditional ores an old axiom of the 1920–1940 era was that 2 tons of ore + 1 ton of coke + .5 ton of limestone and + 4 tons of air = 1 ton of pig iron + .5 ton of slag.

However, the amount of flux varies with the purity and concentration of the iron ore, the fuels and the stone itself. Usually stone is added at a 1 to 1 ratio of flux as CaO equivalent to total acid oxide impurities. Listed below are typical ranges in analyses of iron ore, refined pig iron, and blast-furnace slag. (Of course, wider variances than these also exist.)

Iron ore[a]		Pig iron		Slag[b]	
Fe	55.20%	Fe	92–4%	SiO_2	36.0
Si	8.60	C	4.0–4.5	Al_2O_3	12.4
Al	1.21	Si	1.0–1.5	CaO	41.6
Mn	0.56	Mn	0.5–1.5	MgO	6.2
P	0.08	P	0.11–.22	FeO	0.48
H_2O	8.60	S	0.03–.04	MnO	1.34
				S	1.39

[a] Average composite analysis reported by American Iron Ore Association in 1961 of ore shipped through Great Lakes.

[b] Average analysis reported by National Slag Association.

Before 1930 most metallurgists theorized that high calcium limestone was essential for fluxing pig iron. However, high calcium, magnesian, and dolomitic stone has been successfully used since then, providing the total calcium–magnesium carbonate content is high, generally in excess of 95%.[18] The amount of $MgCO_3$ in the flux depends to a considerable extent on how the slag will be utilized. Most blast-furnace slag is readily marketable. If the slag is to be sold or used for portland cement manufacture (namely, portland slag cement), a high calcium limestone is required, since the raw materials (limestone and/or blast-furnace slag) have strict tolerances of a maximum of 3% $MgCO_3$ equivalent. If the slag will be utilized as construction aggregate, there is a preference for 7 to 15% $MgCO_3$ in the flux. Usually this proportion of $MgCO_3$ is obtained from artificial blends of high calcium and dolomitic limestones. But there are a few magnesian stones of 7 to 10% $MgCO_3$ and low in impurities that provide an ideal balance of calcium and magnesium. The resulting MgO improves the fluidity of the slag and, upon cooling, yields a harder aggregate with up to 10% MgO content. There is some evidence that high alumina in the furnace charge negates the beneficial effect of MgO in slagging by increasing slag viscosity. Another theory is that silicates with more than one base, i.e., calcium and magnesium in the case of dolomitic, tend to fuse at lower temperature than with monobase silicates.

Impurities in stone have a most pronounced effect on fluxing efficiency.[16] Since the objective in pig iron refining is to remove impurities, it is only logical that the vehicle for achieving this—limestone—be as pure as possible within economic limitations. If the raw stone has 2% SiO_2, upon being calcined into lime in the blast furnace, this impurity is doubled to about 4%, since about 44% of the weight of the stone is lost as CO_2 gas. Furthermore, this 4% of SiO_2 will react chemically with about its own weight of CaO, rendering about 8% of the stone unreactive. Thus, effective fluxing value is about four times less than the impurities in the raw stone. Comparing this illustration with a stone of 5% impurity, it would mean that theoretically 12% more flux stone by weight would be required in the furnace charge for the more impure stone even though it contains only 3% more impurities. In practice, generally slightly more than this theoretical amount is required. So in effect a steel mill cannot justify using impure stone unless there is an inducement of an appreciably lower delivered price. For producing certain grades of high-quality iron or steel, it is doubtful if the impure stone can be tolerated at any price. The volume of slag produced generally increases in direct proportion to the extent of impurities in the stone, assuming quality of ore and fuel remain constant.

About half of the United States steel mills will permit no more than a total of 3% noncarbonate impurities, including approximate maxima of 0.1% sulfur and 0.02% phosphorus. There has been increasing stringency in sulfur tolerance due to use of higher sulfur content fuels, resulting from the most advanced mechanized coal mining equipment.

The physical sizes of flux stone charged into blast furnaces with lump iron ore generally range between 8 to 1 in. and 4.5 to 0.5 in., but with more restricted sizing, such as 5×3 in. or 3×1.5 in. or 2×1 in. Usually the precise gradation is dictated by the sizes of the furnace, coke, and iron ore. Fines or soft stone that decrepitates cannot be tolerated.

No other material competes with limestone as *the* flux in blast-furnace operations, but its average consumption factor has steadily ebbed since 1915, interspersed with some plateaus. It has decreased from a high of 0.66 ton of stone/ton of pig iron in 1917 to 0.3 ton in 1963. This is primarily due to use of iron ores of steadily rising assays, more concentrated, and also more recently to new ore beneficiation techniques that have led to agglomerates—pellets sinters, etc.—some of which are, to a greater or lesser extent, self-fluxing. Examples of this proportionate decline in consumption are the following: in the period 1952–1956 flux stone consumption for blast furnaces averaged 29.6 million tons per year when pig iron production averaged 69.2 million tons per year; in contrast, in 1961 stone consumption declined sharply to 18.1 million tons when pig iron ebbed only slightly to 64.9 million tons. This trend is expected to continue as more concentrated ferrous burdens are obtained through importation and new beneficiation methods. The quintescence of high ore purity used to be the renowned Mesabi range, Minnesota ores of 55% iron; in the 1960's 65 to 67% iron contents prevail in Venezuela and some Canadian ores and some agglomerates assay as high as 80% iron. Meanwhile the proportion of the ore burden, as different forms of concentrated ore agglomerates (sinters, pellets, nodules, and briquettes) have steadily grown, until in 1963 it exceeded lump iron ore in tonnage.[17] In fact, it has been prognosticated that ultimately all ore will be charged as agglomerates and that even the most desirable sizes of ore lumps will be pulverized and converted to agglomerates. Depressing the future further for flux stone is the parallel trend of self-fluxing agglomerates, usually in the form of sinters.

Self-fluxing Sinters. These contain pulverized limestone intermixed with iron ore fines and powdered coke which are sintered on strands (traveling grates) into a solid blast-furnace charge.[19] Since the ores have a relatively high concentration of iron, much less limestone is required to make them totally self-fluxing as compared to conventional practice with lump ore. Contributing to this is an apparently more complete reaction of the flux and impurities; 12 to 18% $CaCO_3$ equivalent will generally suffice. Some sinters are only partially self-fluxing with 7 to 10% increments of pulverized limestone, and a small addition of flux stone is supplemented in the furnace, but again the total charge is much less than conventional practice. In 1963 about 24% of total sinter production, or about 11 million tons, were self-fluxing, and this percentage should rise in the future. For such production it is estimated that an average of 15% pulverized limestone is needed or a total of 1.65 million tons. Thus, it appears that former purveyors of lump flux stone will be obliged to install grinding equipment if they hope to salvage tonnage from their declining blast-furnace market.

Both high calcium and dolomitic stone is employed in agglomerates, providing the total carbonate content is high. Arguments propounded in favor of both types by their adherents have been inconclusive or unconvincing. Others are known to be experimenting with various blends of both types, and still others are using hydrated lime and ground quicklime alone or blended with limestone (see p. 351). Smaller percentages are used with the latter two more concentrated sources of lime. It has been empirically established that certain types and concentrations of ore are fluxed more efficiently with different types and forms of liming materials for reasons as yet unknown. The type of flux can also influence increased capacity of sintering strands and enhance the physical quality of the sintered agglomerate. Steel plants have installed elaborate mixing and proportioning equipment for their self-fluxing sinter mixes.

Gradation of the pulverized limestone is also far from standardized. It has ranged from rather uniformly fine material of 15% retained on a no. 200 sieve to a broad gradation, such as 5% + no. 4 mesh, 69% + no. 50 mesh, and 18% − no. 200 mesh.

X-ray diffraction has identified the following mineral phases in the sintering reaction with self-fluxing sinters:

Calcium ferrites—$CaO \cdot Fe_2O_3$, $CaO \cdot Fe_3O_4$, $4CaO \cdot 7Fe_2O_3$
Dicalcium silicate—$2CaO \cdot SiO_2$ (several forms of minerals)
Monticellite (an olivine)—$CaO \cdot MgO \cdot SiO_2$
Iron Monticellite (an olivine)—$CaO \cdot FeO \cdot SiO_2$
Various forms of uncombined iron oxides, like hematite, magnetite, wustite, etc.

Open-hearth Furnaces. In the manufacture of steel, large tonnages of flux stone, usually of the high calcium type, are used for fluxing mainly in open-hearth furnaces for steel.[13a, 15] The United States tonnage involved is only about one-third of the amount consumed in blast furnaces, but it still entails about 6.4 million tons per year (1962). Electric furnaces, most basic Bessemer-type converters (Thomas and related types), and the basic oxygen converter furnaces (LD, LD-AC, LD-Pompey, Kal-Do, OLP, and Rotor types) generally use quicklime, not limestone, almost exclusively as the flux. These latter types, referred to as BOF furnaces, are gradually replacing open-hearth furnaces in the United States and most countries because of improved economy, performance, and lower capital investment. So flux stone is beset again, and this application of lump stone is diminishing gradually, but there will be no *net* loss for limestone per se in view of the increased requirements for quicklime resulting from this steel technological development (see p. 345).

The highest rate of limestone usage in open hearths is in so-called hot-metal plants that depend on molten pig iron from adjacent blast furnaces as the bulk of the ferrous burden along with some ore and often some steel scrap. Such plants will use either as the flux all limestone or 80 to 90% limestone with lime the balance. Invariably all of the *initial charge* is stone, and the lime is added at the end of the heat. With the "cold metal" plants that employ largely steel scrap in the furnace charge, much less limestone and more lime is used. Further information on this joint use of lime and limestone is contained on p. 347. Limestone is charged on top of the ferrous burden rather than under it to prevent it from adhering to the bottom of the hearth and impeding its dispersion throughout the heat.

Open-hearth slag that results is quite different in chemical and physical composition from blast-furnace slag. Regardless of whether limestone or lime is employed there is a much higher ratio of flux to impurities employed (2.5–3:1, lime-silica). The slag is often regarded as too unsound for use as a construction aggregate and has much less value. Fluorspar is commonly added to steel heats to improve slag fluidity without raising the basicity of the slag and inter-fering with the desired lime-silica ratio. In open hearths the average ratio of fluorspar used to burned lime is 1 to 4–5 and for limestone, 1 to 18–24.

The particle size of stone is generally similar to blast-furnace use, tending to be slightly smaller in top size. However, generally a slightly purer stone of lower silica (1.5 to 2% maximum) is specified than in the blast furnace because more intensive refinement is conducted. Also unlike blast furnaces there is a decided preference for high calcium stone over magnesian and dolomitic.

Foundries. Many iron foundries will also do some small-scale fluxing with limestone in cupolas in order to purify the iron a little more for improved or special mechanical qualities that they require. Lime-stone charge is much less than in the blast furnace or open hearth. They purchase about the same chemical quality as required for open hearths and frequently use a stone of $1\frac{1}{2}$–$\frac{3}{4}$ in.

Refractory. Although a preponderance of refractory maintenance material applied to open hearth furnace linings is dead burned dolomite or magnesite (see pp. 350 and 358, Chapter 11), there is raw dolomitic stone also employed by some steel plants as a less costly refractory fettling material.[5] It is often used for routine "drying" of open-hearth bottoms and by some mills for banking doors, usually as a supplement to the more costly refractory materials.

Undoubtedly this stone occupies a partial role as a flux as well as possessing some refractory value since the stone is soon calcined into lime at the high temperatures attained. Since the dolomitic stone does not possess nearly as effective refractory qualities as the dead-burned products, its equivalent rate of consumption must be several times greater. Unlike dead-burned dolomite it has no coalescing agent, so when applied it tends to set unpredictably.

For this purpose dolomitic stone of low silica (1 to 1.5% maximum) is required. High calcium stone is never used for this purpose. A closely graded material of $\frac{3}{8}$ in. to no. 10 mesh is specified. Dust or coarse stone over $\frac{1}{2}$ in. is undesirable and the stone should not decrepitate when subjected to the high heat. In 1962, 332,000 tons of dolomitic limestone were directly consumed by steel plants for this purpose, but in 1963 this tonnage increased sharply to 769,000 tons.

Nonferrous Metals

Smelting. Limestone is also the most widely employed flux for refining metallic *copper, lead, zinc* and *antimony,* in which the stone is charged into the smelters along with the concentrated ores of these metals.[15] Its function is similar to that previously described in the section on pig iron. About the same type of impurities are floated off of the molten metal as slag except that sulfur contents are much greater. Limestone requirements are the same with respect to chemical purity and particle size, but usually high calcium, rather than magnesian or dolomitic limestone is specified. Iron ore is often employed as a fluxing material along with limestone.

In beneficiating a Mexican *manganese* ore that is interbedded with calcium carbonate, the stone is calcined crudely in cupolas; the resulting CaO lumps are hydrated, facilitating separation of the much heavier manganese oxide from the light fluffy hydrate. The hydrate is sold as a by-product.

Alumina. In the red mud sintering (or combination-Bayer) process for alumina manufacture, that utilizes low-grade, domestic bauxite ore, limestone is employed to remove silica.[20, 21]

The low-grade bauxite, which assays about 50% alumina, is intermixed with soda ash and limestone in proportions of about 1.3 parts

Na_2O to 1 part Al_2O_3 (or a 2.24 to 2.0 ratio of sodium carbonate to bauxite) to 0.5 part CaO (or 0.9 part limestone). This limestone proportion is predicated on 12 to 12.5% silica in the bauxite ore; a lime–silica ratio of 2.0–2.2 to 1.0 is employed.

This mixture is ball-milled into a flour of which a minimum of 90% passes a No. 200 sieve. The limestone component calcines into lime in the sintering that follows in a rotary kiln at 2000°F. Water-soluble sodium aluminate and insoluble dicalcium silicate thus are produced. These two reactants are then ground, and the sodium aluminate is leached out as a slurry, filtered, and washed, thereby yielding the sodium aluminate liquor that is used in the digester and is the "key" to the Bayer process that follows. These reactions stated chemically are the following:

$$Na_2CO_3 + Al_2O_3 = 2NaAlO_2 + CO_2\uparrow$$

$$CaCO_3 + heat = CaO + CO_2\uparrow$$

$$2CaO + SiO_2 = 2H_2O + 2CaSiO_2$$

Only high calcium limestone of the same quality as for fluxing is used.

The $NaAlO_2$ is hydrolyzed to $Al(OH)_3$, plus $Na(OH)$. The $Al(OH)_3$ is ignited to Al_2O_3, and the waste caustic is recovered to generate more Na_2CO_3 for the process (above). The Al_2O_3 is then reduced in electric furnaces to the lightweight metal, aluminum. There are, of course, modifications of these procedures.

Chemical Process Industries

In past Bureau of Mines' statistics the alkali, beet sugar, magnesia, and carbide industries have been erroneously included in the limestone chapter. Without exception, these industries only use lime as quicklime or milk-of-lime (hydrated) in their processes. These uses are described in Chapter 11. They were probably listed statistically under limestone, since many plants in these industries manufacture their own lime captively, and, as a result, they either quarry or purchase limestone for their own use. Consequently, compared to the corresponding section on chemical lime in Chapter 11, the direct chemical uses of limestone are not nearly as numerous.

Glass. The manufacture of glass is a poignant example of how the uses of limestone and lime are complementary, interchangeable, and

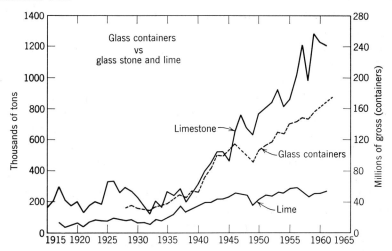

Fig. 5-2. More rapid upward trend of limestone as compared to lime in glass manufacture.

competitive, although the use trend has favored stone since World War II.[22, 23] Since 1946 limestone consumption has quadrupled whereas burned lime has gained very little (Fig. 5-2). The all-time high for glass stone was reached in 1963—nearly 1.5 million tons. Both high calcium and dolomitic types are employed to about the same extent, providing both types are of high purity and uniformity.

Its function in glass, whether applied by stone or lime, is to render the glass more insoluble so that glass can be used unrestrictedly in contact with water and chemical solutions, to improve the mechanical properties of glass by making it less brittle and stronger, to improve the appearance of glass by providing a more enduring luster. Limestone and lime are the lowest-cost fluxing materials for glass. They flux the silica sand, forming chemically fused calcium silicates.

Some general use preferences are evident. A decided trend is to employ high calcium stone for flat or window glass and dolomitic stone or lime for glass containers and tumblers. The reason for the latter preference is that high MgO content glass possesses greater resistance to the etching effect of certain chemical solvents and acids that may cause discoloration in consumer products packed in glass. Furthermore, glass high in MgO has slightly improved resistance to heat shock and appears to promote better fluidity during the melt.

The combined tonnage of lime and limestone are close to soda

ash as easily the third major ingredient of glass; however, if the
lime used to make soda ash by the Solvay Process (p. 356) is included
it would rank second after silica sand. The proportion of lime or
limestone, stated as CaO and MgO, are shown below with the percentages of silica sand, as SiO_2 and soda ash, as Na_2O, illustrating
some of the variation in proportions for different types of *soda–lime*
glass.

SiO_2	68–75%	CaO	5–14%
Na_2O	10–18	MgO	0–10

To these three basic ingredients are added other chemical or mineral
additives to the glass batch to produce greater opacity, decolorization,
heat resistance, and other diverse physical characteristics. A concerted
effort is made to obtain all materials of the same approximate particle
size distribution, which contributes to more intimate mixing and less
segregation and leads to superior-quality glass. Batches of this raw-
material mix are then introduced into glass pots where it is fused
at 3000 to 4000°F to a molten state. The extremely high temperature
quickly calcines the limestone into lime, which reacts with the silica
to form the chemically stable but complex sodium calcium magnesium
silicate, of which glass is primarily composed.

Arguments for using limestone instead of lime are:

1. It is much *cheaper* in material cost—two and one-half to three
and one-half times less costly on an equivalent total oxide basis.
2. The evolvement of CO_2 at usually high pressure as the limestone
is calcined into lime in the glass pot produces a turbulent "boiling"
effect that is regarded as beneficial in improving the homogeneity
of the mix. In any event, soda ash generates more CO_2 bubbling
action than limestone, so considerable turbulence will always be
present.
3. It is easier to handle and store; rain or humidity will not decompose it as in the case of lime.
4. A more homogeneous melt is achieved with a mix containing
coarse particles of Nos. 8 to 10 sieve size.
5. Low-cost production and equal or better quality is claimed for
flat and cheap container glass.

Arguments advanced in the use of lime over limestone are:

1. In high-grade flint glass for containers or tableware, it produces
better color than limestone and greater brilliance and transparency,

even though the equivalent iron contents of both materials is the same. The reason for this is that quicklime is devoid of organic matter and the iron present is in ferric form rather than ferrous, as is the case with limestone. Ferrous oxide adversely affects color more than ferric.

2. Because of No. 1 above there is economy in the use of less decolorizer chemical additives that are comparatively costly.

3. Higher material cost of lime is at least partially compensated for by fuel savings in the glass batch due to extra fuel required for limestone calcination. There is general disagreement on the amount.

4. With glass batch mixes of medium to fine particle gradation there is some evidence that quicklime may melt faster and thereby goes into solution more rapidly for reaction with silica.

5. As a consequence of item 4 some contend that greater capacity results.

Regardless of whether high calcium or dolomitic limestone is used, definite materials specifications exist. There is invariable emphasis on *uniformity* in chemical composition and gradation. Generally, a low iron content is essential, usually less than 0.06% FeO, and for special and optical glass, where a high degree of opacity is required, the FeO tolerance may be 0.01 to 0.02% maximum. Sulfur and phosphorus should also be low but generally present no problem. Very precise gradations are specified, such as 100% passing a no. 10 sieve and 96 to 100% retained on a no. 100 sieve. Other specifications, depending upon the gradation of sand on which a glass plant has standardized, may require discrete gradations between nos. 16–20 and 200 sieve sizes. Most plants cannot tolerate any dust finer than no. 200 mesh or material coarser than no. 8 sieve (and only a small percent at these extremes). Intermediate gradation requirements are frequently specified to assure constancy of grading. Consequently, to supply this requirement a limestone plant must employ more intricate screening procedures than for most commercial uses. If even small amounts of organic matter are present, stone will require washing, although generally wet sieving is essential for the rigorous classification required.

There is no substitute for liming materials in glass when color is important. Blast-furnace slag, which is rich in calcium silicate, has been used successfully in making low-quality brown or dark-colored glass; in fact, the high iron content of slag would be a plus factor for such glass. There is very little slag used for this purpose.

Magnesite and magnesia might be substituted for dolomite, but the cost would be prohibitive.

Pulp and Paper. Limestone is employed in the *sulfite process* for making paper pulp where it is reacted with SO_2 to form the calcium bisulfite pulp-cooking liquor in the Jennsen tower system.[24] When wood pulp is digested in this liquor, all constituents of the pulp except the cellulosic components are either dissolved or removed.

There is a trend to replace limestone with magnesia, ammonia, or soda ash as the alkaline agent to react with SO_2. The principle reason for this is to simplify waste-disposal problems of sulfite paper mills. These other three more costly reagents can be largely recovered and regenerated for reuse in the process, whereas limestone cannot. With steadily increasing pressure from health authorities for stream pollution abatement, it is anticipated that this use of limestone will decline further.

A high calcium limestone is favored over dolomitic in this process. Generally, the minimum $CaCO_3$ content specified is 95 to 97%, and large stone, as much as 8 to 14 in. top size, is preferred for use in the tower.

Extensive use of lime in sulfate and soda pulping processes, including lime recovery systems, is described on p. 354.

Ceramic Pottery. Finely pulverized limestone or whiting is an ingredient of many formulas for glazes and enamels that are applied to pottery and whiteware. In one sense it might be regarded as a type of filler, but its function is more chemical, since when these clay products are burned in kilns, the $CaCO_3$ converts to CaO and is an active fluxing material. Its use is of sufficient importance for the American Ceramic Society to publish in 1928 Specification No. 30, embracing two grades by chemical composition. Both grades specified at least 97% total carbonates, maximums of 2% SiO_2, 0.25% Fe_2O_3, and 0.1% SO_4. One grade limited $MgCO_3$ to 2% and the other to 8%. In fineness, maxima of 1% and 2% for retention on the Nos. 140 and 200 sieves, respectively, were specified.

Mineral Wool. One of the most widely used forms of insulation, "rock" wool (a mineral wool), is commonly made with limestone in the raw-material charge.[25] Since this is a lime-silica reaction, argillaceous limestone or calcareous shale is actually preferred, since in any event considerable silica must be present to react with the lime. Blends of impure limestone, blast-furnace slag, and sandstone are often used. Stone employed alone without blending with other materials is called *wool rock* and should be so impure with large amounts of silica, alumina, and iron that on being sintered in a cupola the ignition loss (mostly as CO_2 from the $CaCO_3$ and $MgCO_3$ components) is only 20 to 30%. This means it would correspond to a total carbonate content of only 45 to 65%.

Layer upon layer of the rock wool charge and coke are fed into the cupola at the rate of 1 part coke to 2–4 parts rock. At 2700 to 2800°F the charge melts, and a small stream trickles from the bottom of the cupola. At this stage it is met by a jet of steam at 100 psi. pressure, creating fine thread which collects into a soft, fluffy mass from which rock wool bats and pellets are made.[13f]

In recent years blast-furnace slag[48] has largely replaced limestone and wool rock as the prime raw material. It contains close to the optimum proportions of lime, magnesia, silica, alumina, and iron. Heat economies are effected since no calcination is needed; there is no loss on ignition, so that a ton of raw material yields about a ton of finished product; better product uniformity is claimed due to less blending. As evidence of this trend, in 1962, 560,000 tons of blast-furnace slag were consumed as against only an estimated 50,000 tons of impure limestone.

Magnesium Insulation. A type of insulation used for pipe and boiler covering is derived from dolomitic limestone and is referred to as *technical carbonate*, magnesia Alba, or just precipitated $MgCO_3$. It is made by the Pattinson process, with various modifications.[5, 6]

Dolomite is calcined and the CO_2 gas emitted is recovered, purified, and compressed. The resulting dolomitic quicklime is slaked in water and then recarbonated with the waste CO_2. This precipitates the insoluble $CaCO_3$ from the soluble magnesium bicarbonate. After $CaCO_3$ is removed by filtration, the remaining magnesium bicarbonate is boiled, which drives off one molecule of CO_2 and precipitates $MgCO_3$. This latter product is compounded with $Mg(OH)_2$ and asbestos fiber and molded into the desired form. Chemically the above reaction is:

$$CaCO_3 \cdot MgCO_3 + heat = CaO \cdot MgO + 2CO_2\uparrow$$

$$CaO \cdot MgO + 2H_2O = Ca(OH)_2 \cdot MgO$$

$$Ca(OH)_2 \cdot MgO + 5H_2O + 3CO_2 = CaCO_3\uparrow + Mg(HCO_3)_2 \cdot 5H_2O$$

$$Mg(HCO_3)_2 + heat = MgCO_3 + H_2O + CO_2\uparrow$$

In addition, technical and U.S.P. grades of $MgCO_3$ have other uses in industry and medicinals.

Industry Fillers

A vast tonnage of about 3.5 million tons of finely divided calcium or calcium-magnesium carbonate is consumed by industry as a *filler* or *extender* in probably over a hundred different products.[14] Specifically, total tonnage consumed annually in 1963 for such purposes is 785,000 tons of limestone and an estimated 130,000 tons derived from shell and marble. To this might be added 457,000 tons of fertilizer filler and 2 million tons of asphalt filler for paving, which will be described later in the agriculture and construction use sections. These latter types are not as finely divided.

WHITING

Some of this industrial filler usage could be classed as standard *pulverized limestone* of nominal No. 100 mesh or No. 200 mesh sizes, but a sizable portion (about 1 million tons) involves finer sizes, known commercially as *whiting*, derived from intense grinding of high calcium or dolomitic limestones, marble, chalk, or oyster shell. Its physical properties are generally analogous to and competitive with precipitated $CaCO_3$, described on p. 360. Just exactly where the *demarcation line* is between pulverized limestone and whiting is a controversial point. But since the most liberal specifications for whiting require at least 90% to pass a No. 325 sieve, this might be regarded as reasonably definitive. Theoretically it would seem that hydrated lime would qualify as an ideal low-cost filler because of its inherent minute size, but it is too chemically active. Chemical inertness is usually a requisite for these carbonate microfillers.

There are so many highly individualized specifications, even for the same ostensible use, that in a sense whiting should not be considered a standard commodity, even by grades. Often it is a tailormade product, expressly for one or a few large consumers, and may be totally unacceptable by many other users, even with price concessions. The consumer establishes his peculiar standards, and when he finds a brand of whiting that satisfies these requirements, confirmed by experimental use, he is loath to change to other sources of supply for fear of impairing the quality or uniformity of his product. Some of the physical and chemical characteristics[13C] demanded in these exacting specifications are:

1. Good white color—high degree of reflectivity.
2. Minute particle size—10 micron to submicron sizes.
3. Special particle size gradations in micron sizes.

4. Particle shape and high specific surface area.

5. Freedom from grit—namely, material retained on a No. 325 sieve.

6. Plasticity and rheological properties.

7. High absorption of oil, ink, and color pigments.

8. Low to zero chemical activity (alkalinity).

9. Freedom from impurities—strict tolerances on SiO_2, Al_2O_3, FeO, Fe_2O_3, $CaSO_4$, H_2O, heat and ignition loss, and even copper and manganese are cited.

10. Bulk density.

11. Specific gravity.

The carbonate fillers encounter considerable competition[13e] from many other types of mineral fillers, the most noteworthy of which are several species of kaolin clay, gypsum, talc, pyrophyllite, mica, diatomite, barite, asbestos fines, silica flour, and siliceous rock dusts. Of all of these mineral fillers, however, the carbonate types (whiting) rank first in tonnage. For some products only the carbonate or siliceous fillers are compatible. In others, like paper and rubber, either whiting or other mineral fillers are compatible and selection is contingent on cost or preference. The exact tonnage of precipitated calcium carbonate that vies for this market is obscure, other than it has grown steadily and is substantial, although far less than the whiting total. Imported British chalk whiting that dominated this field until the 1920's has lost most of its tonnage to domestic limestone (and other carbonate) whiting and precipitated $CaCO_3$, as Fig. 5-3 indicates. European countries still largely use chalk whiting, with Great Britain the dominant producer from its celebrated "white cliffs of Dover."

Only the major industrial markets in which the carbonate fillers dominate or participate heavily are described in some detail below. Statistics on consumption of these markets is not available, but generally about 90% of total whiting is utilized in making paint, rubber, wood putty, paper, and floor covering.

Paint. Some whiting is used as a white pigment filler and pigment extender for such more costly white pigments as titanium dioxide, white lead, and zinc oxide in *all* white paints, oil or water-soluble types.[27] It is used more extensively in the cheaper paints and comprises about 80% of the composition of *calcimine*. Consequently, opacity and a high degree of reflectivity is required for this market. ASTM specification D1199-52T, consolidated in Table 5-4, is a material specification on the use of calcium carbonates as a filler in pig-

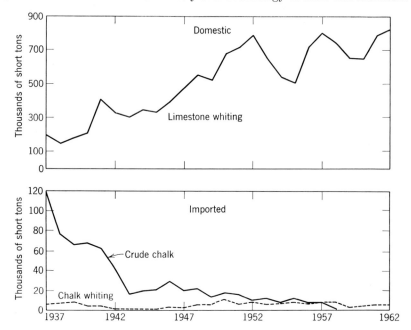

Fig. 5-3. Comparison of U. S. consumption of domestic limestone whiting vs. British imports of crude chalk and chalk whiting. (U. S. tonnage does not include domestic shell and marble whiting and precipitated $CaCO_3$.)

ments and putty. There is still some imported British chalk whiting used for this purpose by some consumers because of its alleged superior reflectivity.

Rubber. Whiting is not used in all rubber products. In fact, there is extensive use of siliceous mineral fillers in various low- and high-quality rubber products that compete with the carbonates. Fillers are utilized with both natural and synthetic rubber elastomers. Most whiting is employed in light-colored rubber or in high-quality rubber products.[26] Some ordinary Nos. 100 to 200 mesh pulverized limestone is used in lieu of whiting as a diluent or filler in low cost rubber products because of economy, but it does not lend as much reinforcement to the rubber as the finer fillers. Carbonate fillers are rarely used in the treads or carcasses of tires, except for white sidewalls. In high-quality rubber products, precipitated calcium silicate is whiting's principal competition. Some special types of whiting are produced expressly for rubber use. These are treated with stearates to enhance their dispersion in the rubber mix.

Table 5-4. ASTM requirements on carbonate fillers for paint and putty
(Spec. D1199-52T)

Type of whiting and grade	Total carbonate	Min. % CaCO₃ H₂O-free	Max. % H₂O volatiles	Fineness requirement	Max. % retained no. 325 sieve	Max. alkalinity[a]
1. P.C. (precipitated calcium carbonate) for paint	—	98	1	majority of particles below 15–20 micron size	1	0.5
2. G.C. (ground high calcium limestone) for filler	—	95	2	largely 10–44 micron sizes	15	0.35
3. G.M. (ground dolomitic limestone) for coarse putty	95	51.6	2	—	30	0.5

Color and oil absorption limits are negotiated between buyer and seller.
[a] Expressed as mg. NaOH per g.

Some specifications for rubber are probably the most detailed and exacting of all applications of whiting. Many differences exist among rubber companies, but the following requirements are common: Minimum of 98% $CaCO_3$; maximum heat loss of 0.2%; maximum alkalinity of 0.03%; maximum of 0.005% copper oxide; maximum of 0.02% manganese; maximum of 0.1 to 1% retained on No. 325 sieve.

Putty. Linseed-oil putty for carpentry and window glazing consists essentially of 85% whiting and 15% linseed oil. Some makers add small amounts of stearates or resins to improve putty plasticity. Whiting requirements cited in Table 5-4 demand very low alkalinity of 0.1 to 0.3 maximum to prevent saponification of the linseed oil, no grit, and rather strict particle size distribution to provide maximum plasticity. There are still some advocates of chalk whiting that contend that the rounder shaped chalk particles provide superior plasticity, but most of this market is supplied by the domestic limestone whiting industry.

Paper. Easily precipitated $CaCO_3$ is the preferred form of carbonate for use in *paper coating*.[28] It seems to possess better printing-ink receptivity and on an average has higher percentages of ultrafine particles in the 5 to 0.5 micron range that contribute most to paper whiteness and gloss. However, there is no monopoly on this use. Other white carbonate and noncarbonate mineral fillers are also employed in paper coating, depending on the type of paper and finish desired, but generally all whiting possesses the highest reflective index. In very high quality white paper some titanium dioxide is employed as a white pigment, which would relegate whiting to the subsidiary role of a pigment extender.

A larger tonnage of mineral fillers is used as an *integral filler*[28] in paper with adhesive agents, like starch and resin, to fill the voids between the fine intermeshed fibers of cellulose. This filler content comprises between 8 and 30% of the weight of paper, exclusive of any coating or ink that may subsequently be applied. Some whiting is employed as a filler but not to the same extent as kaolin clay. Other lesser-used fillers are talc, gypsum, and diatomite. There has been experimental use by a few sulfate pulp producers, which have integrated operations, to use their own waste precipitated $CaCO_3$ from their sulfate process (p. 354) as a paper filler instead of purchased fillers, like kaolin and others. In spite of obvious economy, its future application appears to be limited to a few special situations where this more heterogeneous, lower quality filler can be advantageously used. The substantial savings it potentiates warrant a real research effort.

Estimates on the total amount of *all* mineral fillers used annually as an integral filler and for paper coating range between 1.1 and 1.3 million tons per year, of which carbonates enjoy about 300,000 tons. Paper contains much more than just cellulose.

Floor-covering Materials. Whiting is the preferred mineral filler for such widely used floor-covering materials as asphalt tile, linoleum, vinyl and vinyl–asbestos tiles, and the print paint for printed felt base flooring. All contain substantial percentages of whiting—up to 60% of the product's weight in the case of asphalt tile. Linoleum contains 33 to 40%.

Usually extreme fineness is a requisite, with at least 99% passing a No. 325 sieve. Limestone whiting is the preferred type.

Other Uses. Other important applications of whiting are in *dentifrices*, proprietary *alkalizers*, as digestion aids, and medicinals. Precipitated calcium carbonate dominates these lower tonnage markets since U.S.P. products are required. The solution processes employed in making precipitated $CaCO_3$ develop higher average purity with lower traces of such toxic elements as arsenic and fluorine.

Some whiting is employed in such other varied products as *asphalt roofing material, calking compounds, certain plastics, phonograph records, crayons, baking powder, picture frame moldings, abrasives, white shoe polish, explosives, foundry compounds, white ink, wire insulation coatings, glue, grease, pesticides, special fabrics,* and *chewing gum.* In some of the above products, like pesticides, plastics, and roofing materials, whiting's use is very minor as compared to other mineral fillers, particularly the siliceous types. In other applications above its use and all mineral fillers may be infinitesimal or sporadic.

Portland Cement Manufacture

In volume portland cement manufacture represents the greatest industrial or chemical process use for limestone by a wide margin and is exceeded only by the construction aggregate use category. Limestone is also easily the No. 1 raw material consumed by this huge tonnage industry. Its annual consumption for cement in 1963 was 86.8 million tons.

Before describing the limestone requirements, let us consider its manufacture.[29, 30] Contrary to general belief, cement is strictly a synthetic, manufactured product, made by blending different natural and synthetic materials in order to achieve rather precise chemical proportions of lime, silica, alumina, iron, and sulfate in the finished product, as shown in Table 5-5. Table 5-6 enumerates the raw materials utilized by this industry in obtaining this desired proportion, and it also illustrates great latitude in selection of raw materials.[36] Of the 160-odd portland cement plants in the United States, probably no two will use exactly the same proportionate combination of raw materials. First, these balanced, batched materials

Table 5-5. Typical chemical composition of portland cement

Compound	Oxide composition	Stoichiometric composition	Approx. % content
Tricalcium silicate	$(CaO)_3SiO_2$	Ca_3SiO_5	45
Dicalcium silicate	$(CaO)_2SiO_2$	Ca_2SiO_4	27
Tricalcium aluminate	$(CaO)_3Al_2O_3$	$Ca_3Al_2O_6$	11
Tetracalcium alumino-ferrite	$(CaO)_4(Al_2O_3)\cdot(Fe_2O_3)$	$Ca_4Al_2Fe_2O_{10}$	8
Calcium sulfate	$CaSO_4$	$CaSO_4$	2.5

Table 5-6. List of raw materials used in making portland cement in the U. S. by blending various combinations of these materials

Calcareous		*Argillaceous*	
"Cement rock"	Oyster shell	Clay	Aluminum ore refuse
Limestone	Coquina shell	Shale	Staurolite
Marl	Chalk	Slag	Diaspore clay
Alkali waste	Marble	Fly ash	Granodiorite
Blast-furnace and other slags		Copper slag	Kaolin
Carbide sludge			

Siliceous		*Ferriferous*	
Sand	Quartzite	Iron ore	Iron pyrite
Traprock	Fuller's earth	Iron calcine	Iron sinters
Calcium silicate wastes		Iron dust	Iron oxide
			Blast-furnace flue dust

Sulfurous

Gypsum

Iron pyrite

are ground to a flour of 75 to 90% passing a #200 sieve with either a dry or wet process. The flour or kiln feed is then fed to modern large rotary kilns, 250 to 450 ft. in length, where it is sintered to incipient fusion, forming hard clinkers. Gypsum rock is added as a retarder and the clinker is then ground to a fine gray powder that is finished portland cement.

Limestone Requirements. Over 80% of the raw material for the kiln feed is high calcium limestone or its calcareous equivalent, but due to a much greater loss on ignition during calcination, the limestone component dwindles in weight to 65 to 70% CaO in the finished cement. The other noncarbonate raw material has virtually no loss on ignition. The only rigid requirement is that the limestone should not possess more than 3% $MgCO_3$, since there is a maximum of 5% MgO in all specifications on finished portland cement. This, of course, eliminates all magnesian and dolomitic limestone sources. Frequently the cement industry discovers apparently promising stone deposits only to find that they contain 4 to 6% $MgCO_3$. However, any of the high calcium chemical and metallurgical grades and other relatively impure high calcium stones, low in MgO, easily meet requirements. A few plants are known to utilize substandard stone, providing they are not too far over the MgO maximum, by blending with a chemical-grade stone that contains only 1 to 2% $MgCO_3$. There are important deposits of impure high calcium stone—low in MgO and high in silica, alumina, and iron—that are ideal for cement when there is some constancy in their chemical composition. Large deposits of such stone exist in the Lehigh Valley of Pennsylvania and a few other locations and are known as *cement rock*. These chemically impure, argillaceous stones contain the approximate desired ratio of lime to silica, alumina, and iron, so that little or no blending of raw material is necessary. The savings and resulting simplification in the process is appreciable in eliminating much material handling, blending, and analysis of a combination of materials.

Blast-furnace slags, low in MgO, (derived from high calcium stone), have been utilized for many years as the major raw material by a number of plants in the Chicago, Pittsburgh, and Birmingham steel-producing areas and on a partial basis by other producers.[48] It inherently possesses close to the desired proportion of CaO, SiO_2, Al_2O_3, and FeO and can be used with minimum blending of other materials. Usually a small percentage of limestone supplement is needed.

Other calcareous *industrial wastes* from the alkali, alumina, and copper industries are being increasingly utilized with blending methods. A few German plants have even blended in small portions of *waste hydrated lime* from carbide plants.

About the only commercial use of *chalk* of quantity in the United States is for cement manufacture in Arkansas and Alabama, although it is widely used in Europe. Impure *marls* are rather commonly used in Michigan, Ohio, and Atlantic coastal areas. Oyster, clam, and coquina *shells* are successfully used in coastal plants on the Gulf of Mexico and California. An extra bonus is obtained in utilizing these latter wastes, low grade carbonates, and shell. They are all either precipitated powders or are relatively soft, so that grinding costs are greatly reduced.

Overburden from stone deposits is frequently used as a source of silica, alumina, and iron and is the least expensive of all cement raw materials. Importation of iron wastes (mill scale) is necessary in some cases to provide the small required amount of iron.

However, most of the limestone that is consumed is produced captively by this integrated industry. Some plants located on the Great Lakes receive boatloads of stone and do not operate their stone quarries. Another exception to this would be purchase of nearby high-quality stone, which is used to "sweeten" the mix in the final blending or to enhance the whiteness of special "white" portland cement or improve the color of proprietary masonry cement. Captive production will continue since most plants consume around 0.5 to 1.5 million tons per year, and on this very low cost raw material their production costs are usually far below commercial delivered prices. In fact, their costs in some instances are less than the stone freight and handling costs from a nearby commercial stone source. As a result, this industry in the 1960's produces about 85% of its total stone requirements.

Masonry Cement. Cement companies have one other application for limestone. Most companies make prepared mortars or proprietary masonry cements. A majority of such products contain 35 to 65% pulverized high calcium limestone, most of which they supply from their quarries. Methods of compounding vary, but usually the crusher run-of-the-mine stone is fed into ball mills with portland cement clinker, ground dry, and intimately mixed. Air-entraining agents are invariably added to make the mortar workable. Some companies will substitute 5 to 10% hydrated lime for the equivalent amount of limestone. A minority will use little or no stone and will admix dry hydrated lime, blast-furnace slag, and portland cement clinker or finished cement into a homogeneous mixture. The limestone serves as a filler having no cementitious or plasticity value. For further information on this product, see p. 391. An estimated 1.5 million tons of pulverized limestone is consumed annually for this purpose.

Lime Manufacture

Since the application of limestone for lime manufacture is synonymous to Chapter 8, very little will be mentioned here to avoid repetition.

This is the *third* largest chemical use of limestone—after portland cement and flux stone—with total domestic annual consumption averaging an estimated 25 million tons in the early 1960's. (The author contends that the Bureau of Mines figure is too low due to misallocation of tonnage.) Theoretically, only 1.65 to 1.90 tons of limestone, depending upon its purity and type, is necessary to make one ton of quicklime. However, such an assumption is unrealistic since it would presuppose no loss of stone in manufacture. Making allowance for the inevitable loss of some stone, the industry has standardized its calculation on a reasonable approximate factor of *2 tons of stone for 1 ton of lime.* Actually, some plants, depending upon the physical and chemical characteristics of the stone, will be slightly above or below this average factor.

Again, similar to cement this is primarily a captive use of limestone, in which most lime plants quarry or mine their stone from usually contiguous or nearby deposits. An estimated **83%** of the stone is captive. The remaining **17%** is largely based on water shipments of stone through the Great Lakes to port cities in Ohio, Michigan, Wisconsin, Illinois, and Minnesota. There are commercial purchases of oyster and clam shells for lime burning at some port cities on the Gulf of Mexico, and one company barges stone it produces about 500 miles down the Mississippi river to its captive lime plant. The cost of limestone is such an important factor in the over-all cost of lime that commercial stone is generally uneconomical, almost prohibitive by conventional rail shipment. With this **2** to **1** use factor, transportation charges are doubled in limestone's segment of lime costs.

Miscellaneous Industrial Uses

Coal Mine Dusting. Large amounts of pulverized limestone are sold to coal mines as "rock dust" to improve safety standards.[3] Of all hazards in coal mining, dust explosions can be the most disastrous. This is caused by violent combustion of coal dust particles. It produces an air shock that travels ahead of the flame, stirring up settled dust, and causing the flame to extend the explosion through great distances of shafts and tunnels.

The U. S. Bureau of Mines, in its campaign for improved safety conditions in coal mines, promoted application of mineral dusts to deactivate the combustibility of coal dust. If incombustible dusts are mixed with coal dust up to 50%, the danger of explosions is practically eliminated. The inert mineral dust so dilutes the coal dust and oxygen that continued combustion is rendered virtually impossible.

The requirements and recommendations of mineral dust for this purpose, as set forth by the American Engineering Standards Committee, are generally followed:

Fineness—100% passing a No. 20 sieve; 50% passing a No. 200 sieve.
Silica—not more than 5% preferred.
Combustible content—not more than 5%.
Color of dust—light color preferred.

Relatively low silica is specified to reduce incidence of silicosis through inhalation by miners. Since limestone more easily meets these requirements than other minerals at the lowest cost, it enjoys the largest share of this market. It also tends to be more freely flowing

and will not absorb moisture from the air as readily as some other dusts. Other minerals that are or have been used for this purpose are gypsum, shale, talc, and adobe clay.

It is most efficiently applied by machines to walls and all mine surfaces where coal dust is prone to settle. Safety inspectors frequently spot-check the dust to determine the percentage of incombustibility. If it is below 50%, more pulverized limestone is applied until the desired percent is attained. If methane gas is also present in the mine, a higher minimum than 50% of rock dust is maintained. To reduce chance of combustion, rock dust is increased 10% for each 1% of gas present.

Filter Beds. Sewage plants frequently employ trickling or sprinkling filter beds composed of precisely and closely screened mineral aggregate on which plant effluents are sprayed.[31, 32] The purpose of the aggregate is to provide surfaces for the lodgment of colonies of bacteria which feed upon other bacteria in the effluent and facilitate purification. Large voids at the outset are necessary since formation of bacteria tends to reduce void openings in time.

The size of stone preferred is $1\frac{1}{2}$ to $2\frac{1}{2}$ in. or with higher rate filters, 2×3 in., with very low tolerances for plus or minus sized particles (5 to 10%). A severe soundness test, more rigorous than for any construction aggregate, is specified. Stone should be free of dust and impurities, like pyrite, marcasite, and clay. Fine-grained stone of low moisture absorption is the objective. As a result, this type of stone is relatively high-priced. A filter bed may endure for years with little or no replacement of stone. Most of this tonnage is provided by limestone, but blast-furnace slag has also been used.

Acid Neutralization. Chapter 11 contains information on limestone in acid neutralization[33] of industrial trade wastes and in chemical processing. Tables 11-2 and 11-3 on p. 368 and p. 369 provide a comparison of limestone with lime and other alkalis used for this purpose. No statistics are available on this use of limestone, but it is conjectured that the amount is fairly substantial, probably about 100,000 tons a year.

It has been employed in lump form, as discrete granules of largely #4 to #20 mesh sizes, and as pulverized limestone of No. 100 mesh. The efficiency of its use in lump form is highly questionable since most of the limestone is wasted due to formation of a calcium sulfate shell that impedes neutralization. High calcium stone of comparable size is far more efficient in neutralizing than dolomitic because of greater speed of reaction.

CaCO$_3$ wastes from industrial plants are very competitive to commercial pulverized limestone for neutralization.

Other Uses. Other relatively minor tonnages of industrial limestone are reported for *manufacture of various chemicals, dyes, electric products, ingredient in oil well drilling muds, rayon, rice milling, silicones, and water treatment.* In Europe large tonnages of limestone are consumed in the manufacture of *calcium nitrate,* a widely employed nitrogeneous fertilizer through interaction of limestone and nitric acid, but this fertilizer is not made in the United States.

AGRICULTURE

A vast tonnage of limestone enters the agriculture field as direct application to soil, as a filler and conditioner for mixed fertilizer, in animal mineral feeds, as poultry grit, and in a few miscellaneous minor uses.

Direct Liming

Its direct application in liming soils is by far the largest agricultural application with an average annual consumption of 22.5 million tons between 1957 and 1963. Figure 5-4 protrays its growth over the years. For the past eight years its use has plateaued, except for 1963, but note the big increases in 1936, 1940, 1944, and 1947.[37] The prime reason for its decline since its zenith in 1947 is that the Department of Agriculture has not promoted it as vigorously in connection with its soil conservation program, administered by the Agricultural Stabilization Conservation Service and precursory organizations, and funds for subsidizing liming purchases by farmers have been slightly reduced. Another reason has been the declining farmer equivalent income. Inclement weather caused by droughts or abnormally prolonged rainy periods always adversely affects liming consumption. It still has strong support by (and many adherents among) agricultural and agronomic authorities as an essential soil conservation practice to be nurtured and encouraged. On an average, about 75% of all liming tonnage is subsidized. The same financial stimulus to liming is extended by the governments of some European countries, notably England, France, and West Germany. For further information on liming, its history, and how the United States government administers this program, see p. 383.

Liming is principally practiced in the eastern half of the United

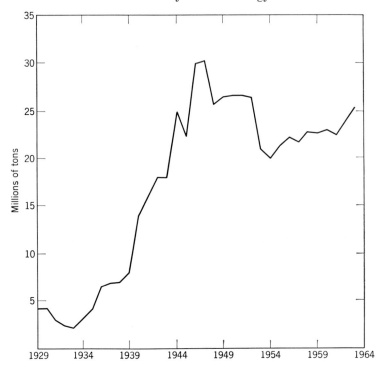

Fig. 5-4. Annual consumption of agricultural liming materials including calcareous materials other than limestone. (From National Lime Association.)

States in a line running roughly east of the 100th meridian from cental Minnesota, through central Nebraska and Kansas, into east-central Texas. West of this line, soils are arid and largely alkaline, and except for a few acid, moist-soil areas in Washington, Oregon, and California, there is virtually no liming. Table 5-7 lists the leading

Table 5-7. Leading agricultural liming states in 1962.

State	Thousands of tons	State	Thousands of tons
1. Illinois	3224	6. Wisconsin	1367
2. Missouri	2330	7. Kentucky	1362
3. Ohio	1883	8. Pennsylvania	917
4. Indiana	1590	9. Iowa	847
5. Tennessee	1400	10. New York	805

Fig. 5-5. Liming application by truck hauled rotary spreader.

under. The presence of lime accelerates decomposition of the sod and foliage and counteracts the acidifying action of decomposition that releases nitric acid.

Heavy or adequate liming applications are generally made every four to six years. However, with sandy soils smaller and more frequent applications are more desirable than with heavier soils. More frequent liming is required in areas of high rainfall, which causes greater loss of calcium and magnesium through leaching and erosion. The intensity of cropping is also a factor. The most dependable determinant on how often and when to lime is systematic pH testing.

Acid-loving Plants. Certain plants do not require much calcium and magnesium and tend to thrive best in acid soils. These are called acid-loving plants, most noteworthy of which are such plants as blueberries, azaleas, most evergreen shrubs, and dogwood trees. Other crops, like corn, oats, and timothy, are tolerant of a moderate degree of acidity but will also grow satisfactorily in neutral and faintly alkaline soils. Potatoes are unique in that optimum growing conditions demand a precise acid pH range of 5.0 to 5.5. Yields of potatoes are adversely affected above and below this range. Peanuts require soils of high soluble calcium but thrive best under moderate acidic conditions. Since liming would elevate pH levels too high for optimum peanut growth, gypsum (calcium sulfate, a neutral salt) is applied instead of carbonate liming materials.

Overliming. Rates of liming should not exceed recommended optimum conditions since the beneficial results obtained usually become progressively less. Heavy liming of an acid soil beyond a pH of 7 is wasteful, although many plants will thrive in a pH range of 7 to 8, but they will be no more productive (usually less) than in a pH range of 6 to 7. To reduce alkalinity of high pH soils or increase acidity of soils for acid-loving plants, sulfur or alum is utilized to lower the pH.

Testing Soils. Experienced agronomists can often visually detect Ca and Mg deficiencies in plants, but even the most experienced rely on soil tests. Depending upon the degree of accuracy desired, pH determinations can be estimated with litmus paper or more precisely by titration with dye indicators, the colors of which alter with changes in pH, or most accurately with the potentiometer or pH meter apparatus. Agricultural county agents provide valuable soil-testing service and perspicacious liming recommendations for farmers.

Total Neutralizing Power. The neutralization efficiency of a liming material is measured by determining its total neutralizing power (TNP) by titration with standard acid solution in the laboratory. An arbitrary value of 100 is given for 100% pure calcium carbonate. Thus, high calcium limestones may have a TNP value ranging between 80 and 98, depending upon the extent of impurities. However, pure magnesium carbonate possesses a higher TNP value, since it has a lower molecular weight but the same neutralizing power per molecular unit as $CaCO_3$. The factor of $MgCO_3$ as $CaCO_3$ equivalent is 1.19 to 1.0. In a pure dolomitic stone of a 60 to 40 ratio of $CaCO_3$ to $MgCO_3$, the TNP is 108. Consequently, dolomitic limestone of comparable purity invariably possesses a higher theoretical TNP value than high calcium stone. A liming material with a TNP of 80 would require theoretically a 20% heavier application for equivalent neutralizing power or would be worth about 20% less in price than one with a TNP of 100. Relative neutralizing values of various liming materials are depicted in Table 5-11.

Most states, in their liming laws or specifications, and the federal government, in its purchases and subsidies of liming materials under its Agricultural Conservation and Stabilization Administration (formerly AAA and PMA), recognize the TNP principle. So by establishing an arbitrary TNP value of, say, 85 or 90, depending upon the quality of available material, bonuses or penalties on bid prices are

Table 5-11. Relative neutralizing values of pure liming materials[a]

Liming materials	Relative neutralizing values	Pounds of liming materials equivalent to 1 ton	
		Calcium carbonate	Calcium oxide
Calcium carbonate	100	2,000	3,570
Magnesium carbonate	119	1,680	3,000
Calcium oxide	178	1,120	2,000
Magnesium oxide	250	800	1,430
Calcium hydrate	135	1,480	2,640
Magnesium hydrate	172	1,160	2,070
Dolomite	108	1,850	3,330
Dolomitic hydrate	175	1,145	2,040

[a] Assumes all liming materials are 100% pure.

employed to evaluate which material, high, low, or intermediate quality, provide the most TNP for the dollar.

Fineness. However, in addition to TNP, fineness and particle gradation is another important consideration in evaluating liming materials. Although there is disagreement among agronomists on the relative value of coarse versus fine liming materials and the degree of fineness to yield optimum results per dollar, there is general accord that neutralizing efficiency rises as fineness is increased. Stone particles of No. 100 mesh dissolve and react with soil acids in usually two to four weeks, whereas No. 10 mesh or limestone meal require six to twelve months, or longer. If soil tests before planting reveal that liming in needed, the coarser fractions are of little or no value in the ensuing crop. Eventually the coarse material will react, but one or two crops will suffer in the meantime. The benefit from the fine material eventuates with the *first* crop. As a result, the faster-acting pulverulent fractions behave more like a fertilizer; the coarse resembles more a long-range soil conditioner. Middle Western agronomists tend to disparage this theory by contending that the coarse material provides delayed neutralization reaction and liming value for one to three years hence, an asset in four-year liming programs. Since generally the prices of limestone accompany increases in fineness, they believe that farmers will be more amenable to purchase their full requirements of the lowest-price (coarse) material. Unfortunately

Table 5-12. Influence of limestone fineness on alfalfa yield in pot cultures and on pH, exchangeable Y, Ca, and Mg in Canfield silt loam[a]

Particle size, screen	pH	Yield of alfalfa per pot	Exchangeable ions per 100 gm. soil		Average relative efficiency	pH	Yield of alfalfa per pot	Exchangeable ions per 100 gm. soil			Average relative efficiency
			Hydrogen decrease	Calcium increase				Hydrogen decrease	Calcium increase	Magnesium increase	
			Calcitic limestone[b]					*Dolomitic limestone*[b]			
		gm.	*me.*	*me.*			*gm.*	*me.*	*me.*	*me.*	
No lime	4.96	4.60	7.26	2.48		4.96	4.60	7.26	2.48	0.56	
<100	6.45	9.51	3.24	6.78		6.56	8.39	3.22	4.77	3.14	
			Relative values					*Relative values*			
No lime	0	0	0	0	0	0	0	0	0	0	1
4–8	−3	16	7	2	5	5	10	6	8	10	8
20–30	44	60	51	61	54	35	49	38	37	37	39
40–50	62	72	76	90	74	53	89	61	56	65	65
60–80	91	99	90	103	96	80	82	92	90	96	84
<100	100	100	100	100	100	100	100	100	100	100	100
Screenings[c]	52	72	53	24	50	40	69	43	41	38	46
Meal[c]	62	77	61	37	59	44	81	50	31	50	51
Ground[c]	79	85	82	88	84	67	102	76	84	75	81

[a] Reference 34 at end of chapter.
[b] Calcitic and dolomitic limestones applied at 3 tons per acre.
[c] Agricultural limestone grades.

these materials are so slow to act that the long-range results are often not visible to the farmer, causing him to be apathetic about liming. Table 5-12 presents research results by the Ohio State Agricultural Experiment Station[34] on the relative speed of neutralization influenced by fineness of limestone particles. There are many varied state specifications based on particle size requirements, ranging from limestone meal or no. 8 mesh to no. 100 mesh. Extra credit in bids is allowed for greater fineness by some procurement agencies.

Types of Liming Materials—Limestone provides 97.5% of all agricultural liming material; usually it is of the high calcium, magnesian, or dolomitic type, but it also includes some shell and marble fines. The remaining 2.5% is divided among such more concentrated and expensive liming materials as ground burnt and hydrated lime, described in detail on p. 382, and other less concentrated materials and calcareous wastes, such as blast-furnace slag, marl, lead mine chats, beet sugar and paper mill refuse, and water softening sludge. Actually any calcareous material can be used for liming, providing the material is ground or preferably pulverized and assuming a sufficient amount is applied.

Most dolomitic limestone that is used is finely ground since coarse material is too slow in reacting to be of any practical value. Dolomite of equal fineness with high calcium stone is invariably slower reacting. For this reason some buyers will not recognize dolomite's greater comparative neutralizing power and will equate both types at parity. In an academic sense there should be no competition between calcium and magnesium. Both elements are essential plant nutrients and both possess compensating inherent individual advantages that dictate the preferential use of one or the other under various circumstances or sets of conditions. County agents make impartial recommendations on which type to use (assuming both are economically available).

Fertilizer Fillers

Fertilizer compounders frequently add pulverized dolomitic limestone up to about 250 lb./ton of mixed fertilizer.[40] It serves as a filler or diluent, but unlike other siliceous fillers that are inert and valueless, the dolomite contributes Ca and Mg as plant nutrients and counteracts acidity stemming from acidic nitrogen chemicals. By proper proportioning of the fertilizer ingredients, limestone can create a so-called *physiologically neutral fertilizer*.[41] In general, 4 lb. of dolomite will counteract acidity of 1 lb. of ammonium sulfate, and 2.5 lb. are needed for ammonium nitrate and urea.[42] It also helps prevent bag rot caused by traces of free acid in the superphosphate component of fertilizer and conditions the fertilizer so as to reduce caking. Its use paralleled the steady growth of mixed fertilizers, but has ebbed in the past five years, since more concentrated fertilizers, which require less filler, are being produced.

High calcium limestone is rarely employed in mechanical fertilizer mixtures, since, being more reactive than dolomitic, there is some danger that it may release ammonia from the fertilizer. Also an excessive calcium content tends to reduce the availability of phosphorus in the fertilizer. Furthermore, there are agricultural areas that are frequently deficient in magnesium, so dolomite helps to alleviate this problem. Great fineness is not desired since stone of fine, sandlike consistency (No.s 15 to 80 mesh) tends to make the fertilizer more free-flowing and approximates the gradation of the other plant foods. However, this source of liming never replaces direct liming, since the amount applied with fertilizers is only a small percent of average liming requirements, as can be concluded from the liming rates in Table 5-10 on p. 108. It is just a minor supplementary source of lime.

Cal-Nitro is a well-established nitrogeneous fertilizer, a mixture of ammonium nitrate and high calcium limestone. *Calcium nitrate* is widely employed in Europe and is chemically precipitated by reacting nitric acid with high calcium limestone. For some unaccountable reason calcium nitrate has never been produced in any quantity in the United States. Except for a tendency to cake, it is theoretically the most ideal type of nitrogeneous fertilizer, being neutral to faintly alkaline and possessing 100% active plant nutrients. *Calcium cyanimide* is discussed on p. 356.

Mineral Feed. There is a steadily expanding application of high calcium limestone as an ingredient in mineral feed supplements, which are used to fortify organic feeds for livestock and poultry.[44] In tonnage it has soared from only 25,000 tons in 1929 to 618,000 tons in 1963 (692,000 tons in 1962). This scientific feeding in animal husbandry promotes more rapid and healthier growth. Since bone structure is almost entirely composed of calcium and phosphorus, it logically follows that these elements are among the principal constituents added to mineral feed supplements, usually second to salt in quantity.

The principal sources of calcium for mineral feeds are derived from finely pulverized high calcium limestone, bone meal, and dicalcium phosphate, in that order. The latter two materials are the prime source of phosphorus. Usually calcium is added at a higher ratio than phosphorus by about 1.3–1.5 to 1.0. The importance of adding sufficient calcium to proprietary mineral feeds is safeguarded in many areas by state laws. A minimum of 35% available calcium is frequently specified. There are also specifications on the particle size of limestone, such as 95% passing a No. 100 sieve or even finer, substantially No. 200 mesh material, as well as tolerances on such toxic impurities as arsenic and fluorine. Such untoxic impurities as silica and alumina are generally regarded as somewhat deleterious. As a result, only limestone of high purity qualifies for this use of 98 to 98.5% minimum $CaCO_3$ content. Dolomitic limestone is not employed, since little (if any) value is attached to its magnesium content. A small amount of magnesium is provided in some supplements in the form of epsom salt ($MgSO_4$).

The three principal supplements—salt, calcium, and phosphorus —are either added by the farmer to his organic feeds and mashes or are sold to the compounders of mineral feed supplements, who usually add other minerals, such as iron, copper, manganese, iodine, magnesium, and various trace elements to the mixture. Quite a number of limestone producers perform this compounding service themselves and actually market mineral feeds.

Poultry Grits. Granular forms of shell and limestone are consumed by hatcheries as poultry grits.[3] The grits lodge in the gizzard, enabling poultry to grind its feed intake properly. A uniform size gradation, without fines and coarse aggregate, is required. $\frac{3}{8}$, $\frac{1}{4}$, and $\frac{1}{8}$ in. sizes predominate. The purity of the carbonate is unimportant; in fact, siliceous aggregates, primarily granite, are used to some extent. So it is the physical size and shape of the aggregate that is important. Rounded or spherical granules are preferred to flat or elongated ones. There is a decided preference for carbonate types of poultry grits, presumably since there is evidence that the calcium nutrient is absorbed by the poultry, abetting egg formation. This predominance is evidenced by the following consumption figures in 1962 with oyster shells an obvious favorite:

	Thousands of tons
Oyster shell	587
Limestone	161
Granite	197
Total	945

Some English poultry authorities contend that absorption of too much calcium through limestone or shell assimilation can be injurious, and they urge partial substitution of insoluble siliceous grits for carbonates. A proportion of 1 to 3–4 granite to carbonates is advocated with about 2 oz. of this mixture fed per chicken per week. Other theories exist.

Miscellaneous Uses. In recent years some direct liming of *forest saplings* and *orchard trees* has been initiated in the United States. Such liming is much more extensive in Germany, where it has been employed for many years. In fact, a 229-page book has been written expressly on this—*Der Wald Braucht Kalk* (*The Forest Needs Lime*). Frequently, pulverized limestone is applied to the floors of *dairy barns* as a mild sanitizing agent and is often used jointly with lime whitewash. Cows are less prone to slip on a floor covered with limestone fines than on a wet whitewashed surface. A very small amount of pulverized limestone has been used with hydrated lime in restoring boggy, turbid *lakes and ponds,* which is described on p. 386.

There is increasing evidence that liberal liming programs may benefit mankind in still another way. In *radioactive fall-out,*[43] strontium 90 is particularly dreaded, since it is readily absorbed by plants from the soil over a long period of time and is ultimately assimilated by livestock from pastures and by humans from their food intake. It tends to accumulate in the skeletal structure and is considered a potential cause of bone tumors and leukemia. It has been repeatedly proven in tests by the Department of Agriculture and Atomic Energy Commission that plants absorb much less strontium 90 from well-limed soils than from soils deficient in calcium and that the ratio of strontium 90 to the calcium cation is markedly reduced in soils containing a high exchangeable calcium level. The reduction in absorption of strontium 90 varies greatly depending upon conditions, but reports range from about 25 to 80%.

CONSTRUCTION

Disregarding its established usage as a dimension stone in building, which is not included in this book, essentially limestone's role in construction lies in its many applications as a coarse and fine mineral aggregate, embracing countless sizes and gradations for all forms of portland cement concrete work, portland cement and bituminous concrete highway pavements, road "metal" for highway base courses, surface courses for unpaved and bituminous macadam roads, ballast for railroad beds, rip rap, fills, and for various building materials and structural purposes. A much smaller but also important tonnage is consumed as a filler in finely pulverulent form.

Competitive Aggregates. Although construction easily constitutes its largest use category, limestone encounters considerable competition from other mineral aggregates as Table 5-13 clearly indicates.[1] *Sand and gravel* combined exceeds limestone, which ranks a strong second; other stone comprising *trap rock, basalt, granite, sandstone, quartzite, miscellaneous crushed and broken stone,* and *blast-furnace slag* offer formidable competition, although limestone consumption averages nearly 75% of the total *stone* aggregate production. Consequently, its application as an aggregate is not indispensible. There is no construction use that limestone dominates 100%.

A comparison of the average prices per ton in 1961 for all of these species of aggregate is also contained in Table 5-13. Note that

Table 5-13. Comparison of all construction aggregates (1961 U. S. Bu. Mines figures in thousands of tons)

Construction Use	Lime-stone^a	Granite	Basalt and trap rock	Sandstone and quartzite	Other stone types^d	Total stone^f	Blast-furnace slag	Gravel	Sand
Average price/ton	$1.31	$1.46	$1.51	$1.36	$1.13	$1.41	$1.81	$1.11	$1.11
Concrete aggregate and road "metal"	273,306^b	36,658	50,456	14,688	15,289	390,398	15,400	423,000	234,000
Railroad ballast	4,260	1,291	1,561	771	1,734	7,883	2,841	3,232	517
Rip rap	9,138	3,051	8,935	4,061	1,900	27,085	**	—	—
Stone sand	1,693	1,555	65	*	*	3,527	—	—	—
Fill material	266	660	1,719	*	1,572	4,325	—	30,119	26,924
Asphalt filler	2,130	—	*	*	*	f	—	—	—
Roofing granules	*	**	**	*	*	2,114	376	—	—
Concrete products aggregate	**	—	*	—	—	f	2,800^e	—	—
Terrazzo	397^c	—	—	*	*	407	—	—	—

^a Includes shell and marble.

^b Includes 14,698 of shell.

^c This amount is marble; there is also an indeterminate amount of limestone.

^d Includes volcanic rock, chert, flint conglomerates, limestone chats.

^e Largely lightweight aggregate made from slag.

^f Does not include in subtotals indeterminate amounts for asphalt filler, concrete products, stucco, cast stone, tile, patching plaster, etc. This unallocated tonnage is listed at 16,388.

* There is an indeterminate amount used.

** There is an indeterminate amount used, believed to be substantial.

sand and gravel has the lowest cost, which along with its extremely widespread availability (it occurs in or near most communities), probably accounts for its preeminence. Generally, however, limestone is considered superior to gravel, since its sharp, angular, chunky shape provides a stronger interlocking action and greater density with the fine aggregates than the smooth, rounded shape inherent to sand and gravel. Thus, this disparity in prices may be further explained.

Values of some other siliceous stone products, notably trap rock, basalt and granite, exceed limestone. On an average these aggregates are usually appreciably harder and possess more abrasive resistance than limestone as well as possessing the same desirable sharp, angular, interlocking qualities. As a result, they are preferred by some engineers, but they are not as extensively produced as limestone and sand and gravel. The other types of stone listed in Table 5-13 on an average are inferior to limestone as aggregates.

Blast-furnace slag[48] is the most costly aggregate per ton, but this is deceptive because of its lower specific gravity and bulk density (70 to 87 lb./ft.^{3}), it covers a given area with much less weight than other aggregates, fully compensating for its higher price per ton. Its hardness, strength, and soundness are sufficiently established, in spite of its porous, vesicular surface, so that it is only required to meet a 70 lb./ft.^{3} density requirement. However, as of 1964 the demand for this aggregate was greater than the supply, due primarily to decreased production as a result of technological changes in blast-furnace practice involving the use of higher iron ore concentrates and less flux stone, as described on p. 82.

Relatively smaller tonnages of shell and waste spalls from marble operations are another carbonate source of aggregate. Most of the shell aggregate is consumed with sand in coastal areas on the Gulf of Mexico where, except for shell, there would be a great dearth of local coarse aggregate. Recent indications point to possible fairly imminent depletion of existing off-shore shell beds, and conservation regulations are being proposed by one state (Texas) to restrict the use of shell for seaboard industrial consumers, like portland cement, lime, and chemical plants that for economy reasons have employed it instead of inland sources of limestone.

Gradation. Table 5-14 displays a large number of the major aggregate sizes for the principal construction applications under the Simplified Practice Recommendation (R163-48) on "Coarse Aggregate Crushed Stone, Gravel, and Slag," promulgated by the National Bureau of Standards, Department of Commerce. These are the same sizes depicted in Table 4-2 on p. 69. This table does not include

Table 5-14. Various construction uses for standard aggregate gradations

Use	Size number and nominal size[a]																						
	1	1-F	2-F	2	24	3	357	4	467	5	56	57	6	67	68	7	78	8	89	9	10	G1	G2	G3
	3½ to 1½	3½ to 2	3 to 1½	2½ to 1½	2½ to ¾	2 to 1	2 to No. 4	1½ to ¾	1½ to No. 4	1 to ½	1 to ⅜	1 to No. 4	¾ to ⅜	¾ to No. 4	¾ to No. 8	½ to No. 4	½ to No. 8	⅜ to No. 8	⅜ to No. 16	No. 4 to No. 16	No. 4 to 0	1½ to No. 50	1½ to No. 8	1½ to No. 4
Water-bound macadam:																							
Coarse aggregate	x			x		x																	
Filler																					x			
Bituminous macadam, penetration method:																							
Coarse aggregate	x			x		x																	
Choke										x	x		x	x			x							
Seal																		x	x	x				
Bituminous plant mixes, base or surface courses:[b]																							
Base, open mix					x	x		x			x													
Base, closed mix							x		x			x		x										
Binder course												x		x										
Surface course• coarse grading										x	x	x		x										
Surface course• fine grading																	x	x	x	x				
Seal																	x	x	x	x				
Bituminous road mix:																							
Mixing course														x										
Choke																		x	x	x				
Seal																		x	x	x				
Drag leveling course:																							
Leveling course														x	x		x							
Seal																		x	x	x				
Bituminous surface treatment										[c]x	[c]x		x	x	x	x	x	x	x	x				
Seal for airport construction																			x	x				
Portland cement concrete	x			x		x	x	x	x			x		x		x								
Railroad ballast:																							
Stone or slag					x	x		x			x	x												
Gravel																						x	x	x
Roofing																		x						
Sewage trickling filter media		x	x																					

[a] In inches, except where otherwise indicated. Numbered sieves are those of the United States Standard Sieve Series.
[b] For plant mixes the aggregate should consist of appropriate sizes selected from table 4-2 combined with suitably graded fine aggregate.
[c] Bottom course of multiple surface treatment.

numerous individual specifications and very broad gradations that may have size distributions ranging from $2\frac{1}{2}$ in. to no. 325 mesh, etc. It was conceived as a means of standardizing sizes and to disparage special, odd sizes. Oftentimes the contractor will purchase two or more of these sizes for blending his own peculiar gradation requirement. In the succeeding use descriptions, reference should be made to this table on gradation requirements, since the sizes will not be repeated.

Some *definition* is needed to define coarse and fine aggregate. Generally, material that passes a no. 4 mesh sieve is considered fine aggregate; that which is retained on the same screen is coarse.

Specifications. There are various national specifications on construction aggregates, the most noteworthy of which are by the American Society of Testing Materials (ASTM), American Association of State Highway Officials (AASHO), U. S. Engineers, and the Federal Specification Board that are regarded generally as guide or referee specifications, often of minimal quality. Frequently, local or state specifications are much more stringent, particularly if more exacting gradations can be obtained locally at reasonable cost.[45, 47] State highway departments are continually evaluating local aggregates in their own laboratories. In fact, a vast amount of research and testing has been conducted to determine the optimum sizing and proportioning of aggregates for maximum density, strength, durability, and soundness (and in portland cement concrete with the minimum amount of the more expensive cement binder). In many construction uses the gradation that produces the lowest percentage of voids and maximum density is the most desirable, but in other applications gradations with narrow limits yield high voids and low density. There is, however, considerable disagreement among engineers on their evaluation of limestone with other aggregates and their gradation and design theories and criteria. In order to stimulate competition among the aggregate industries, most specifications are written to include all or most of the available mineral aggregates that have demonstrated reasonable performance in the field and in laboratory test. Yet occasionally individual project specifications may be restrictive to only one or two specified aggregate that, in the opinion of the engineer, have exhibited superiority for a given use.

ASTM specifications and methods of test in particular, along with a few of the other more significant tests, are described on pp. 460–71. These and most standards are predicated on the following general aggregate requirements.

Requirements. In addition to *gradation* requirements there are numerous physical and some chemical tests that are often imposed.[46] Aggregate should be reasonably *clean*. In case of coarse aggregates this would mean free of stone dust and fines as well as clay, silt, and soil. Stone fragments should be *hard, strong,* and *durable,* more *cubical in shape* and not laminated or containing incipient cracks. Flat and elongated fractions, even though hard, are undesirable, since they interlock poorly and contribute to excessive voids. With portland cement concrete, in particular, aggregates should also be *chemically sound. Excessive impurities* in the stone of soluble sulfides, alkali salts, and organic matter are objectionable since they can produce an adverse reaction with cement, leading to concrete of poor or suspect durability. Impurities, such as silica, alumina, and iron, are generally not deleterious and are aften regarded as beneficial, since they may contribute hardness to the stone. Stone *color* is usually unimportant except for a few special building products, ornamental facing effects, or where high reflectivity is desired. The most important tests are enumerated as follows:

1. *Dorry hardness*—measures resistance to abrasion.
2. *Los Angeles abrasion*—most widely used measurement of abrasion.
3. *Paget impact*—measures toughness and resistance to impact.
4. *Specific gravity*—measures density.
5. *Absorption*—measures porosity and weatherability after obtaining specific gravity.
6. *Compressive strength*—determines average strength value under compression in psi.
7. *Cementing value*—determines suitability of stone in concrete.
8. *Sodium sulfate and magnesium sulfate*—measures soundness.
9. *Alkali reactivity*—determines whether stone is reactive with alkalis contained in portland cement.
10. *Freezing and thawing*—determines soundness of porous stone under simulated winter weather conditions.

Descriptions of most of above tests are found in Chapter 13.

Portland Cement Concrete. The physical properties of aggregate employed in portland cement concrete for all purposes is important since they influence its workability, strength, and durability. Hardness, resistance to abrasion, and soundness are vital to concrete.[50] In addition, gradation is extremely important. Generally the most desirable gradation for concrete is so-called straight-line grading, which provides equal percentage proportions for all standard sizes.

Any deviation from this norm would be to specify higher percentages of fine aggregate of #10 mesh and finer, thus providing an ellipse toward the end of the straight-line gradient.

Most limestone that is reasonably hard provides a highly satisfactory coarse and fine aggregate, with the latter derived from limestone sand. Coarse limestone aggregate is also compatible with siliceous sand. Primarily, coarse fractions serve as a filling material (in addition to imparting strength) in concrete and have the objective of reducing to a minimum the more costly portland cement. In so doing, it tends to reduce shrinkage in concrete. Most engineers strive to introduce the maximum amount of coarse aggregate in concrete commensurate with the necessary degree of workability (or consistency) that is required. Too much coarse aggregate produces harsh working concrete.

Top sizes of aggregate and mix proportions vary greatly, depending upon application—dams, highways, walls, foundations, footings, revetments, curbs, sidewalks, reinforced concrete, bridges, concrete products, etc. A few typical concrete designs are enumerated below:[49]

		By Volume (cu. yd. or cu. ft.)		
Portland cement	*Fine aggreg.*	*Coarse aggreg.*	*Water (gal.)*	*Compressive strength, psi*
1	1 to 1.5	2 to 3	4.5	5,500–6,000
1	1.5 to 2	3 to 3.5	5.5	4,500–5,000
1	2 to 2.5	3.5 to 4	6.75	3,000–4,000

Proportioning of concrete is generally determined by the water—cement ratio theory (proportion of cement to mixing water), arbitrary volume proportions, or compressive strength and a combination of these factors.

Stone is used by contractors, who operate concrete batching plants or mixers at the job site or by commercial central-mix plants (ready-mixed concrete), who use transit mix trucks for delivery.

Road Stone. This is a general term embracing uses of coarse, fine, and graded aggregate in constructing or maintaining road bases and subbases and as an aggregate in various types of bituminous pavement surfaces.[51] In all cases it is added to the soil and/or aggregate in situ or the bitumen (asphalt or tar) to improve gradation, stability, strength, and durability of the road or road layer. Figure 5-6 shows a few typical highway profiles which distinguish between the various sections that comprise a road. Limestone may be employed in one or all of these highway layers, described briefly as follows:

1. *Macadam-type Bases*

a. Traffic-bound macadam. Aggregate is spread on an unpaved road surface to a depth of several inches. Part of the aggregate is windrowed to both sides of the roadway. Traffic action then embeds the aggregate gradually into the base or subbase (subgrade). The windrowed material is periodically bladed onto the road, so that the traffic compacts more aggregate into the road. This method is extensively employed on secondary and tertiary roads, detours, and temporary haul roads.

b. Bank-run macadam. Coarse aggregate is spread on the road subgrade to a depth of 4 to 8 in. Then, after fine aggregate (screenings) is added to fill the voids, water is applied and the aggregate is densified with compaction equipment.

c. Dry macadam. Same procedure is followed as in bank-run macadam except that no water is added. Screenings generally comprise about 25% of the aggregate that is applied.

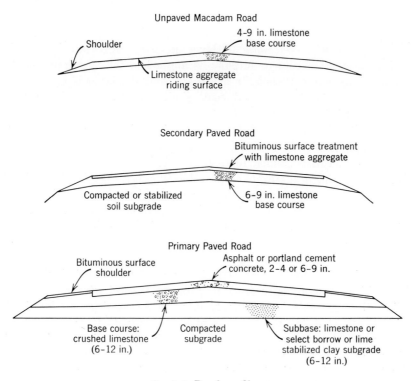

Fig. 5-6. Road profiles.

d. Water-bound macadam. One or more layers of graded aggregate is spread on a well compacted subgrade or subbase. Some fine aggregate is added to fill voids, and the base is watered and compacted. Usually this base is ultimately covered with a bituminous wearing surface in stage construction and represents the highest type of unpaved road.

2. *Bituminous Surfaces*

a. Road-mix. Bitumen and graded aggregate are mixed in place on the road base, followed by compaction. This is usually considered the lowest quality paved surface, since in road mixing bitumen and aggregate it is difficult to obtain homogeneity in the mix, so that this surface is prone to ravel and disintegrate.

b. Surface treatment. Bitumen, usually hot asphalt, is spread on the road base or worn-out bituminous surface. One inch of coarse aggregate is then spread uniformly over the surface, impregnating into the asphalt. Usually, multiple treatments are applied as stage construction. When the second asphalt coat is applied, the second application of aggregate is finer in size than the first coat. This is also called an inverted penetration treatment. If distress occurs to the surface, a third or fourth such coat may be applied.

c. Penetration macadam. A layer of coarse aggregate is spread, and a penetration coat of bitumen is applied to the aggregate. Voids are filled with fine ("choke") aggregate, and the layer is compacted with rollers. Often, a second coat, similar to the first, is applied up to about 3 in. thickness.

d. Bituminous or asphalt concrete. This is also called a "hot mix" application, since usually 6 to 7% of asphalt, the optimum amount that can be added for stability, is heated and mixed with a dense, well-graded aggregate in portable-type central mix plants. The mixture is applied to the road base or onto old pavement and is compacted by rolling. It is 1 to 4 in. thick, depending upon the character of road, volume of traffic, etc. Generally this is considered to be the highest-quality bituminous pavement and is most commonly employed on primary roads and parking areas.

e. Bituminous Cold Mix. This is another form of asphalt concrete but utilizes a more open-graded aggregate instead of dense graded. When the aggregate is completely dried in rotary driers, it is mixed with the heated bitumen in a central mix plant. It is applied to the road at a lower temperature than the "hot mix" method, 2*d* above, and compacted similarly. It is widely used for maintenance, patching "pot-holes," and resurfacing distressed road sections.

f. Sheet asphalt. This is similar to 2*d* except no coarse aggregate is used. Instead, only fine aggregate and mineral dusts are employed. It provides a dense, smooth form of asphalt concrete that is applied extensively to municipal streets.

3. *Miscellaneous Highway Uses.* Often 2 to 3 in. of fine aggregate is applied as a blanket underneath a graded aggregate base to improve *drainage* ("French" drains). After underground water and sewage laterals or drainage conduits are installed, the accepted practice in *backfilling* is to add 3 to 6 in. of graded aggregate on top of the conduit and compact thoroughly.

For all highway purposes hardness and resistance to abrasion are the all-important requirement of aggregate. For this reason many soft, porous limestones are totally unsuitable as coarse aggregate, but such material might be employed as fine aggregate unless they are extremely soft, like chalk. A soft, corralline Florida limestone that is largely unsuitable as a coarse aggregate is extensively used as a fine aggregate in stabilizing and hardening sandy and silty soils as road bases for flexible road pavements. In all instances, however, finely pulverized limestone can be employed as mineral fillers and dusts since strength is not a factor.

In bituminous construction a prime consideration is whether the aggregate has a stronger affinity for water than for asphalt. In asphalt concrete many aggregates, often the siliceous types, after prolonged exposure to moisture, are displaced from their adhesion to the asphalt binder, resulting in pavement disintegration and raveling. This is called *stripping*. Such aggregates are classified as *hydrophilic* (they attract water). Other aggregates that exhibit a positive adhesion for bitumens and resist displacement by water are the *hydrophobic*. Basic aggregate, like limestone, tends to qualify for the latter type, more so than gravel and siliceous stones, which are often hydrophilic in their reaction. As a result, the stripping propensity of a given aggregate and bitumen are usually predetermined by laboratory test before use of the materials is approved. However, many limestones have a compensating disadvantage in that they tend to polish to a slippery smoothness under prolonged traffic action, thus contributing a safety hazard in wet weather. Generally, siliceous stone, like granite and basalt, resist the abrasive (polishing) effect of repeated wheel loads better than most limestones. However, a minority of limestones are eminently satisfactory.

As a result, considerable laboratory testing is performed on materials used in asphalt concrete, but discordant opinions result in evaluation of test results.

Asphalt Filler. In design of asphalt concrete mixes, tests frequently reveal that greater density and stability are achieved if the percentage of voids that exist in the aggregates can be reduced to the lowest practical minimum. As a result, mineral dusts in amounts ranging from 1 to 10% (about 4% average) are added to the aggregate mixture. These fines are largely Nos. 200 to 325 mesh sizes, and they should be chemically inert in contact with bitumen for use as filler.

Finely pulverized limestone is the most commonly used mineral filler for this purpose. Specifications vary somewhat on the fineness specified but generally adhere to minimum criteria established by ASTM of 100% passing a No. 30 mesh and a minimum of 60% passing a No. 200 sieve.

Other competitive mineral dusts are siliceous stones, like granite, diatomaceous earth, asbestos fines, kaolin clay, portland cement, and talc. In 1962 limestone dominated this market with 3.4 million tons of the estimated 5.5 million tons consumed. Inexplicitly, limestone declined to about 2 million tons in 1963. This use for limestone has grown steadily from only 177,000 tons in 1932 and 709,000 tons in 1953. Whiting would be too costly to employ for this purpose, and it is doubtful if its finer particle size would be much more beneficial. For further information see p. 423 on the application of hydrated lime as an asphalt additive.

Limestone Sand. Limestone sand that simulates silica sand in gradation and particle shape is increasingly utilized in construction as a fine aggregate in concrete, mortar, and road construction. Its extent of use, however, will never jeopardize silica sand's domination of this market. Furthermore, due to considerable regrading necessary in its processing, it cannot compete cost-wise with silica sand in most locations. It is produced in a size distribution of 98 to 100% passing a No. 8 sieve with only 0 to 2% passing a No. 100 mesh. The advent of air entraining agents in portland cement concrete has enhanced limestone sand's workability characteristic, some of which with normal concrete is regarded as harsh-working.

Small amounts of white marble sand have been added to special lime *plaster and stucco-finish coats* and even silica sand–float finish coats to lend more body, strength, and sheen to the finish.

Railroad Ballast. It has the same general aggregate requirements that exist for road stone in the maintenance and construction of railroad beds. The American Railway Engineering Association has prepared a specification for railroad ballast. A gradation of $2\frac{1}{2}$ to $\frac{3}{4}$ in. is stipulated, but they have relaxed this provision to permit

5 to 10% of the material to be $\frac{3}{8}$ in. and in some instances as low as No. 4 mesh. In abrasive resistance the AREA requires stone to meet a maximum of 40% wear in the Los Angeles abrasion test. Proportionately there is much more stone than gravel used for this purpose. In 1963 limestone consumption was 4.9 million tons.

Rip Rap. This consists of heavy irregular stone fragments, some of which may be as large as boulders. For *breakwater* work "one-man" stones of up to 15 tons in weight are employed. Rip rap is principally used in river and harbor work in the construction of spillways for dams, shore protection bulkheads, jetties, piers, and docks. Land-based "dry rock" walls offer another outlet. Any dense, sound type of rock may be used. Since no crushing or grinding is employed, it is usually the lowest cost broken stone product, unless its size and shape are specified by the consumer. Its principal competition would be much more costly poured portland cement concrete. Annual consumption of limestone rip rap is 9 to 10 million tons.

Fill. In 1963, 338,000 tons of limestone were employed as a fill material in elevating swampy or low-lying areas to the desired grade in highway, railroad right-of-way construction, in development of industrial plant areas, and in urban redevelopment projects. Usually the fill is mechanically compacted to provide a more stable, stronger foundation than ordinary earth fills. Yet some stone fills are permitted to consolidate by normal settlement over a period of years. Some gradation is needed in the fill to obtain adequate density. Stone would represent the most costly fill material.

Roofing Granules. There is an indeterminate amount of limestone and shell used in roofing granules and believed to be relatively small in the total of mineral aggregates consumed for this purpose.[3d] Of the total of 2.7 million tons of stone products employed as granules, limestone and shell comprise only an estimated 10 to 15% of the total. Siliceous rock, notably granite, basalt, trap rock, and sandstone, constitute the balance. Blast-furnace slag also enjoys extensive use, but more as a cover material for built-up roofs (about 350,000 tons per year in roofing). Siliceous stone's dominance is largely due to its superior weathering qualities, being completely insoluble.

White crystalline limestone or marble chips of $\frac{3}{8}$ to $\frac{1}{4}$ in. are favored as the final coat on certain types of flat *built-up roofs* on residential homes because of aesthetic considerations and higher reflectivity. Some limestone is employed in other types of built-up roofs for factories and apartment buildings, where aggregate is imbedded thickly into a bituminous surface to minimize weathering and particularly to provide melting resistance against the destructive effect of the sun's ray. Gradation of such aggregate is nominally ¾ in. to No. 8 mesh, according to ASTM specification D1863-61T ("Mineral Aggregate for Use on Built-up Roofs").

Granules for composition roofing (asphalt shingles and related types) are supplied from siliceous stone aggregates and have an average gradation of No.s 12 to 40 mesh.

Stucco. Limestone chips usually of ⅜ to ⅛ in. are employed competitively with other aggregates in the finish coats of textured exterior plaster or stucco, and are called stone or pebble dash stucco. The tonnage is indeterminate and relatively very insignificant.

Concrete Products. Coarse and fine aggregate and mineral filler derived from limestone are commonly consumed by concrete products manufacturers in most of the burgeoning, multifarious products of this rapidly growing industry, which involves masonry units (block and brick), pipe, large conduits, coping, curbstones, flagstones, floor units, lintels, monuments, mantles, manhole covers, posts, silos, vaults, wall slabs, bird baths, and many other products.[50] No statistics are available on the quantity of limestone consumed, but it is appreciable, probably over 5 million tons for all types of aggregate and filler. However, all other types of aggregate, including lightweight types, are also widely employed. There appears to be no clear-cut preference for any particular aggregate. Literally hundreds of concrete mix designs are employed in order to comply with local building codes, architect specifications on strength, density, and aesthetic effects.

Gradation of aggregate is very important. For this reason, and to obtain maximum density and watertightness, limestone flour of No. 200 mesh is frequently added to the other aggregates as a void filler and densifier. Often such filler is also added to enhance whiteness in the product.

Precast Concrete. This is another new and growing industry that fabricates large prestressed beams and ornamental wall-facing panels of reinforced and prestressed concrete to meet many individual, specified, complex designs. It also consumes limestone aggregate in increasing amounts. Uultramodern office buildings, hotels, apartments, factories, and institutions of the 1950's and 1960's are often constructed almost entirely with such products along with structural steel and have replaced much unit masonry and other facing materials.

In many of the ornate exterior and interior wall-facing panels, chips of white limestone and marble are embedded rough-cast into the concrete, producing varied architectural effects.

REFERENCES

General References on Stone Uses

1. U. S. Bureau of Mines, *Minerals' Yearbook* (1915–1963), chapters on "Stone" by various authors.
2. A. Searle, *Limestone and Its Products,* E. Benn (London), 709 pp. (1935).
3. O. Bowles, *The Stone Industries,* McGraw-Hill (New York), 519 pp., 2nd Ed. (1939).
4. O. Bowles and N. Jensen, "Limestone & Dolomite in the Chemical & Processing Industries," *U. S. Bu. Mines I.C. 7169,* 15 pp. (Mar. 1941).
5. O. Bowles, "Limestone & Dolomite," *U. S. Bu. Mines I.C. 7738,* 79 pp. (1956).
6. S. Colby, "Occurrences & Uses of Dolomite in U. S.," *U. S. Bu. Mines I.C. 7192,* 21 pp. (1941).
7. P. Hatmaker, "Utilization of Dolomite & High Magnesian Limestone," *U. S. Bu. Mines I.C. 6524,* 18 pp. (Sept. 1931).
8. H. Kriege, "Symposium on Mineral Aggregates," *ASTM Bull.* (1948), pp. 205–219.
9. J. Lamar, "Uses of Limestone & Dolomite," *Ill. State Geol. Surv. Circ. 321,* 41 pp. (1961).
10. J. Thoenen, "The Stone Industry—A 50-Year Review," *Crushed Stone J.* (Dec. 1954), pp. 6–24.
11. G. Whitlatch, "Limestone & Lime," *Tenn. Dept. of Conserv. Markets Circ. 10,* 38 pp. (Apr. 1941).
12. A. Goldbeck, "Crushed Stone," *Industrial Minerals & Rocks,* AIME (1949), pp. 245–293.
13. AIME, *Industrial Minerals & Rocks* (1960)—the following chapters:
    a. J. L. Gillson et al, "The Carbonate Rocks," pp. 123–203.
    b. C. F. Clausen, "Cement Materials," pp. 203–233.
    c. W. W. Keys, "Chalk & Whiting," pp. 233–243.
    d. J. A. Brown, "Granules," pp. 443–454.
    e. A. B. Cummins, "Mineral Fillers," pp. 567–585.
    f. K. M. Ritchie, "Mineral Wool," pp. 595–605.
14. H. Wilson and K. Skinner, "Properties & Preparation of Limestone & Chalk for Whiting," *U. S. Bu. Mines Bull. 395,* 160 pp. (1937).

More Specific Limestone

15. O. Bowles, "Metallurgical Limestone," *U. S. Bu. Mines Bull. 299,* 40 pp. (1929).
16. F. Crockard, "A Single Formula for Evaluating Iron Ores," *Blast Furnace Steel Plant* (March 1953), p. 295.
17. AIME, *Proc. Int'l Symposium on Agglomeration* (1961).
18. R. Raddant, "Limestone, Unsung but Vital Rock," *Iron Age* (May 21, 1953), p. 79.

19. K. Hass, G. Bitsianes, and T. Joseph, "Calcium Ferrites in Relation to Sintering of Iron Ore," *AIME Proc.* Blast Furn. Conf. (1960).
20. Anon., "8 Processes Vie in Alumina Race," *Chem. Eng.* (Oct. 1954), p. 130.
21. D. Blue, "Raw Materials for Aluminum Reduction," *U. S. Bu Mines I.C.* 7675 , 11 pp. (Mar. 1954).
22. Anon., "A New Lime Glass," *Glass Ind.,* V. 17, 8 (Aug. 1936), p. 279.
23. J. Jones, "Dolomite & Its Uses in Glass Making," *Glass,* V. 14 (1937), p. 184.
24. O. L. Cook, *Paper Tr. J.* **92**, 17 (1931), p. 55; **94**, 17 (1932), p. 37; **96** (June 22, 1933), p. 35.
25. O. Bowles, "Industrial Insulation with Mineral Products," *U. S. Bu. Mines I.C. 7263,* 17 pp. (1943).
26. R. Ecker, "Dispersion of Fillers in Rubber," *Rubber Chem. & Tech.* **27** (1954), pp. 899–919.
27. A. Lukens, "How to Choose Extender Pigments," *Am. Paint J.* **37**, 40 (1953), p. 60.
28. H. Roderick and A. Hughes, *Paper Tr. J.* **110** (Feb. 1940), pp. 104; **113** (Oct. 1941), p. 39.
29. R. Bogue, *The Chemistry of Portland Cement,* Reinhold (New York) (1955).
30. F. Lea, *Chemistry of Cement & Concrete,* E. Arnold (London) (1956).
31. American Society of Civil Engineers, "Filtering Materials for Sewage Treatment Plants," *Manual of Eng. Prac.,* 13 (1937).
32. J. Lamar, "Limestone for Sewage Filter Beds," Ill. State Geol. Surv. *R. I. 12* (1927).
33. R. Hoak, C. Lewis, and W. Hodge, "Basicity Factors of Limestone & Lime," *I.&E. Chem.,* **36** (Mar. 1944), p. 274.
34. Symposium on Agricultural Liming (eight articles, various authors) *Soil Science* **73**, no. I, 75 pp. (Jan. 1952).
35. National Lime Association, *100 Questions & Answers on Liming Land,* 32 pp., 4th Ed. (1960).
36. C. Whitaker et al, "Liming Soils for Better Farming," *U. S. Dept. of Agric. Farmers' Bull. 2032* (1951).
37. R. Boynton, "Survey of Liming Material Use," *Rock Prod.* (Oct. 1952), p. 90.
38. F. Parker, "Importance of Ca & Mg to the Nitrogen and Fertilizer Industry," *Proc. Am. Soc. of Agron.* (Nov. 1931).
39. J. Hester, "Magnesium Deficiency," *Farm Chem.* (May 1958).
40. R. Boynton, "Ca & Mg in Mixed Fertilizer" *Chemistry & Technology of Fertilizers,* Reinhold (New York, 1960), pp. 446–453.
41. K. Beeson, W. Ross, "Preparation of Physiologically Neutral Fertilizer Mixtures," *I. & E. Chem.* (Sept. 1934), p. 982.
42. F. Keenan, W. Morgan, "Rate of Dolomite Reactions in Mixed Fertilizers," *I. & E. Chem.* **29**, no. 2 (Feb. 1937), p. 197.
43. Anon., "The Limestone Industry's Interest in Atomic Fall-Out," *Pit & Quarry* (May 1962), p. 168.
44. L. Maynard, "Meeting the Mineral Needs of Farm Animals," Cornell *Ext. Bull. 350,* 30 pp. (Apr. 1936).
45. U. S. Bureau of Public Roads, "Aggregate Gradation for Highways," 26 pp. (May 1962).

46. D. Woolf, "Results of Physical Tests on Road Building Aggregates," *U. S. Bu. Pub. Rds. Bull.* (1953).
47. J. Gray, "Specifications for Mineral Aggregates," *Crushed Stone J.,* **36,** no. 2 (June 1961), p. 3.
48. G. Josephson, *Iron Blast Furnace Slag, U. S. Bu. Mines Bull. 479* (1949).
49. National Ready-Mixed Concrete Association, "Control of Quality of Ready Mixed Concrete," *Publ. 44,* 5th Ed., 51 pp. (Oct. 1962).
50. R. Blanks, H. Kennedy, *The Technology of Cement & Concrete,* **I.** John Wiley (New York, 1955).
51. L. Hewes, *American Highway Practice,* **1, 2,** John Wiley (New York, 1942).
52. L. Dow, "Crushed Stone Sand," *Rock Prod.* (Feb. 1932), p. 63.
53. D. Knill, "Petrographical Aspects of the Polishing of Natural Roadstones," *Crushed Stone, J.* **36,** no. 2 (June 1961), p. 13.

For further references see list following Lime Uses (Chapter 11), some of which also pertain to limestone.

CHAPTER SIX

Theory of Calcination

It is now appropriate to consider the principal chemical property of limestone—its thermal decomposition—in which co-products, lime and carbon dioxide, are formed. It is due to this characteristic that lime manufacturing, another auxiliary industry, was created by a process generally known as calcination . . . as well as most of limestone's direct uses as a raw material or reagent in chemical and metallurgical thermal processes (Chapter 5).

This is probably the most basic and apparently the simplest of all chemical reactions. But while it is theoretically very prosaic (many erudite chemists are even disdainful of it because it is so elementary), there are many complexities attendant to this reaction. In spite of incontrovertible scientific data delineating calcination, this process still remains to some extent a *technique* or an *art* that only an experienced lime burner fully comprehends. There are *numerous variables* that require considerable trial-and-error methods for optimum performance and delicate empirical (often impulsive) modifications for operating efficiency. As evidence of this, there have been high-calibre engineers with imposing technical degrees and fortified by extensive experience in other pyrochemical processes who have encountered complex operating difficulties in lime burning. Such formally trained engineers are simply no match, at least at the outset, for the veteran lime burner, who may have no formal education. The latter is like a "French" chef; the former is a novice cook. The writer is familiar with several instances of large chemical-process companies, esteemed for their research and engineering acumen, who manufacture in captive plants surprisingly poor quality lime with mediocre efficiency and, most pathetic of all, do not seem to know the difference. They have simply regarded the product and process as mundane and have devoted scant study to it. Such oversimplification is costly.

Stated chemically with molecular weights, this reversible reaction for both high calcium and dolomitic quicklime is diagrammed as follows:

$$\overset{100}{CaCO_3} \text{(high calcium limestone)} + \text{heat} \rightleftharpoons \overset{56}{CaO} \text{(h.c. quicklime)} + \overset{44}{CO_2}$$

$$\overset{100}{CaCO_3} \cdot \overset{84}{MgCO_3} \text{(dolomitic limestone)} + \text{heat} \rightleftharpoons$$
$$\underset{56}{CaO} \cdot \underset{40}{MgO} \text{(dol. quicklime)} + \overset{88}{2CO_2}$$

There are three essential factors in the kinetics of limestone's decomposition.

1. The stone must be heated to the dissociation temperature of the carbonates.

2. This minimum temperature (but practically a higher temperature) must be maintained for a certain duration.

3. The carbon dioxide gas that is evolved must be removed.

Dissociation Temperature. Values of $CaCO_3$'s dissociation temperature, developed by Johnston[1] and Mitchell,[2] are still generally recognized and have been well authenticated by other researchers.[3] For calcite it is 898°C (1648°F) for 760 mm. pressure (1° atm.) for a 100% CO_2 atmosphere. The temperature of dolomite, however, is not nearly so explicit. Magnesium carbonate dissociates at a much lower temperature of 402 to 480°C (756 to 896°F), to which a majority of investigators subscribe. The range in values probably stems from the different testing conditions and purities of the magnesite used. The lowest value reported was possibly catalyzed by the presence of moisture. This 402°C isotherm curve is contained in Fig. 6-1.[4] Another investigator hypothesizes that decomposition occurs in three stages at 402°C, 438°C, and 480°C.[5, 6]

Since the proportion of $MgCO_3$ to $CaCO_3$ differs in the many species of dolomitic and magnesian limestone, the dissociation temperature naturally also varies and is much more difficult to calculate. Differences in the crystallinity of the stone also appear to add to the disparity of data. The $MgCO_3$ component of dolomite decomposes at higher temperatures than natural magnesite. Azbe[7] and others have detected some dissociation of dolomite at about 510°C, but no appreciable decomposition is attained until 590°C; then dissociation progresses rapidly. Others[8] found that the commencement of dissociation ranged from 500 to 750°C for three dolomites (500°C for a dense, fine crystalline type; 650°C for "fairly" crystalline; and 750°C for "highly" crystalline, with complete dissociation near 800°C). A good

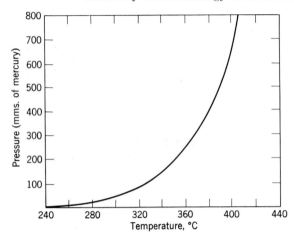

Fig. 6-1. Dissociation pressures of $MgCO_3$.

average value for complete dissociation at 760 mm. pressure in a 100% CO_2 atmosphere is 725°C; the $CaCO_3$ component of dolomite would, of course, adhere to the above higher value representing dual-stage decomposition. The latter was concluded by Linzell and his co-workers[9] (and many others). As a result of these differences in dissociation points, the MgO is usually necessarily hard-burned in varying degrees before the CaO is formed due to retention at relatively high temperature. The degree of hard-burning can be alleviated by cooling the quicklime immediately after the $CaCO_3$ is calcined and by calcining at minimum and constant temperatures, but for longer duration. Even if the CaO constituent is soft-burned, the harder-burned MgO component influences a denser quicklime of lower reactivity than a comparably calcined high calcium lime.

Dissociation always proceeds gradually from the outside surface inward. Usually the depth of penetration moves uniformly inward on all sides of the stone, like a "growing" veneer or shell. Actually a certain amount of exterior or surface dissociation of the carbonate molecules can occur at lower temperatures than the above under favorable conditions, such as low concentrations of CO_2 with low partial pressures.[3] Azbe[10] has detected traces of surface dissociation as low as 1335°F with high calcium stone. But in order for dissociation to penetrate into the interior of the limestone, higher temperatures are necessary and further elevated for dissociation to occur in the center or core of the stone (in practice, generally well in excess of the above dissociation points). The larger the diameter of the stone, the higher the temperatures required for dissociation of the core due to increasing internal pressure as the CO_2 gas forces its escape. Thus, with the same stone purity the difference between dissociation temperatures of the surface and core may be 3 to 700°F, depending primarily on the stone's diameter.

The above dissociation temperatures and their rates are not rigid values since they vary with the counteracting CO_2 pressure and concentration, as shown in Fig. 6-2.[1, 2] If temperature and pressure are in equilibrium, regardless of their values, dissociation is static. But if there is a minute change in one of these variables, such as a decrease in CO_2 pressure or concentration or an increase in temperature, dissociation immediately proceeds with evolution of CO_2 gas and the simultaneous formation of oxides. E.g., in Fig. 6-2 if the dissociation pressure of $CaCO_3$ is only 380 mm., corresponding to 50% CO_2 concentration, then the dissociation temperature is reduced to 848°C. *In all cases* there is a definite relationship between CO_2 pressure, concentration, and temperature.

However, the sensitive reversible nature of this reaction can also manifest itself here—*recarbonation.*[7] This can occur when large lumps of limestone are calcined. As the higher-temperature heat inflow pene-

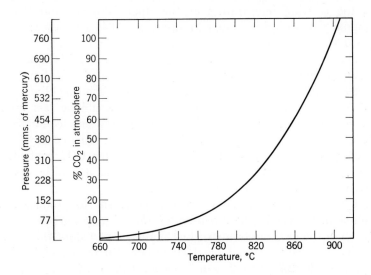

Fig. 6-2. Influence of CO_2 concentration and pressure on dissociation temperature of $CaCO_3$.

trates well into the lump near its center, dissociation in an atmosphere of pure CO_2 will start to exert considerable pressure in excess of the atmospheric pressure up to as high as 100 psi in extreme instances. The temperature rises as the pressure increases and causes the already calcined surface to be overburned. This tends to shrink the stone, occluding or narrowing the pores and fissures through which the CO_2 must escape, thus generating more pressure. If such lime is discharged into the cooler before all of the core is calcined, there is still a faint diffusion of CO_2 from the red-hot core which can be reabsorbed on the surface of the cooling lime. This, of course, vitiates the value of the resulting quicklime. So the greater the pressure of CO_2, the higher is the dissociation temperature, as Fig. 6-2 indicates. With pressures exceeding $1°$ atmospheric, the dissociation temperature as explained in Fig. 6-2 increases correspondingly. Rapid and continuous evolution of CO_2 is a major objective in lime burning.

Heat Consumed. Attainment of the theoretical minimum dissociation temperature, of course, requires considerable heat. This can be calculated from the specific heat of high calcium or dolomitic limestone (p. 24) by the following formula: specific heat of limestone $\times$ 2000 lb. $\times$ net gain in $°F$ from atmospheric temperature.

Assume that the specific heat is calculated for $CaCO_3$ at 0.255, that the starting temperature of the stone is $50°F$, and that the minimum dissociation temperature is $1648°F$; then: $0.255 \times 2000 \times 1648 - 50$ (or 1598) = 814,980 Btu/ton of limestone. But 44% of the limestone is volatilized as CO_2, so that theoretically 1.79 ton of limestone is required to yield 1.0 ton of quicklime; then: $814,980 \times 1.79 = 1,458,814$ Btu/ton of high calcium quicklime.

Calorific requirements for dolomitic stone would be less because of its lower dissociation point (about $1337°F$ for a pure equimolecular double carbonate). Based on the above formula, same constants, and a factor of 1.90 to allow for dolomite's greater CO_2 content, the theoretical kinetics would be 656,370 Btu/ton of dolomitic stone, or 1,247,103 Btu/ton of dolomitic quicklime.

But there is another, larger heat requirement to produce a ton of quicklime. This is the *retention of the dissociation temperature* until all of the CO_2 has been expelled. Values reported for this requirement have varied between 2.5 and 3.0 million Btu/ton for high calcium quicklime, but 2.77 and 2.60 millions of Btu/ton for high calcium and dolomitic quicklimes, respectively, appear to be the most accurate values. For magnesian lime the value is closer to the high calcium value. All values are predicated on theoretically 100% pure carbonates. As impurities increase in limestone, this thermal requirement is reduced correspondingly, since there is less carbonate to decompose.

For further information on heat consumption, refer to the examples of heat-balance studies on p. 254.

Decrepitation and Preheating. In attaining these dissociation temperatures free moisture in the stone is, of course, volatilized and a small amount of organic matter that exists in most limestone is ignited. This adds slightly to the heat requirement given above. Thus, at the point calcination commences the stone is more porous and may contain a myriad of micropores and crevices (fissures) interspersed throughout its crystal lattice.

Hedin[11] experimentally calcined different types of limestone and pure calcite in the laboratory and discovered microscopic physical changes in some calcitic limestones at 800°C, below the dissociation point. Generally this heat caused expansion in the crystal matrix of the limestone. In the highly crystalline limestones the heat produced stresses in the individual crystals, causing them to fracture. He found that the degree of decrepitation of these crystals maximized with crystals of the largest lineal dimensions; with some very large crystals the heat completely disintegrated them to dust. In contrast, small crystals of 2.5 mm. or less, prevalent in fine-grained stone, generally resisted the temperature stresses without cracking. Another factor that he revealed, which contributed to resistance of the expansion stresses, was the presence of a network of minute fissures separating the small crystals. These fissures behaved like expansion joints as the heat caused the crystals to dilate, thus maintaining the integrity of the matrix. The above findings are generalizations, since there was a minority of inexplicable contradictory results, e.g., one fine-grained limestone's crystals disintegrated into a powder along its intercrystalline boundaries, whereas cretaceous (chalky limestones) of very minute crystals did not crack or decrepitate.

This research by Hedin proves what every experienced lime production man knows: that certain kinds of limestone cannot be successfully calcined into lump or pebble quicklime. These types may largely dissociate into an oxide, but they will decrepitate into small fractions down to dust, rendering the product unsalable for many uses and seriously complicating the calcination process. Large crystalline forms are most prone to behave like this, but there is no sinecure to observe, and experimental calcination of the stone is judicious before a plant investment is made. This is a *physical* oddity; the chemical analysis of the stone has little or no influence on this characteristic.

Foster [12] also observed that the $CaCO_3$ crystal lattice in preheating undergoes thermal expansion prior to calcination with gains frequently 5 to 10% longer in the matrices' linear dimensions. A residual effect of this expansion is an increase in the stone's porosity.

Rate of Heating. A general axiom among lime manufacturers, regardless of the type or quality of limestone, is that the higher the burning temperature and the longer the calcining period, a harder-burned quicklime results that has high shrinkage, high density, low porosity, and low chemical reactivity; that the converse prevails at lower burning temperatures and/or shorter burning duration, yielding the desirable soft-burned, highly reactive limes of low shrinkage and density and high porosity. This empirically conceived theory has been authenticated by many researchers.[15, 14, 16]

While Murray[13] generally adheres to this theory, he discovered some flagrant exceptions in his calcination studies with forty-three commercial limestones. An example of such a discrepancy is given as follows with two high calcium stones:

	Stone 1	Stone 2
Calcination No.	21-11	19-1
Maximum temperature	1765°F	1850°F
Retention time	46 min.	474 min.
Activity coefficient (ΔT5)	12.0	33.4
Porosity	38.4%	46.6%
Shrinkage	26.2%	15.1%

Two other stones were both calcined for nearly four hours but at temperatures that varied by 550°F, the highest being 2400°F. Yet these two stones had practically identical lineal shrinkage and reactivity values.

Such a paradox defies explanation, and it cannot be attributed to laboratory error. It simply underscores the fact *again* that there are still some unexplained incongruities involving the divergent physical properties of limestone. *There is an optimum calcination temperature and rate of heating for every limestone* that can only be determined by experimentation.

Murray hypothesizes that the *rate of heating* (both during preheating and calcination) has the greatest influence on lime quality—its shrinkage, porosity, and reactivity—more than maximum temperature or retention time. He advocates gradual, rather than shock preheating, and then a gradual increase in calcination temperature up to the point that dissociation is complete, avoiding then, if possible further retention time. This is opposed to the theory of calcination for a *constant time* and at a *constant temperature*. Based on the writer's experience, Murray's theory is the most fruitful to pursue.

Rate of Dissociation. Hedin also conducted a series of calcination experiments with seven different limestones, ground to a uniform size of 15 to 20 mm. edge lengths and calcined at fixed temperatures of 1000°C in 100% CO_2 of atmospheric pressure until dissociation was complete, as determined by measuring the loss of weight on ignition (CO_2). The results, depicted in Fig. 6-3,[11] show a wide variance in the rate at which CO_2 is liberated. The two slowest stones (No.s 2 and 3) are dense, coarse crystalline limestones, while the next slowest (No. 4) is finely crystalline. These stones in relation to the others might be classed as difficult to burn. Of the stones that burn rapidly No.s 6 and 7 are cretaceous, fine-grained limestones, and the one most easily calcined, No. 5, is a very fine-grained, compact stone, possessing organic impurities.

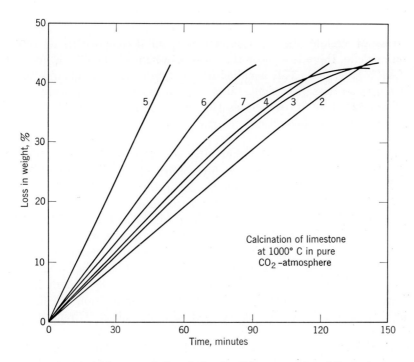

Fig. 6-3. Variable rates of dissociation for different species of limestone.

Hedin theorizes that the coarse, crystalline limestones, because of their dense microstructure, release CO_2 slowly, since the gas has difficulty penetrating through the dense crystal lattice. It only occurs under considerable sinking pressure, namely, that the CO_2 pressure within the crystal would have to be much higher than the external pressure. However, the No. 2 dense stone did not decrepitate during preheating, whereas the No. 3 related stone did. The cracks caused by decrepitation of the latter stone accelerated its decomposition more than the former stone, which was lacking in avenues for the gas to escape.

The explanation for the rapid calcination of the No. 5 stone is that the lattice inclusions of organic impurities are consumed by the heat leaving a loose lattice structure of high porosity, providing many avenues for the escape of the gas.

Thus, more rapid burning is usually achieved with the most porous stone or stone with natural cracks or fissures that are either open or are filled with moisture or carbonaceous matter. These pores provide more surface area for the evolution of CO_2. Stated another way, the slowly dissociating stone (large crystalline types) do not conduct heat as readily as the more porous, fine-grained types.

Loss of Weight. In the complete thermal decomposition of 100% pure calcite, there is a theoretical loss of weight as the CO_2 is evolved of 44% (Fig. 6-4). This is called *loss on ignition*. With dolomite there is a greater loss in weight, since a pure magnesium carbonate will release 52.2% of its weight as CO_2 gas. So the greater the $MgCO_3$ component is in dolomitic or magnesian limestone, the greater is the weight loss.

Complete calcination does not infer zero CO_2 in the lime. Linzell,[9] Knibbs,[3] and others have repeatedly confirmed that zero CO_2 is virtually impossible to obtain, even in the laboratory. Even if the limestone is hard-burned so that no possible carbonate core can exist in the interior of the limestone fragment, there will be at least slight surface absorption of CO_2, due to *recarbonation* from the CO_2-rich combustion gases enveloping the lime in the kiln. With effective combustion practices these gases are diluted and vented off rapidly, so that recarbonation may be reduced to as low as 0.1 to 0.2% CO_2 content in the lime. Azbe[10] has reported an instance of recarbonation as high as 42% in a highly inefficient vertical kiln operation. Circumstances that are conducive to excessive recarbonation in the kiln are excessive calcination temperatures, uneven distribution of combustion gases that envelop the stone in the burning zone and a resulting disparity in

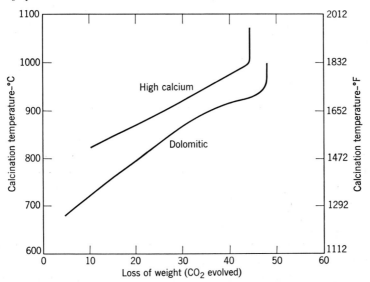

Fig. 6-4. Loss of weight with increases in temperature over dissociation point (¼ in. stone calcined under partial CO_2 pressure corresponding to 25% CO_2 concentration).

heat exchange, exuding CO_2 gases in the cooler with insufficient draft to vent the CO_2 up the kiln shaft with exhaust gases, and inordinately long retention of calcined lime in the cooler (usually a combination of these factors). Such heavy recarbonation dilutes the concentration of the lime oxides, rendering the product unsalable or unusable for many purposes.

Core. The other source of CO_2 that exists in lime stems from incomplete calcination or underburning. Usually it is practically nonexistent in superior quality quicklime or manifests itself as a small 10 to 40 mesh grain or miniature rice-sized kernel of $CaCO_3$ that is lodged in or near the center of the limestone fraction and which has resisted dissociation. It is commonly referred to as *core*. In large-lump quicklime of poor quality the core may be the size of an acorn. Chemically it is not deleterious but is regarded by many consumers as an impurity since it vitiates the availability of the lime. Some core, usually 0.25 to 2%, is present in soft-burned quicklimes. The producers, who are striving to obtain a reactive lime of low shrinkage, may intentionally discharge the lime to the cooler just before dissociation would be complete to avoid producing a hard-burned material—or some fractions may be underburned due to uneven heat exchange in the kiln.

Core of 5 to 8% is commonly found in typical commercial limes in Europe and other foreign countries. Such percentages are generally unacceptable in the United States; 3 to 5% is considered high, and many domestic manufacturers restrict core content to an average of only 0.5 to 1%, including the slight CO_2 contamination that results from recarbonation as mentioned above.

A semiburned lump of limestone with 25 to 50% core ($CaCO_3$) or the core itself (assuming that the outer shell of lime is removed by hydration) may be recharged into kilns and calcined into lime. There still persists in some areas the erroneous belief that core can never be calcined.

Maintenance of Temperature. If the minimum dissociation temperature was precisely maintained, calcination would be hopelessly retarded under most conditions, probably with some limestones never to 100% completion. As a result, in practice to accelerate this reaction, higher temperatures are applied of 1900 to 2450°F for $CaCO_3$ and usually 150 to 200°F less for dolomitic limestone. Ideally, these temperatures are maintained until all of the CO_2 from the limestone fragments is expelled. The rate of expulsion of the CO_2 accelerates rapidly as the temperature is elevated above the dissociation point as Fig. 6-4 demonstrates for both high calcium and dolomitic limestone.[15]

High versus Low Temperatures. An increase in temperature exerts a much greater influence on the dissociation rate than temperature retention. Mather[18] discovered that an increase in temperature of only 50°F exerted as much influence on the calcination as extending burning time from two to ten hours. In fact, the rate of calcination can be so materially shortened with some limestone at high temperature that greater thermal efficiency is achieved. The optimum temperature for maximum calcination efficiency (capacity and Btu consumption) varies with different stone and can only be determined with exactitude by experimentation. However, such a temperature must be balanced against the optimum temperature for maximum quality, which are usually disconsonant. Generally these two temperatures differ by 2 to 500°F, and the dilemma on which to adopt is often resolved through a compromise intermediate temperature. For high calcium and dolomitic, respectively, 2100 and 1950°F might be regarded as a median demarcation line between high and low temperatures in lime burning. Approximate maximum and minimum practical calcining temperatures would be, depending on the stone's crystallinity and lineal dimension: high calcium, 2450–1850°F; dolomitic, 2250–1725°F.

The above maximum for high calcium is in agreement with Bleininger and Emley,[19] but the latter concluded that the maximum for dolomitic should be 250°C less than high calcium, or only 1050°C, about 100°F less than the above. In excess of these maxima, overburning will result; below these minima, underburning will usually occur. For optimum quality the lowest range of temperatures is generally indicated. In calcining impure limestones low temperatures are essential to achieve even slight to moderate chemical reactivity.

Shrinkage. Murray[13] studied the shrinkage characteristics of forty-three commercial limestones (high calcium) of widely varying bulk density, and crystal structure when calcined at four different temperatures ranging from 1750 to 2450°F. All stones were completely calcined as proven by their losses on ignition. Calcination was consummated in an electric muffle furnace. As revealed in Table 6-1, shrinkage of all stones increased as the calcination temperature was elevated. The percentage of accretion in shrinkage, however, varied with the different stones. A few stones at the lowest temperature (1700 to 1750°F) even expanded initially before eventually shrinking. Coinciding with the stone number designation in Table 6-1 is a brief petrographic summary of each stone type on whether the crystals were fine, medium, coarse, or oolitic. Strangely, there was only partial correlation between crystal size and shrinkage, which is in agreement with Hedin. A preponderance of coarse crystal and oolitic stones were among those displaying the greatest shrinkage. A majority of the fine- and medium textured stones exhibited the least shrinkage. However, there were some disconcerting exceptions. In attempting to explain these inconsistencies, complete chemical, spectrographic, and petrographic (thin sections) analyses were made on the forty-three limestones. Although these analyses varied widely, there was almost no correlation with chemical composition, qualitatively or quantitatively, with the shrinkage results, except with Na_2O content (see p. 154). A total of twenty-five elements, most of which were traces, was found in these stones. The shrinkage of the quicklime was calculated from the bulk specific gravity of the limestones and allowing for the loss on ignition by the following formula (see p. 145):

Murray also found a definite relation with optical calcite on shrinkage during calcination and its degree of reactivity (slaking rate as determined by ΔT_5, described on p. 484). Calcitic quicklimes of low shrinkage and high porosity had high reactivity, and the converse

Table 6-1. Shrinkage of high calcium limestones at rising temperatures

No.	Stone crystal[a]	\multicolumn{4}{c}{Volumetric shrinkage after one hour at}
		1750°F	2050°F	2250°F	2450°F
6A	Fine	+17.3	+6.4	2.6	15.7
7A	Medium	+2.6	+1.8	5.3	13.8
7B	Medium	+2.9	0.7	6.2	20.5
10	Coarse	—	3.2	9.6	22.6
11A	Medium	+0.8	2.0	7.0	30.1
6B	Medium	7.3	6.9	14.1	18.6
13	Medium	—	0.5	13.5	26.9
19	Coarse	+6.6	2.0	20.1	36.3
17A	Fine	+1.6	7.3	33.5	26.8
11B	Medium	1.4	7.7	29.8	32.1
11C	Fine	5.3	15.5	24.9	31.7
15	Oolitic	4.5	14.5	28.1	38.8
26	Medium	0.0	21.5	30.4	39.3
24A	Coarse	9.5	20.3	27.5	—
1	Fine	3.9	22.8	33.2	36.6
12A	Coarse	3.3	25.1	30.5	39.2
17B	Coarse	1.3	20.0	31.8	46.0
12B	Fine	5.9	21.9	30.6	43.5
24B	Coarse	7.4	24.0	37.4	42.2
28	Coarse	5.6	27.1	37.0	41.4
2	Medium	18.6	22.8	35.5	37.7
25B	Coarse	4.9	26.7	38.2	45.0
24C	Medium	16.1	21.8	40.6	39.4
20A	Medium	11.0	32.0	40.5	38.1
22	Oolitic	15.5	24.6	40.5	42.5
14A	Fine	22.4	26.6	34.8	42.2
23	Fine	16.8	27.4	39.6	44.2
25A	Coarse	13.1	29.0	43.1	43.9
20B	Medium	11.8	32.0	44.6	41.3
4	Medium	23.9	31.1	32.0	45.6
16	Oolitic	12.7	33.6	41.2	46.5
14B	Fine	24.4	35.9	30.6	44.1
8	Coarse	18.5	39.0	42.1	41.9
5B	Medium	14.2	39.0	43.7	44.9
18A	Coarse	17.7	38.3	40.9	46.5

Table 6-1. (Continued)

No.	Stone crystal[a]	Volumetric shrinkage after one hour at			
		1750°F	2050°F	2250°F	2450°F
20C	Medium	24.1	34.6	43.4	44.5
3	Coarse	24.8	31.0	44.6	47.7
27	Medium	9.0	42.6	49.5	47.6
21A	Oolitic	24.1	35.5	44.5	45.1
18B	Coarse	33.6	40.6	39.0	42.4
21B	Oolitic	24.0	43.6	46.7	46.3
5A	Oolitic	25.7	38.2	46.5	51.0
9	Oolitic	25.3	39.5	45.7	52.6

[a] Limestone characterized as to type of crystallinity.

$$S = 100 \ \frac{100/Ds - \dfrac{100 - L}{DL}}{100 \ Ds}$$

S = shrinkage (%)
Ds = bulk density of stone
DL = bulk density of quicklime
L = loss on ignition of stone

applied to the quicklimes with high shrinkage that were induced by high temperature and long calcining retention periods.

Murray[23] also studied the influence of varying calcination atmospheres on the shrinkage of quicklime using a heating microscope with temperatures up to 1400°C. Six different high calcium limestones were exposed to three different atmospheres: 100% Co_2, 30% CO_2 plus 70% N_2, and 100% N_2. Inexplicably, the diverse atmospheres exerted a considerable influence on shrinkage with three of the stones and much less on the remaining three. Nitrogen atmosphere yielded the least shrinkage for three of the stones and 100% CO_2 for the remaining three.

Dolomites also exhibit considerable variation in shrinkage but tend to shrink generally less than high calcium stones at ordinary calcination temperatures of up to 1400°C. Studies of six dolomites in the heating microscope in CO_2 and N_2 atmospheres yielded consistent shrinkage results. All six stones displayed the least shrinkage in a CO_2 atmosphere and the greatest in N_2. Unlike the high calcium stones, the quantitative chemical analyses of the dolomites offer a plausible explanation for their shrinkage behavior. The degree of lineal shrinkage increased proportionately with the total impurity contents of the dolomites; the greater the impurity content, the greater the shrinkage.

Porosity and Density. The factors of porosity, density, and pore size distribution are interrelated and exert a profound influence on such standard measurable properties of quicklime as reactivity, available lime, and the particle size distribution and surface area of the resulting hydrated lime. A porous lime usually has low shrinkage; a dense lime of low porosity has high lineal shrinkage. When calcination temperatures are increased or constant temperatures in a medium to high calcination range are retained, there is diminution in porosity (the converse of Table 6-1 on shrinkage).

Data on porosity measurements obtained from forty-one calcinations of calcite under diverse temperature-time combinations is contained in Table 6-2, based on Murray's research.[16] In a series of tests in a muffle furnace on ten high calcium limestones, Fischer[21] observed that retention time of one to four hours had little or no effect on porosity, surface area, or reactivity at mild calcination temperatures of 1750 to 1950°F. The bulk densities of these quicklimes remained constant. He also discovered that shock calcination (i.e., sudden application of high temperature without preheating) invariably densified the resulting oxide, usually abnormally, except at very low temperature (1780°F). Staley and Greenfeld[22] obtained somewhat similar results with a high calcium stone of 98.5% $CaCO_3$ content, as indicated in Figs. 6-5 and 6-6 on the effect of time and temperature on surface area.

Murray also studied porosity distribution with a specially designed porosimeter and concluded that the type of porosity that contributes the most to high surface area are pores of the lowest average pore sizes, measured in angstrom units. He speculates that if two quicklimes have the same percentage of porosity, but one has an average diameter of pores of 1000 Å and the other 40 Å, the internal surface area of the lime with the largest pores will have only 4% of the surface area of the smaller.

Generally Murray was able to obtain higher porosity, surface area, and reactivity with calcination experiments with Iceland Spar than with most limestones, which indicates the adverse effect of impurities on reactivity and quality.

Hedin's explanation of increased density and shrinkage is that the CaO molecular crystals that form through dissociation are simultaneously uniting into progressively larger crystallites.[14] In tests with

Table 6-2. Properties of quicklime as affected by temperature and time of calcination

Calcination no.	Max. T of bed °F	Retention time min.	Time to reach 1700°F min.	Activity coefficient ΔT_5	Bulk specific gravity	Porosity %	Volumetric shrinkage %
19-1	1850	474	182	33.4	1.78	46.6	15.1
19-2	1850	230	88	35.2	1.74	47.8	13.1
19-3	1834	185	72	35.2	1.73	48.1	12.6
19-4	1857	103	39	36.3	1.73	48.1	12.6
19-5	1850	97	37	36.3	1.73	48.1	12.6
19-6	1760	82	34	35.5	1.72	48.4	12.1
19-7	1727	74	31	36.3	1.72	48.4	12.1
19-8	1707	67	29	33.2	1.72	48.4	12.1
21-1	2020	474	163	30.2	1.80	46.0	16.0
21-2	2018	230	79	34.3	1.76	47.2	14.1
21-3	2020	103	35	33.6	1.78	46.6	15.1
21-4	2023	95	33	30.5	1.87	43.9	19.1
21-5	2023	92	32	30.9	1.86	44.2	18.7
21-6	2020	82	28	26.0	1.89	43.3	20.0
21-7	2000	76	27	26.6	1.87	43.9	19.1
21-8	2000	67	23	22.3	1.92	42.4	21.3
21-9	1853	63	24	11.4	2.10	37.0	28.0
21-10	1780	49	20	14.2	1.96	41.2	22.9
21-11	1765	46	19	12.0	2.05	38.4	26.2
23-1	2218	474	146	31.5	1.81	45.7	16.5
23-2	2220	474	146	33.2	1.79	46.3	15.5
23-3	2223	308	95	33.8	1.79	46.3	15.5
23-4	—	230	71	31.9	1.81	45.7	16.5
23-5	2229	127	39	7.9	2.21	33.7	31.6
23-6	2225	103	36	8.4	2.20	34.0	31.3
23-7	2218	82	25	11.4	2.17	34.9	30.3
23-8	2204	67	21	6.4	2.30	31.0	34.3
23-9	—	66	20	6.7	2.34	29.7	35.4
23-10	2158	48	15	5.1	2.35	29.4	35.7
23-11	2184	47	15	5.5	2.47	25.8	38.8
25-1	2362	474	136	20.8	1.88	43.5	19.6
25-2	2397	308	86	30.2	1.83	45.0	17.4
25-3	2400	230	65	28.7	1.85	44.5	18.3
25-4	2396	103	29	7.2	2.40	27.9	37.0
25-5	—	100	28	9.6	2.21	33.6	31.6
25-6	2412	95	26	3.6	2.61	21.6	42.1
25-7	2402	82	23	2.2	2.63	21.6	42.1
25-8	2397	74	21	2.0	2.71	18.6	44.2
25-9	2407	67	19	2.6	2.65	20.4	42.9
25-10	2402	49	14	2.0	2.63	21.0	42.7
25-11	2341	48	14	2.4	2.67	19.8	43.5

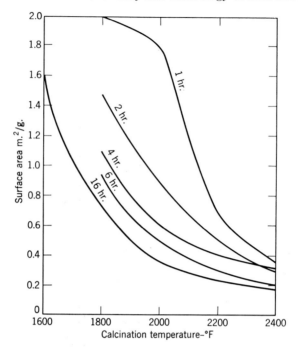

Fig. 6-5. Relation of surface area to calcination temperature.

calcite through microphotography, he has calculated that these crystals at 900°C, the start of dissociation, are only about 0.1μ; at 1000°C they are 1μ, due to coalescence of these crystallites. At 1100°C a still larger coherent mass of crystallites is formed that become steadily more strongly bonded together with increasing temperature. Figure 6-7 illustrates this enlargement of crystals, and the accompanying explanation by Hedin elucidates the cause of shrinkage and densification. Fischer's[21] findings on the formation of crystallites coincided with Hedin's.

Mayer and Stowe, adhere to this same crystallite theory and have observed through X-ray the inexorable build-up of the crystallites as calcination temperatures are raised. They estimate that the average CaO crystal is increased about a thousandfold (from 0.1μ to 100μ) between initial calcination at minimum dissociation temperature and at sintering temperature of about 3000°F, when the lime is completely dead-burned. However, with dolomite they observed that the accretion of the CaO crystallites is more rapid than its MgO component, re-

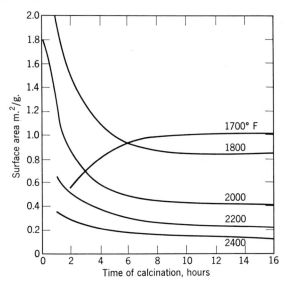

Fig. 6-6. Relation of surface area to duration of calcination.

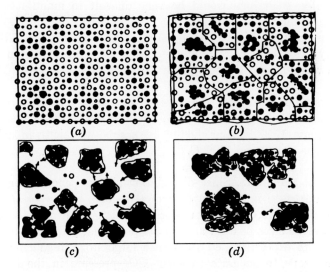

Fig. 6-7. Graphic display of how lime oxides gradually form much larger crystallites as calcination temperature increases. Gradual agglomeration of small crystals into steadily larger crystallites is illustrated sequentially in (a) through (d).

sulting in much larger ultimate CaO crystals. Their explanation of this phenomena is that the CaO molecules possess lower "bond energy" than their MgO counterparts and, as a result, migrate more easily throughout the matrix. They also commented that there is no lineal change in the apparent density of the original stone fragment as these crystallites accrete, up to the dead-burned stage when the physical size of the particle shrinks perceptibly. Before the dead-burned stage, the only change in the molecular geometrics of the lump is immeasurable—namely, the corresponding reduction in pore space and surface area. Mayer and Stowe's research was predicated on two chemical-grade limestones, high calcium and dolomitic types from northern Michigan.

The true specific gravity of CaO is 3.34 to 3.40, but this only occurs when theoretically pure quicklime has been extremely dead-burned, yielding optimum shrinkage, densification, and zero porosity—virtually impossible to obtain commercially. Murray[16] experimentally produced in the laboratory limes of 3.30 specific gravity, but generally values in the 2.5 to 3.0 range would be considered very hard-burned, unreactive lime. Azbe[10] claims the ultimate in calcination should be highly porous, reactive lime of only 1.45 to 1.65 specific gravity, which would indicate a porosity of over 50%. Practically, such lime would be very difficult to manufacture with any consistancy and economy.

A summary of Murray's on apparent specific gravity (bulk density) of forty-three high calcium limestones reveals a range from 1.72 to 2.71; twenty-five of the forty-three samples ranged between 1.72 and 1.96.

Murray observed that the general shrinkage pattern also occurs with pure calcite (Iceland spar). Figures 6-8 and 6-9 display the approximate 15% lineal reduction of a cube of calcite in a heating microscope at 500 and 1400°C.

Effect of Stone Size. Since dissociation always penetrates gradually from the surface into the interior of the stone, the *larger stone sizes* are more difficult to calcine uniformly and require more time. Large cubical stone sizes of 6 in. and up are particularly difficult to calcine. In order to expel the CO_2 from such large stone, high temperatures are necessary to generate sufficient CO_2 pressure in the interior of the crystal lattice for the escape of the gas. Frequently these high temperatures (2300 to 2450°F range) will overburn the surface layer of the stone, causing excessive shrinkage, which narrows and closes the pores and fissures. Pores are also clogged by the slagging effect

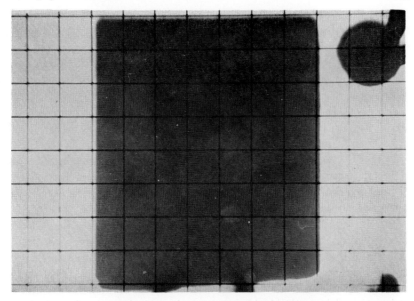

Fig. 6-8. Specimen of a calcite cube as seen in heating microscope at 500°C.

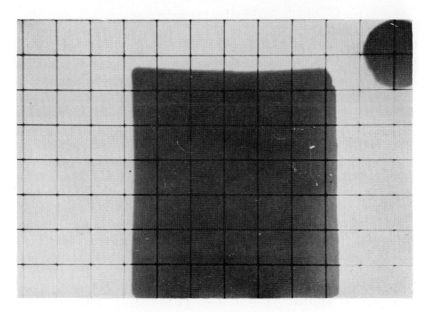

Fig. 6-9. Same specimen as Fig. 6-8 after being dead-burned at 1400°C, causing visual shrinkage.

of impurities (mainly SiO_2 and secondarily Al_2O_3) in the stone that
are fluxed with lime, forming occlusions. Due to these stifled dissociation conditions, recarbonation of the surface and/or incomplete dissociation of the core (center of the stone) may occur, thus producing a dense, highly unreactive lime of low surface area, but one paradoxically underburned or uncalcined in the center.

Usually this condition is aggravated when the *stone gradation* is *broad*, such as 3×9 in. or 4×10–12 in. The smaller sizes tend to calcine faster and at lower temperature than the larger fractions, since the CO_2 gas has a shorter distance to travel in the smaller diameter. Temperatures that calcine these smaller sizes adequately without overburning will only calcine the outer shell of the larger diameter stone. Or if the large stone is effectively calcined, then necessarily the smaller stone is usually excessively hard burned due to the long retention at elevated temperature for the small stone. The same problem, although not as critical, occurs with small stone of a broad or irregular gradation, $\frac{1}{4}$–$\frac{1}{2} \times 2$ in. or no. 10 mesh $\times 1$ in., etc. Consequently, restricted gradations of stone, regardless of size, are much easier to calcine. (Desirable ranges of gradation are described on p. 224 and p. 231.)

Another advantage of smaller stone is that it provides greater surface area per ton of kiln feed for heat transference.

Azbe[26] equates the "calcining effort" of small and large stone in terms of their relative heating and calcination times as being *directly proportional to the square of the thickness* (or average diameter for irregularly shaped stone). "Calcining effort" is defined as a composite of time, temperature, and heat transfer. This formula means that it would require about four times as long for stone A of 4 in. thickness to be calcined as stone B of 2 in., assuming both are cubicly shaped, since the thicker fraction contains eight times as much volume to calcine and the heat transfer has twice as far to penetrate. Stated another way, the small stone absorbs heat twice as fast as the large, but the calcining effort is only one-fourth. *Thickness* of the stone rather than width or length (in irregularly shaped stone) or total volume in cubic inches is the main criterion in the influence of the stone's size or calcination rate or effort, e.g., comparing stone C that is cubicly shaped and $2\frac{1}{2}$ in. in thickness with slab-shaped stone D of same thickness but with a dimension of $2\frac{1}{2} \times 7\frac{1}{2} \times 10$ in., they both have the same distance for heat to transfer to the core, in spite of the fact that stone D has about eleven times as much volume in cubic inches. But the calcining effort, according to Azbe, is proportionately much less for its size—only about twice as much as stone C.

A widespread particle size distribution in kiln feed also interferes with combustion. The small stone accumulates between the voids formed by large stone in shaft kilns, thus impeding the draft and the flow of combustion flame and gases. Such resulting poor circulation leads to an uneven distribution of heat in the calcining zone.

The ideal stone to calcine for optimum quality, uniformity, and thermal efficiency, of course, would be small size ($-\frac{1}{2}$ in.) and be of uniform size and shape. Production of such kiln feed would be economically impossible. However, probably the greatest deterrent to high, uniform quality among lime manufacturers is the use of stone gradations that are too broad and in some cases of an erratic size distribution. Attaining more narrow and consistent gradations can potentially improve over-all calcination performance more than any other single factor.

Effect of Calcination on Ionic Spacing. Hard- and soft-burned quicklimes and their hydrates, produced from the same limestones, were examined under X-ray. The crystal lattice parameters of the limes were calculated in angstroms and were found to be identical down to 4 to 5 minus decimal points.[20] This, of course, proves that the crystals' ionic spacings are identical and that hard-burning does not shrink the molecular structures, only the spaces between the intercrystalline matrix. When subjected to X-ray diffraction, sharp, fine X-ray lines indicate clear, perfect crystals of impressive purity. In contrast, broad, indistinct X-ray lines indicate imperfections in the crystal caused by impurities.

Effect of Salts. For many years some lime manufacturers in several countries have experimented with small additions of sodium chloride to the limestone or to the fuel preparatory to calcination. The salt is added either dry (0.2 to 1.0%) or the limestone is soaked in or doused with brine. Contradictory results have been reported; some claim benefits from improved lime quality and even in superior fuel economy. The Japanese in particular have reported encouraging results, some of which border on the incredible.

Murray[16, 23] experimented with salt additions in the laboratory. Comparing both 10% sodium chloride and sodium carbonate solutions separately, limestone particles were soaked for sixteen hours and then calcined in a laboratory rotary kiln. He observed a definite trend of lower shrinkage. The limestones that were prone to high shrinkage contracted less; those with a low shrinkage tendency were unaffected. The efficacy of the NaCl and Na_2CO_3 solutions averaged out about equal; some limes were enhanced more by one or the other of these two salts. Murray also observed that limestones containing the most Na_2O as an impurity generally exhibited less shrinkage than those stones with only a trace. Salt additions to some dolomites yielded even more profound changes, as determined by differential thermal analysis, in comparing the same stone with and without salt.

Consequently, there is enough evidence to indicate possible improvement for some limestones, and some experimentation by individual companies would appear justified since there is a virtual absence of adverse results reported.

Influence of Stone Impurities. A description of limestone impurities is given on pp. 17–18 and in Table 2-2, p. 19. Since lime is employed primarily as a chemical reagent, it is only logical that the purest stone usually provides the best quality of lime. High purity also generally enhances lime's physical or rheological properties for structural uses. That is why lime producers, at least in the United States, are constantly striving to obtain the purest stone deposits available. In addition to *stone* the other source of impurities is derived from the *fuel*. For a majority of lime uses the *quantity* of impurities is more critical than the *quality*. Not only will impurities dilute the availability of the lime, but they will also adversely affect its reactivity, particularly when calcined at high temperatures of 2200°F and above.

The absorption of impurities from fuel occurs on the most reactive portion of the lime particle—its surface. Hard-burned lime of sluggish reactivity does not possess as strong an affinity for the acid oxide impurities or iron as soft-burned. If silica in the ash is in contact at high temperature with a lump of lime whose surface is partially hard- and soft-burned, the most reactive CaO molecules will react with the SiO_2 first; the hard-burned CaO molecule reacts more slowly.

With integral impurities in the stone at low temperatures of 1650 to 1700°F, relatively little silica and other impurities combine with the lime. But as the temperature increases, the uncombined acid oxide impurities are increasingly absorbed, forming various complex calcium compounds, such as monocalcium and dicalcium silicates, calcium aluminates, dicalcium ferrite, etc. A slagging effect occurs that tends to occlude the pores in the quicklime and mute its reactivity. This means that when the lime is slaked, water encounters more difficulty in penetrating the interior of the lime because of its clogged pores, thereby retarding hydration. As a result, lime abnormally high in impurities may react like a hard-burned or recarbonated lime with water, even though overburning did not occur. Consequently, the inevitable presence of some degree of impurities interjects still another *variable* in lime burning.

Although it is usually much smaller in quantity than the other major impurities, sulfur is probably the most baffling to the lime manufacturer because of its sporadic, inexplicable accretions in the lime. Instances have occurred, fortunately usually short-lived, where the total sulfur in the lime exceeded by an appreciable amount the total sulfur present in both the stone and fuel, in effect representing absorption of sulfur from sources unknown. Often this mysterious sulfur contamination may ebb to a plausible total as suddenly as it appeared. This is one area of lime production that needs some fundamental research to explain the unpredictable phenomenon of sulfur build-up. Sulfur is present in all fuels in varying degrees, but unlike solid fuels it is the main impurity present in gaseous and liquid fuels. It is generally absorbed from the fuel gases as SO_2 on the outside surface of the lime particle, forming sulphites and sulphates (SO_3 and SO_4), although sulfides (S) are also formed. When this element is an integral part of the stone, it behaves similarly.

Another old axiom of the lime industry is that impurities in the limestone are doubled when it is calcined into lime. This is misleading, as explained on p. 81, since additional available lime is also dissipated as it fluxes (combines) with the acid oxide and iron impurities. In practice, the concentration of lime is vitiated by about four times its percent of total impurities as a result of this fluxing effect during calcination. So for 2% total impurities in the stone there is a loss of about 8% free CaO or oxides in the quicklime. Such loss is minimized at lowest calcination temperature levels.

Murray and his co-workers[13] analyzed forty-five different high calcium limestones, most of which were presumably employed in United States commercial lime manufacture. He also analyzed six dolomites.[23] The principal impurities are summarized as follows for both types, showing the range in values:

High calcium

SiO_2	0.10 to 2.89%	Na_2O	0.00 to 0.16%
*R_2O_3	0.13 to 0.92%	SO_3	0.00 to 0.05%
K_2O	0.00 to 0.21%	MgO	0.12 to 3.11%

Dolomitic

SiO_2	0.02 to 0.50%	H_2O and organic	0.03 to 0.80
Fe_2O_3	0.10 to 0.44%	Mols CaO	.518 to .548
Al_2O_3	0.27 to 1.05%	Mols MgO	.456 to .541

* Includes primarily Al_2O_3.

Murray also examined twenty-five of the high calcium limestones spectrographically for twenty-five metallic elements. His findings for other than calcium, magnesium, silicon, and iron are enumerated as follows:

Aluminum was present in all samples; five samples had 0.35 to 0.60%; the remainder were less than this range.

Barium was present in all samples; quantity was only a trace except for one sample.

Boron was found in only trace amounts in three samples.

Chromium was present in twenty samples; of these, three possessed more than 10 ppm.

Cobalt was present in only nine samples; of these, two contained more than 10 ppm.

Copper traces were present in all samples.

Lead traces were present in fifteen samples.

Manganese was present in all samples in amounts greater than 10 ppm.; one sample had over 0.1%.

Mercury traces were found in only four samples.

Molybdenum traces were found in eight samples.

Nickel was present in all but three samples (twenty-two stones in all); it existed as a trace in all samples except one, which was $<0.01\%$.

Phosphorus traces were found in all stones; only two contained an amount between 0.001 to 0.01%.

Potassium was present in all samples, usually as a trace or less than 0.1% except for one sample which was 0.2%.

Rubidium traces were present in seventeen samples.

Silver traces were present in thirteen samples.

Sodium was present in all samples; all except three samples contained a trace or less than 0.1%; the latter three contained slightly over 0.1%.

Strontium was present in all samples in amounts ranging between 0.01 to 0.1%.

Tin was present in all samples as a trace.

Titanium was present in twenty-three samples (all but two); in all of these except one, the amount ranged between a trace and 0.1%; the latter was slightly over 0.1%.

Zinc was present in twenty-three samples (all but two) and occurred in amounts ranging between a trace and 0.1%, except for one sample which was slightly over 0.1%.

In some of the limestones and dolomites analyzed above, the impurities were rather uniformly disseminated throughout the stone;

in others it was concentrated in localized sections of the stone, making it difficult to secure a representative sample. As cited previously on p. 143, Murray was unable to explain shrinkage, reactivity, and surface area discrepancies in the different limes burned from these stones as a result of impurities. The sole exception was Na_2O (see p. 153).

In general, limestone for lime manufacture in the United States will range from 95 to 99% total carbonates. Most of it will assay at least 96%, and over 50% of the stone employed for commercial lime is in the 97 to 99% range. Lower average purities prevail in foreign countries—90 to 95% carbonate ranges are common. Some United States dolomites are extremely pure, analyzing 98.7 to 99.3% in total carbonates.

Effect of Steam. A wide difference of opinion exists on the efficacy of calcining in an atmosphere of steam. Research has indicated that steam tends to reduce the dissociation temperature of some stones by 5 to 15%, thereby catalyzing the rate of dissociation slightly. A patented calcination process[24, 25] for calcining under superheated steam was developed based on the theory that more highly reactive lime could be produced in this heat-tempered atmosphere. Somewhat the same effect as a steam blast might be simulated in using wood as a fuel (see p. 248) or in calcining a porous limestone that has a 7 to 15% moisture content. Numerous manufacturers who have experimented with steam, however, contend that no visible benefits were gained, so that its virtue is inconclusive. Knibbs[3] appears to favor the use of steam generally and claims that some steam is almost invariably present in most commercial kilns, derived from the stone, solid fuels, and atmosphere. Thus, steam injection would simply be an artificial supplement to naturally generated steam. As a result, it appears that the application of steam *might* produce moderately successful results with certain types of stone or with some kiln designs. Those producers probably enjoying the greatest benefit from steam are the dolomitic shaft-kiln producers, who have found steam injections of aid in tempering kiln heat in order to produce the soft-burned lime required for finishing hydrated lime. Steam generation, of course, adds to the thermal cost of lime burning.

Dead-burned Lime. Soft-burned, hard-burned, and intermediate calcines have been described in the preceding sections, but no mention as yet has been made of dead-burned lime, also referred to as *double-burned* and *sintered* lime. By elevating the calcination temperature to a 2800 to 3000°F+ range, a hard-burned lime (usually the product of 2250 to 2600°F temperature) undergoes still further shrinkage, increased specific gravity, reduced porosity, less surface area, and virtual elimination of any chemical reactivity, e.g., the lineal shrinkage of such dolomitic types approximates 55%; bulk density—120 lb./cu. ft.; specific gravity—3.25 (true specific gravity for dolomitic is 3.48 to 3.65); porosity—8%.[7] Values for corresponding high-calcium types would be quite comparable, except that it would still possess a faint degree of reactivity, more than dead-burned dolomite.

Chemically, the product is the same as quicklime, a true oxide of calcium or calcium–magnesium, but it is practically as inert and stable as its limestone antecedent. This material is usually devoid of CO_2, except possibly traces from recarbonation, since no core is present due to the severity of the sintering, and most surface recarbonation that would have occurred during calcination would be recalcined at the high temperatures. Further information on its manufacture follows in Chapter 8.

Recapitulation. From the foregoing it should be abundantly clear that there are numerous critical variables in lime burning which even individually can exert a profound or disastrous effect on lime quality, but most complicated of all is the fact that each stone has its own peculiarities, the exact limitations and optimum conditions of which can only be ascertained by tedious trial-and-error methods. Otherwise, results are usually not predictable. These variables are summarized as follows in the approximate order of their importance:

1. Stone quality.
    a. Physical characteristics (type of crystallinity).
    b. Tendency to decrepitate.
    c. Quality and quantity of impurities.
2. Stone size and gradation.
3. Rate of calcination.
4. Calcination temperature.
5. Calcination duration.
6. Chemical reactivity.
7. Shrinkage characteristic (density and porosity).
8. Surface area and spacing of crystallites.
9. Quality and type of fuel.
10. Possibility of recarbonation.

These seemingly anomalous conditions encompassing lime manufacture are not surprising if thin sections of different limestones are studied comparatively through photomicrographs. Vast differences in geologic origin and physical structure are clearly revealed. Figures 6-10 through 6-17 are examples of thin sections made by Murray

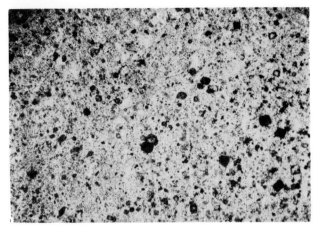

Fig. 6-10. Stone 11C—fine-grained.

and his co-workers at M. I. T. in which the stone number designations correspond to the numbers given in Table 6-1. Two figures each of fine, medium, coarse crystalline, and oolitic limestones (high-calcium types) are shown. The same pronounced differences also apply to dolomitic types. Brief details of each stone are enumerated as follows:

Stone 11 C (Fig. 6-10). This is predominantly fine-grained crystal-

Fig. 6-11. Stone 23A—fine-grained.

Fig. 6-12. Stone 6B—medium grain size.

line of Devonian age with the largest crystal cleavage faces up to 1 mm. across. It consists of fossil fragments and recrystallized calcite in a chert matrix (SiO_2 measures 2.69%). It is buff-colored.

Stone 23 A (Fig. 6-11). This is fine-grained crystalline and of probably Oligocene age. Essentially it was derived from a calcareous mud, but some fossil debris and small quartz grains are present. There is some manganese staining, and the color ranges between reddish-brown to buff, with a small amount of white.

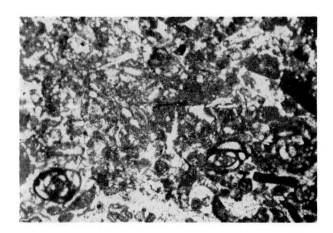

Fig. 6-13. Stone 20B—medium grain size; cretaceous.

Fig. 6-14. Stone 24A—coarse-grained.

Stone 6 B (Fig. 6-12). This averages as medium grain size (some fine grains) of Silurian age. The stone is clastic with brachiopod fragments and coral molds that have been consolidated in calcareous mud. Recrystallized calcite rhombs are clearly visible; it is largely gray and has some white. It is quite pure with only 0.66% SiO_2 and 0.39% R_2O_3.

Stone 20 B (Fig. 6-13). This is fine- to medium-grained crystals

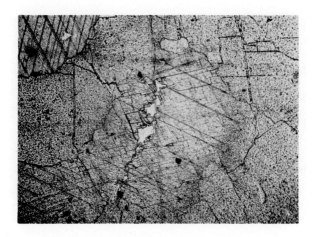

Fig. 6-15. Stone 25B—very coarse-grained.

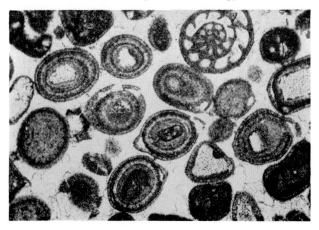

Fig. 6-16. Stone 16A—oolitic.

of Cretaceous age. It consists of fossil fragments and calcite masses consolidated in a calcareous mud. Fossil debris cemented into calcite is visible. It is gray in color and rather pure.

Stone 24 A (Fig. 6-14). This is a massive, coarse-grained, highly crystalline stone of probable Eocene age. It contains some fine grains and consists of brachiopod shells and other fossil debris with recrystallized calcite. It is light buff in color and of high purity.

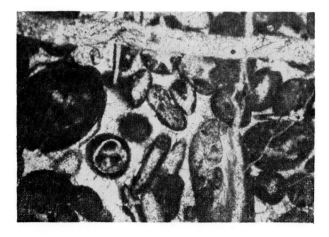

Fig. 6-17. Stone 21B—oolitic.

Stone 25 B (Fig. 6-15). This is very coarse-grained crystalline of Mississippian age. Cleavage faces of the crystals range from 1 to 10 mm. across. Marked cleavage and polysynthetic twinning is visible. It is gray in color and very pure, only 0.14% SiO_2, 0.28% R_2O_3, and 0.12% iron.

Stone 16 A (Fig. 6-16). This is composed entirely of oolites, well cemented together in a crystalline calcite matrix. Fossil fragments are brachiopods and crinoids. Occasional cleavage masses of calcite measure 1 mm. across. Entire foraminifera casts are present. It is dull white in color.

Stone 21 B (Fig. 6-17). This is composed of tightly packed oolites, well cemented, with recrystallized calcite and some fossil fragments and is of the late Paleozoic age. Visible in picture are quartz crystals traversing the oolites. It is gray in color and very pure with only 0.10% SiO_2.

The only possible classification of these diverse limestone types and possibly another thousand relatively pure species, based solely on their calcination behaviors, would be the following four very general and arbitrary categories:

1. Those that fracture and decrepitate readily during preheating and low calcination temperatures.

2. Those that will yield a porous, reactive lime under most calcination conditions and which are difficult to overburn.

3. Those that will yield a dense, unreactive lime of low porosity even under the mildest calcining conditions.

4. Those that yield a porous, reactive lime under mild temperature conditions and a denser, less porous lime under harder burning conditions.

Category No. 1 (above) is, of course, virtually useless for lime manufacture. Number 3 is very little better since a porous, reactive lime is preferred for most lime uses. Number 2 is a desirable type of stone since it can be utilized for most uses. However, for all-around versatility, category No. 4 is slightly preferable over No. 2 because through adjustments in time-temperature conditions a tailor-made lime can be made by a skillful operator for *all* purposes. The flexibility offered by such a stone, assuming equal chemical quality, is becoming an increasingly decided advantage in modifying a lime's characteristics to satisfy the often rigorous and fickle demands of many consumers. The key to such a lime is the kiln, its design, and operating methods, which are embraced in Chapter 8.

REFERENCES

1. J. Johnston, *J. Am. Chem. Soc.* **32** (1910), p. 938. Also quoted widely in other articles and books by Emley, Bleininger, Knibbs, Searles, and others.
2. J. Mitchell, *J. Am. Chem. Soc.* **123**, (1923), p. 1055.
3. Marc and Simek, *Z. Anorg. Chem.* **82**, (1913), p. 17.
4. N. V. S. Knibbs, *Lime and Magnesia*, E. Benn (London, 1924), pp. 85–104.
5. Brill, *Z. Anorg. Chem.* **45**, (1905), p. 279.
6. Brill, *Z. Anorg. Chem.* **57**, (1907), p. 721.
7. V. J. Azbe, *Rock Prod.* (Sept. 1944), p. 68.
8. Bole and Shaw, *J. Am. Ceram. Soc.* **5**, (1922), p. 817.
9. H. K. Linzell et al, "Lime: the Loss in Weight of Limestone as a Function of Time and Temperature of Burning," *Proc. Am. Inst. of Chem. Engs.* (1926).
10. R. Hedin, "Changes Occurring in the Limestone during Heating before Calcination," *Swed. Cement & Conc. Res. Inst. Bull. 23*, 34 pp. (1954).
11. R. S. Foster, M.I.T. research report to the Nat'l Lime Assn. (Unpublished, 1946).
12. V. J. Azbe, "ASTM, Symposium on Lime," 15 pp. (1939).
13. J. A. Murray, et al, *J. Am. Ceram. Soc.* **37**, 7 (1954), pp. 323–328.
14. R. Hedin, "Investigations of the Lime Burning Processes," *Swed. Cement & Conc. Res. Inst. Bull. 32* (1960). Condensation of the above by Nat'l. Lime Ass'n, "Structural Processes in the Dissociation of $CaCO_3$", Azbe Award #2 (1961).
15. J. Murray, H. Fischer, and D. Sabean, *Proc. ASTM* (1950), p. 1263.
16. J. A. Murray, Research Rept. to Nat'l Lime Assn. on "Summary of Fundamental Research on Lime" (1956).
17. R. T. Haslam and E. C. Hermann, *I. & E. Chem.* **18**, (1926), p. 960.
18. K. W. Ray and F. C. Mathers, *I. & E. Chem.* **20**, (1928), pp. 415–419.
19. A. V. Bleininger and W. E. Emley, *J. Am. Ceram. Soc.* **13**, (1911), pp. 618–638.
20. R. Mayer and R. Stowe, "Physical Characterization of Limestone and Lime," Azbe Award #4A, Nat'l Lime Assn. (1964).
21. H. C. Fischer, *J. Am. Ceram. Soc.* **38**, nos. 7, 8 (1955), pp. 245–251, 284–288.
22. H. R. Staley and S. H. Greenfeld, *I. & E. Chem.* **41**, no. 3, "Surface Areas of Quicklimes" (1949).
23. J. A. Murray, M.I.T. research reports to Nat'l Lime Ass'n (unpublished, 1955).
24. W. H. MacIntire and T. B. Stansell, *I. & E. Chem.* **45** (Jul. 1953), pp. 1548–1555.
25. W. MacIntire, U.S. Pat. 2,155,139 (to Am. Zinc), "MgO & $CaCO_3$ from Dolomite" (1939).
26. V. J. Azbe, "Theory and Practice of Lime Manufacture" (Part II), *Rock Prod.* (Mar. 1953), pp. 102–104.
27. H. C. Lee, Bull. *Ohio State U. Eng. Sta. News* **19**, no. 2 (Apr. 1947).
28. S. Brunauer, D. Kantro, and C. Weise, "Surface Energies of CaO & $Ca(OH)_2$," *Bull. of Port. Cement Assn.* (1955).

CHAPTER SEVEN

Definitions and Properties
of Limes

Nomenclature

Before a delineation of physical and chemical characteristics of lime, it is appropriate at the outset to define lime, the first derivative, manufactured product of limestone. Many of the following definitions[1] are repetitious and overlap, but these terms are widely employed in the industry and among its consumers:

Agricultural hydrate conveys a relatively coarse, unrefined form of hydrated lime that is mainly used for neutralizing soil acidity and for purposes where high purity and uniformity are unnecessary.

Air-slaked lime contains various proportions of the oxides, hydroxides, and carbonates of calcium and magnesium which result from excessive exposure of quicklime to air that vitiates its quality. It is partially or largely decomposed quicklime that has become hydrated and carbonated.

Autoclaved lime is a special form of highly hydrated dolomitic lime, largely utilized for structural purposes, that has been hydrated under pressure in an autoclave.

Available lime represents the total *free lime* (CaO) content in a quicklime or hydrate and is the active constituent of a lime. It provides a means of evaluating the concentration of lime.

Building lime may be quick or hydrated lime (but usually connotes the latter), whose physical characteristics make it suitable for ordinary or special structural purposes.

Calcia is the chemical compound—calcium oxide (CaO).

Carbide lime is a waste lime hydrate by-product from the generation of acetylene from calcium carbide and may occur as a wet sludge or dry powder of widely varying degrees of purity and particle size. It is gray and posseses the pungent odor of acetylene.

165

Chemical lime is a quick or hydrated lime that is used for one or more of the many chemical and industrial applications. Usually it possesses relatively high chemical purity.

Dead-burned dolomite is a specially sintered or double-burned form of dolomitic quicklime, which is further stabilized by the addition of iron, that is chemically inactive and is employed primarily as a refractory for lining open-hearth steel furnaces.

Fat lime connotes a pure lime (quick or hydrated), distinguishing it from an impure or hydraulic lime; it is also used to denote a lime hydrate that yields a plastic putty for structural purposes.

Finishing lime is a type of refined hydrated lime, milled in such a manner that it is suitable for plastering, particularly the finish coat. Putty derived from this hydrate possesses unusually high plasticity.

Fluxing lime is lump or pebble quicklime used for fluxing in steel manufacture—or the term may be applied more broadly to include fluxing of nonferrous metals and glass. It is a type of chemical lime.

Ground burnt lime refers to ground quicklime used for agricultural liming.

Hard-burned lime is a quicklime that is calcined at high temperature and is generally characterized by relatively high density and moderate to low chemical reactivity.

Hydrated lime is a dry powder obtained by hydrating quicklime with enough water to satisfy its chemical affinity, forming a hydroxide due to its chemically combined water. It may be high calcium, magnesian, dolomitic, or hydraulic.

Hydraulic hydrated lime is a chemically impure form of lime with hydraulic properties of varying extent that possesses appreciable amounts of silica, alumina, and usually some iron, chemically combined with much of the lime. It is employed solely for structural purposes.

Lime is a general term which connotes only a *burned* form of lime, usually quicklime, but may also refer to hydrated or hydraulic lime. It may be calcitic, magnesian, or dolomitic. It does *not* apply to limestone or any carbonate form of lime (although it is often erroneously used in this vein).

Lime putty is a form of lime hydrate in a wet, plastic paste form, containing free water.

Lime slurry is a form of lime hydrate in aqueous suspension that contains considerable free water.

Lump lime is a physical shape of quicklime, derived from vertical kilns.

Magnesia is the chemical compound, magnesium oxide (MgO), that is an important constituent in dolomitic and magnesian limes.

Mason's lime is a hydrated lime used in mortar for masonry purposes.

Milk-of-lime is a dilute lime hydrate in aqueous suspension and is the consistency of milk.

Pressure hydrate is synonymous to autoclaved lime and is the most common variety of ASTM designated Type S hydrated lime.

Pebble lime is a physical shape of quicklime.

Quicklime is a lime oxide formed by calcining limestone so that carbon doxide is liberated. It may be high calcium, magnesian, or dolomitic and of varying degrees of chemical purity.

Refractory lime is synonymous to dead-burned dolomite, an unreactive dolomitic quicklime, stabilized with iron, that is used primarily for lining refractories of open-hearth steel furnaces.

Roman lime is synonymous to hydraulic lime, but of the more impure and highly hydraulic type.

Run-of-kiln quicklime is unclassified quicklime as discharged from a kiln.

Slaked lime is a hydrated form of lime, as a dry powder, putty, or aqueous suspension.

Soft-burned lime is a quicklime that is calcined at relatively low temperature. It is characterized by high porosity and chemical reactivity.

Spray lime is a specially milled dry hydrated lime of very fine particle size of at least 98% passing a No. 325 mesh screen.

Type S hydrated lime (also called *special* hydrated lime) is an ASTM designation to distinguish a structural hydrate from a *normal* hydrated lime, designation *Type N,* that possesses specified plasticity and gradation requirements. It may be dolomitic or high calcium and is more precisely milled than Type N hydrates.

Unslaked lime is any form of quicklime.

Vienna lime is a special fossiliferous pulverized dolomitic quicklime used for buffing precision metal parts.

Whitewash is synonymous with milk-of-lime, a dilute lime hydrate suspension.

Physical Properties of Quicklimes

Color. Generally quicklime is white of varying degrees of intensity, depending on its chemical purity. The purest types of quicklimes are the whitest. Other less pure or improperly calcined types may have a slight ash gray, buff, or yellowish cast. The quicklime is invariably whiter than its derivative, limestone.

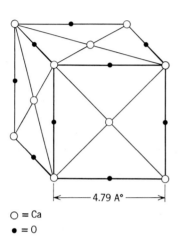

○ = Ca
● = O

Fig. 7-1. Crystal structure (unit cell) of calcium oxide.

Odor. It possesses a faint but distinctive odor that is difficult to define, except it is slightly "earthy" and pungent in aroma, but not offensive.

Texture. All quicklimes are crystalline, but the crystallite conglomerates vary greatly in size and spacing in their matrices. Some appear to be amorphous, but they are microcrystalline.

Crystal structure. X-ray diffraction reveals that a pure calcitic oxide crystallizes in the cubic system as depicted in Fig. 7-1. The edges of the cube are 4.797 Å in length, with calcium atoms located midway in between.

Magnesium oxide possesses the same cubic crystal lattice of CaO, except that the MgO crystal is slightly smaller and denser, with edge lengths of 4.203 Å. This accounts for the slightly higher average density of dolomitic quicklime.

The crystal lattice (AX type) is similar to the cubic arrangement for NaCl and is tabulated as follows:

Crystal	Coordination number	a_0	Sum of ionic radii*	Expt.
CaO	6	4.797 Å	2.39 Å	2.40 Å
MgO	6	4.203 Å	2.05 Å	2.10 Å

* Calculated ionic radii for coordination number 6 are as follows: Ca^{++} = 0.99 Å, Mg^{++} = 0.65, and O^{--} = 1.40.

Porosity–density. The degree of porosity of commercial quicklime varies widely in percent of pore space from 18 to 48%, with an average value of about 35%, depending on the structure of the limestone, temperature, and severity of calcination (see Table 6-2 on p. 147). Dead-burned dolomite has much lower porosity of 8 to 12%.

Specific gravity. The *true* specific gravity of pure calcium oxide is 3.34, but this presupposes zero porosity, a condition that is impossible to achieve in manufacture. Values have been reported at 3.40, and lower, but 3.34 appears to be a generally recognized average value. Commercial limes may range as low as 3.0; pure dolomitic oxides may range as high as 3.5 to 3.6.

The *apparent* specific gravity varies similarly, from 1.6 to 2.8 (p. 150). Average values for commercial oxides are 2.0 to 2.2. Values for dolomitic quicklimes average about 3 to 4% more than the above. Dead-burned has the highest value of all—an average of 3.2.

Bulk density. The same variance pertaining to specific gravity is prevalent as well as the added variable of the different physical size and gradation of the quicklime particle. The range in values in lb./cu. ft. is 48 to 70, with an estimated average of 55 to 60 lb./cu. ft. for commercial quicklime of pebble size.[2] Lump size would be about 10% lower, and ground or pulverized would be 12 to 15% greater than this average value. (The larger the particle and the more restricted the gradation, the lower the bulk density.)

Values for dolomitic would average 3 to 4% greater than for high calcium.

Hardness. Hard-burned and sintered dolomitic quicklime lies between 3 and 4 on the Moh's scale. Ordinary quicklime is variable, but usually between 3 and 2. The same broad divergence in hardness and strength in limestones is manifest in their derivative limes (p. 22).

Coefficient of expansion. The only values reported are 145×10^{-7} between 300 and 700°C and 138×10^{-7} between 0 and 1700°C[3]. As a result, this data probably only represents the magnitude of this measurement; certainly a variance would exist with commercial quicklimes.

Electrical resistivity. 71×10^8 ohms/cm. at 15°C, declining to 91 ohms at 1466°C, has been calculated. The presence of nitrogen depresses values.

Refractive index. The pure calcitic oxide is 1.83 and the value of commercial quicklime ranges between 1.70 and 1.82. A value of 1.736 for pure MgO means that dolomitic quicklime has a slightly lower value than CaO.[4] Both types possess slight refractive properties.

Luminescence. All lime oxides are very luminescent at high temperatures in the calcining range of 900°C and higher; hence, origin of the term "limelight."

Thermal conductivity. It has been estimated at 0.0015 to 0.002 cal./cu. cm./sec./°C temperature difference, but this value may be undependable.

Heat of fusion. It is doubtful if this has ever been accurately measured; 28,000 cal./mole has been estimated as the approximate value.

Melting point. Recognized values for CaO are 2570°C (4658°F) and MgO are 2800°C (5072°F), with dolomitic oxides intermediate.[5]

An eutectic mixture of about 50% $CaCO_3$ and 50% CaO is reported to melt at 1240°C under high pressure of 30,000 mm.[6]

A most recent investigation[7] of the system CaO–MgO, involving X-ray diffraction and optical methods, that may comprise the most authoritative data, reveals a maximum solid solution of MgO in a CaO lattice of 17% weight and a maximum solid solution of CaO in the MgO lattice of 7.8% weight, both at temperatures of 2370°C. In both instances, the extent of solid solution is higher than that reported by other investigators. Melting point for the eutectic 67% CaO and 33% MgO is 2370°C; for 100% CaO–2625°C; for 100% MgO-2825°C. Figure 7-2 shows the phase equilibrium diagram, calculated by these investigators.

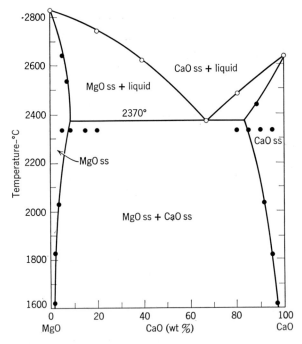

Fig. 7-2. Phase equilibrium diagram for system, CaO·MgO. Solid circles are data points of solid solution, and open circles represent liquidus points.

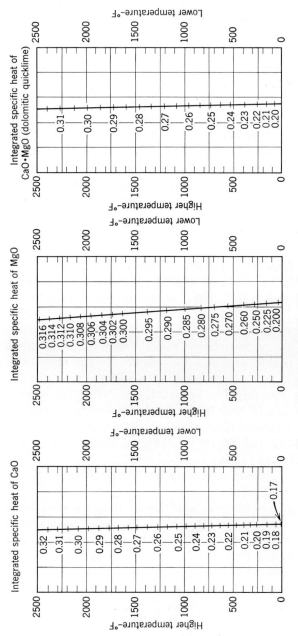

Fig. 7-3. Specific heats for CaO, MgO, and dolomitic quicklime.

Heat of combination. Same value for Heat of Formation below.

Boiling point. Values for CaO are 2850°C (5162°F) and for MgO are 3600°C (6512°F), with dolomitic oxides intermediate.[8]

Specific heat. Figure 7-3 graphically reveals the range of mean values for high calcium and dolomitic quicklimes and MgO at temperatures from 0 to 2500°C.[9] They encompass:

> For high calcium—from 0.17 to 0.32
> For dolomitic   —from 0.185 to 0.319
> For MgO   —from 0.195 to 0.316

These values increase gradually as the temperatures rise.

Earlier investigators developed slightly different values than the above—0.182 to 0.232 for CaO and 0.197 to 0.243 for CaO.MgO (dolomitic).[10, 11] The former values are preferred.

Transition point and heat of transition. It has been variously estimated at between 400 and 415°C and 425 and 430°C, with heat absorption of 280 cal./mole.

Heat of formation. 151,900 cal./mole is generally recognized for CaO; for MgO (periclase) 143,750 cal./mole is the value at 298°C, but this latter value rises gradually to 174,050 cal./mole at 2000°C.[12]

Angle of repose. There is some variance in values with different quicklimes and with different particle sizes and gradations, but 50 to 55° for pebble sizes is a reasonable average magnitude for this measurement.

Solubility. See values of hydrated limes on p. 176, since quicklime is converted to a hydrate before dissolution occurs.

Heat of solution. Heat of solution for CaO has been measured between 844.72 and 847.08 cal./g.[13] Solubility of MgO is so slight that this value may be immeasureable.

Surface energy. Surface energy of CaO at 23°C has been measured at 1310 ± 200 erg/cm.[2]

A value that is 330 erg<cm.[2] higher than the corresponding CaO value has been calculated for MgO. Brunauer[13] estimates this value at a minimum of 1400 erg/cm.[2], about 300 units higher than the 1000 to 1040 erg/cm.[2] for MgO calculated between 0 to 298° K by another source. The higher value is preferred.

Surface entropy. A value on surface entropy of 0.28 erg/cm[2] degree[-1] has been estimated for CaO.[14]

Physical Properties of Hydrated Limes

Color. All dry hydrates, except those that are quite impure, are extremely chalky in color, invariably whiter than their derivative

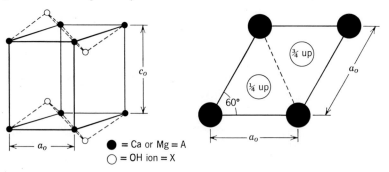

= Ca or Mg = A
= OH ion = X

Fig. 7-4. Crystal structure (unit cell) of calcium hydroxide.

quicklime. Overburning of quicklime may yield a faint yellowish cast in the resulting hydrate. Dark flecks of coarse particles are simply impurities, usually silica.

Individual pure hydroxide crystals are clean and colorless.

Odor. Same approximate aroma exists as with quicklime (p. 168).

Form. It occurs as a fine powder. Fineness varies in size, but it may be of microcrystalline or colloidal size (submicron). For this reason, many people erroneously regard some forms of hydrate as amorphous. But X-ray reveals a definite crystal structural pattern for even the finest hydrates.[15] Coarse hydrate particles are clearly crystalline in appearance.

Crystal structure. The crystal is a hexagonal-shaped plate or prism with perfect basal cleavage, but the physical particle is of varying size since the microscopic crystallites agglomerate in varying degrees. Hedin describes the crystals in solid state as forming a "ditrigonal scalenohedral belonging to the hexagonal-rhombohedral system."

The crystal lattice (AX_2 type) of the unit cells is similar to the molecular arrangement for $Cd\ I_2$, depicted as follows and in Figs. 7-4 and 7-5 for both $Ca(OH)_2$ and $Mg(OH)_2$:

Crystal	Coordination number A	Coordination number X	Edge length of base	Height	Acute angle of base
$Ca(OH)_2$	6	3	3.5844 Å	4.8962 Å	60°
$Mg(OH)_2$	6	3	3.11 Å	4.74 Å	60°

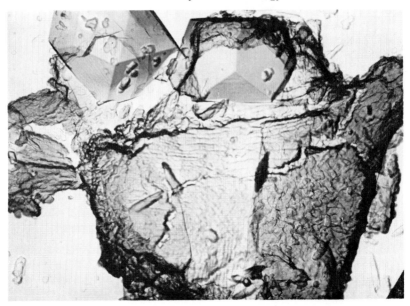

Fig. 7-5. Microphotograph clearly showing the single large crystals of $Ca(OH)_2$ and its characteristic rhombohedron shape (50,000 X).

Specific gravity. The range in specific gravities for different commercial hydrates are as follows:

High calcium	2.3 to 2.4
Highly hydrated dolomitic	2.4 to 2.6
Normal hydrated dolomitic	2.7 to 2.9

The latter contains 25 to 35% MgO on an average. (Hedin calculated a value of 2.244 g./cm.[3] for pure reprecipitated $Ca(OH)_2$ from X-ray diffraction data.)

Apparent density. Values of 0.4 to 0.55 g./ml. have been reported.

Bulk density. As determined by the Scott Method, a range in bulk density would be 25 to 40 lb/ft.[3] for commercial hydrates, with about 35 lb./ft.[3] an average value.[2] Degree of air entrainment affects bulk density values.

Angle of repose. There is a very wide range in angle degrees for different commercial hydrates, depending upon particle size, moisture content, degree of aeration, and particularly the extent of the electrostatic charge. It may range from 15 to 80°, but 70° is considered most common by some authorities.

Hardness. It is intermediate between 2 and 3 on the Moh's scale, with 2.5 an average value reported for pure $Ca(OH)_2$ crystals.

Refractive index. Values for both $Ca(OH)_2$ and $Mg(OH)_2$ are similar, being 1.574 and 1.545 for the former and 1.559 and 1.580 for the latter. Crystals are uniaxially negative. Maximum index of double refraction is 0.029.

Coefficient of expansion. It is 3.34×10^{-5} along principal axis and 0.98×10^{-5} normal to it. Data is scant and of only academic interest.

Specific heat. The values would depend on the specific heats of the two reactants, CaO and MgO (see p. 171), and water that forms the hydroxide, all of which are well established. But the data reported for hydrates is sparse and discordant. An intermediate value between quicklime and water of 0.27 at 0°C, increasing to 0.37 at 400°C, appears reasonable.[16] Value for dolomitic hydrate is believed to be about 5% higher.

Heat of formation. This is also synonymous to the *heats of hydration* and *reaction.* The most dependable values for commercial hydrates appear to be:[17, 18]

$Ca(OH)_2$	15,300 cal./g. mole or
	488 Btu/lb. of quicklime or
	27,500 Btu/lb. mol.
$Mg(OH)_2$	8000 to 10,000 cal./g. mol or
	14,400 to 18,000 Btu/lb. mol

$Ca(OH)_2 \cdot Mg(OH)_2$—intermediate between the above values.

Melting and boiling points. See quicklime (p. 170), since the hydroxides are converted to oxides before melting occurs.

Heat of solution. For $Ca(OH)_2$ a value of 2790 cal./mole at 18°C is probably the most accurate. Value for dolomitic hydrate is unknown, but it is probably about 30 to 45% less than the above values. For pure $Mg(OH)_2$, its solubility is so slight that its heat of solution would be nearly zero.

Vapor pressure. Measurements on the vapor pressure incident to the dissociation of $Ca(OH)_2$ into oxides and water range between 510 to 547°C at 760 mm. pressure, with 540°C the most plausible value.[19]

This figure is also the *decomposition point.*

Values for $Mg(OH)_2$ are lower, but very discordant; 345°C (653°F) at 760 mm. pressure is probably the best value,[19] but figures as low as 180°C have been reported.

Dolomitic hydrates are intermediate between the above.

Diffusion Constant. For a concentrated $Ca(OH)_2$ solution at 25°C temperature, the diffusion constant was reported by Hedin[15] as $1.931 \cdot 10^{-5}$ cm.$^2 \cdot$ sec.$^{-1}$ (An interpolated value of 1.922 on the same basis has also been reported.) As the concentration increases, this value decreases gradually until at a concentration of 1.038 g.$\cdot$ CaO/l. the diffusion constant is $1.385 \pm 0.027 \cdot 10^{-5}$ cm.$^2 \cdot$ sec^{-1}.

Solubility of $Ca(OH)_2$. The solubility of calcium hydroxide has been measured many times by different investigators, and the values are in reasonably good agreement, with a maximum spread of about 10%. Of these values, probably those developed by Haslam and his co-workers on commercial hydrates and pure $Ca(OH)_2$ are the most reliable, and they tend to be intermediate among the mass of solubility data reported.[20]

The magnitude of solubility of $Ca(OH)_2$ is 1.330 g. CaO/l. of saturated solution at 10°C in distilled water (or 0.13%). At 0°C solubility increases to 1.4 g. CaO/l. Figure 7-6 presents the results of Haslam's findings. Table 7-1 provides solubility data in tabular form expressed as CaO equivalent or as $Ca(OH)_2$. Contrary to limestone, solubility of lime hydrate is in inverse proportion to temperature, decreasing as temperature rises. Data is sparse above 100°C.

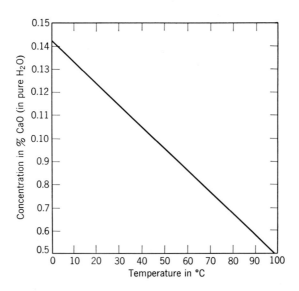

Fig. 7-6. Solubility of pure CaO at different temperatures.

Table 7-1. Solubility of lime expressed as CaO or Ca(OH)$_2$ at different temperatures g./100 g. sat. sol.

t°C	CaO	Ca(OH)$_2$
0	0.140	0.185
10	0.133	0.176
20	0.125	0.165
30	0.116	0.153
40	0.106	0.140
50	0.097	0.128
60	0.088	0.116
70	0.079	0.104
80	0.070	0.092
90	0.061	0.081
100	0.054	0.071

In spite of the fact that Ca(OH)$_2$ is about one-hundred times more soluble than CO$_2$-free calcium carbonate (p. 25), it still is only regarded as very slightly soluble. Haslam[21] calculated that commercial hydrated limes are about 7% more soluble on an average than chemically pure Ca(OH)$_2$ made from Iceland spar and speculated that this slight deviation is due to traces of highly soluble alkalis (K$_2$O and Na$_2$O) that exist in commercial hydrates. Another investigator, Bassett,[22] calculated that a freshly slaked quicklime with fine microparticles was about 10% more soluble than a coarse, crystalline type of hydrate. Unlike Haslam, Bassett and Herold[23] reported values at higher temperatures, which are listed below in g. of CaO equivalent per 100 g. sat. sol.:

°C	Bassett	Herold
90	0.0591	
99	0.0523	
120		0.032
150	0.246	0.018
190		0.009

EFFECT OF INORGANIC SOLUTIONS.[24] The solubility of Ca(OH)$_2$ is affected by some salt and inorganic chemical solutions in varying degrees, depending upon the concentration. Most salts increase Ca(OH)$_2$ solubility by 10 to 15% in 0.1 to 0.2% salt solutions. Generally, increases in temperature still depress solubility of the hydrate, as in pure water. The effect of these extraneous solutions on the solubility of Ca(OH)$_2$ is summarized as follows:[21, 25]

1. *Calcium chloride* increases solubility slightly in increasing concentrations up to 30%. An exception prevails in that rising temperatures increase solubility in concentrated $CaCl_2$ solutions. This anomaly results from the formation of a new solid phase at a $CaCl_2$ concentration of 17% for the system: CaO, $CaCl_2$, and H_2O. The reverse occurs with dilute $CaCl_2$ solutions. Highest $Ca(OH)_2$ concentration reported is 0.4922 g./100 c.c. in 30% $CaCl_2$ solution at 100°C. At concentrations above 30%, solubility decreases.

2. *Calcium iodide* causes a slight, steady decrease in solubility up to a 59% CaI_2 solution at 25°C.

3. *Ammonium chloride* elevates solubility gradually until a NH_4Cl concentration of 4.4% is reached at 25°C, tripling $Ca(OH)_2$ solubility to 4.42 g./l.

4. *Calcium nitrate* increases solubility slightly (a maximum of about 100%) in concentrations rising to 45% $Ca(NO_3)_2$ at 25°C; then solubility ebbs as a new solid phase occurs.

5. *Calcium sulfate* in very dilute concentrations depresses solubility. At 25°C, a 0.2% $CaSO_4$ concentration reduced $Ca(OH)_2$ solubility to 0.0062 g./100 c.c. sat. sol.

6. *Ammonium hydroxide* contributes no change of consequence (a slight increase) in solubility with NH_4OH concentrations up to 2%; then a very slight decrease in solubility occurs with added increments of this salt.

7. *Sodium and potassium chlorides* increase solubility at a slow rate up to 20% NaCl and KCl concentrations at three temperatures—0, 15, and 99°C. Total increase in solubility is only about 25%; rising temperatures lowered solubilities uniformly in the same proportion as with pure water.

8. *Lithium chloride* increased solubility gradually up to a 10.37% LiCl concentration, doubling original $Ca(OH)_2$ solubility.

9. *Barium chloride* increased solubility steadily up to 1.59% $BaCl_2$ concentration, more than doubling original $Ca(OH)_2$ solubility.

10. *Strontium chloride* reacted the same as barium chloride with concentrations up to 2.23%.

11. *Sodium and potassium bromides* yield no change of consequence; slight gains in $Ca(OH)_2$ solubility occur up to 3 to 4% bromide concentrations and then slight declines occur.

12. *Sodium iodide* yielded only inconsequential changes.

13. *Sodium hydroxide* in minute fractional concentrations causes a steady, sharp decline in $Ca(OH)_2$ solubility at four temperatures: 20, 50, 70, and 100°C. For all of these temperatures 0.8% NaOH solution renders lime almost totally insoluble. Even 0.1% NaOH

concentration drastically curtails solubility. Na_2CO_3 solutions act similarly.

EFFECT OF ORGANIC SOLUTIONS. Several soluble organic substances produce a dramatic change in calcium hydroxide's solubility characteristic, markedly increasing solubility to a far greater extent than any inorganic chemical. The most important of these organics is sugar, which, judging from the abundance of available data, must have intrigued research chemists. Glycerol and phenol solutions also greatly increase lime's dissolution.

Glycerol at 25°C yields a steady increase in solubility up to a glycerol concentration of 69%. $Ca(OH)_2$ solubility at this maximum level is 3.55 g. CaO/100 c.c. or a total increase of about twenty-two times as compared to pure water.

Phenol at 23°C produces an even steeper rise in solubility up to a 30% phenol concentration, 8.67 g. CaO/100 c. c. phenol solution, or an increase of seventy-five times is reported.

Sugar exerts the greatest influence on lime solubility, as Table 7-2 demonstrates.[24] Saturated solutions of $Ca(OH)_2$ rise rapidly and

Table 7-2. Effect of increasing sugar concentrations on the solubility of lime[a]

25°C			80°C		
G./100g. sat. sol. sugar	CaO	Solid phase	G./100g. sat. sol. sugar	CaO	Solid phase
0.0	0.122	$Ca(OH)_2$	0.0	0.071	$Ca(OH)_2$
2.1	0.242	$Ca(OH)_2$	4.90	0.117	$Ca(OH)_2$
4.2	0.461	$Ca(OH)_2$	9.90	0.189	$Ca(OH)_2$
6.6	0.750	$Ca(OH)_2$	14.75	0.230	$Ca(OH)_2$
8.6	1.11	$Ca(OH)_2$	19.50	0.358	$Ca(OH)_2$
11.8	1.86	$Ca(OH)_2$	24.60	0.548	$Ca(OH)_2$
15.4	2.76	$Ca(OH)_2$	29.70	1.017	$Ca(OH)_2$
21.1	4.53	$Ca(OH)_2$			
27.2	6.72	$Ca(OH)_2$			
31.4	8.39	$Ca(OH)_2$			
35.2	9.8	$Ca(OH)_2$ + Saccharate			
35.0	10.1	Saccharate			
36.2	9.8	Saccharate			
43.7	8.84	Saccharate			
53.2	7.87	Saccharate			
68.3	4.08	Saccharose			

[a] Data indicates that lime is progressively covered with a layer of saccharate, which impedes eventual dissolution of lime and at very high sugar concentrations causes the solubility of lime to retrogress.

steadily as the increment of sugar is increased up to 35% sugar solution. At this concentration and 25°C an increase of nearly 10.1 g. of CaO/100 c.c. can be dissolved—one-hundred times over a pure $Ca(OH)_2$ saturated solution. Beyond this sugar content, the free, undissolved lime hydrate is progressively coated with a layer of saccharate, a solid phase reactant, that steadily decreases the CaO solubility. Sugar solutions parallel pure lime solutions in that solubility also is in an inverse ratio to rising temperatures, decreasing as the temperature increases. Between 25 and 80°C solubilities decline between 8 and 10 times at various sugar concentrations.

In an exhaustive study of the literature on lime solubility, Haslam concludes:

1. MgO and such impurities as silica, alumina, and CO_2 (as $CaCO_3$) do not affect the solubility of $Ca(OH)_2$, but they may retard its rate of solution.

2. Lime solubilities in other inorganic salt and organic solutions are generally predictable in dilute concentrations, based on semi-empirical relationships. Thus, solubility decreases sharply in alkaline solutions and increases in acid salt solutions. Marked increases also occur with some organic substances when lime forms a new chemical compound with the dissolved organic material.

3. Agitation increases both the rate of solution and degree of solubility. With agitation, 25% more solubility was obtained in one day than in seven days without stirring the solute.

4. Dissolution of K_2O and Na_2O, as impurities in commercial limes, occurs immediately on contact with water—well in advance of the CaO dissolution.

Solubilities of many calcium and magnesium salts are contained in Appendix A at end of the book.

ELECTROLYTIC CONDUCTIVITY. Ringquist developed a vast amount of data on the electrolytic conductivity of calcium hydroxide solutions and established a direct relationship of conductivity to the concentration of $Ca(OH)_2$ solutions at temperatures between 0 and 100°C. From his data, the concentration of a $Ca(OH)_2$ solution at a given temperature can be accurately ascertained by obtaining the conductivity value of the $Ca(OH)_2$ electrolyte, which varies from 0.560 to 11.805 in ohms^{-1}/cm.$^{-1}$/10^3. Variance in conductivity depends on temperature and concentration of the solution as displayed in Table 7-3 covering a representative portion of Ringquist's data. Through interpolation, similar data has been calculated for each degree of temperature between 0 and 100°C.

Table 7-3. Electrolytic conductivity measurements of Ca(OH)$_2$ solutions of varying concentration and temperature[a]

Temperature °C	Concentration in grams of CaO per litre					
	0.1217	0.3050	0.6090	0.9130	1.2170	1.4900
0	0.560	1.400	2.640	3.915	5.090	6.070
5	0.650	1.600	3.030	4.465	5.790	6.910
10	0.735	1.800	3.420	5.005	6.485	7.750
15	0.820	2.000	3.805	5.540	7.175	8.590
20	0.905	2.200	4.195	6.075	7.870	9.415
25	0.990	2.395	4.565	6.610	8.550	10.210
30	1.075	2.590	4.940	7.140	9.225	10.995
35	1.160	2.780	5.310	7.670	9.900	11.740
40	1.240	2.975	5.670	8.190	10.555	×
45	1.325	3.160	6.030	8.690	11.195	
50	1.410	3.350	6.390	9.190	11.805	
55	1.490	3.540	6.740	9.685	×	
60	1.575	3.725	7.075	10.165		
65	1.660	3.910	7.410	10.640		
70	1.740	4.085	7.745	11.095		
75	1.820	4.255	8.070	×		
80	1.900	4.425	×			
85	1.980	4.590				
90	2.060	4.745				
95	2.140	4.895				
100	2.220	5.045				

[a] Values are expressed in ohms^{-1}/cm.$^{-1}$/10^3. × indicates that solid phase was reached.

Similar data has not been developed for Mg(OH)$_2$, but one value at 18°C is given of 80 × 10^{-6}.[27]

Solubility of Mg(OH)$_2$. There is a paucity of discordant solubility data for magnesium hydroxide. Remy and Kuhlman[28] report 0.016 g. MgO/1. at 18°C in pure water; other values are much less: 2 to 6 × 10^{-4} at 18 to 45°C, but such data probably refers to periclase, chemically inert MgO. Knibbs[27] theorizes that a value of 0.01 g./l. is a reasonable approximate. For freshly slaked, chemically reactive milk of magnesia, this is its probable magnitude of solubility at temperatures around 15 to 30°C. Therefore, it appears that Ca(OH)$_2$ is about one-hundred times more soluble than Mg(OH)$_2$;

even $MgCO_3$ is about ten times more soluble. Values on the effect of temperature are erratic, but solubility is believed to decline with rising temperatures. Total insolubility, measured to five decimal places, has been reported at 178°C:

One source reports[29] a sharp increase in $Mg(OH)_2$ solubility in concentrated ammonium chloride and nitrate solutions at 29°C:

		g./l.			*g./l.*	
Salt conc.	NH_4Cl	4.13	20.86	NH_4NO_3	6.09	14.69
Sat. col.	$Mg(OH)_2$	1.43	4.55	$Mg(OH)_2$	1.45	2.43

In very dilute concentrations of NH_4Cl solutions of less than 1%, Fredholm[30] added $Mg(OH)_2$ and measured the Mg^{++} content of the saturated solutions at 18°C. The calculated equivalent $Mg(OH)_2$ solubility values were 0.0472, 0.0698, and 0.1237 g./l., with values rising as the fractional salt concentrations were increased.

There is some evidence that solubility is increased with combined NaCl–NaOH solutions and then decreases to complete insolubility, based on the following paucity of data of questionable dependability:[31]

NaCl g./l.	125	125	140	160
NaOH g./l.	0.8	0.4	0.8	0.8
MgOH g./l.	0.7	0.03	0.45	nil

It appears that there is no confirmed data on dolomitic hydrate, so that its solubility must necessarily be subject to some speculation. Haslam[20] concluded that the solubility of $Mg(OH)_2$ was too minute to be of any consequence in the presence of the more soluble $Ca(OH)_2$. Yet his experimentation indicated that nearly the same percent of CaO was dissolved in saturated solutions of both dolomitic and high calcium hydrates, indicating there is little, if any, difference in their respective solubilities. Thus, it is believed that the presence of MgO or $Mg(OH)_2$ does not affect the solubility of the $Ca(OH)_2$ component and only acts as a minor diluent.

Chemical Properties of Quick and Hydrated Lime

Table 7-4 delineates the gravimetric percentages of the critical constituents of all limes, quick and hydrated, high calcium and dolomitic, both as elements and compounds.[32] (More detailed comparable information is also contained in Appendix A.)

Table 7-4. Gravimetric percentages of critical constituents of limes

| Lime substances | Percents of elements | | | | Percents of compounds | | | |
	Ca	Mg	O	H	H_2O	CaO	MgO	Alkali oxides (CaO + MgO)
CaO	71.47		28.53			100.00		100.00
Ca(OH)$_2$	54.09		43.19	2.72	24.31	75.69		75.69
MgO		60.32	39.68				100.00	100.00
Mg(OH)$_2$		41.69	54.85	3.46	30.88		69.12	69.12
CaO·MgO	41.58	25.23	33.19			58.17	41.83	100.00
Ca(OH)$_2$·MgO	35.03	21.26	41.95	1.76	15.75	49.01	35.24	84.25
Ca(OH)$_2$·Mg(OH)$_2$	30.27	18.36	48.33	3.04	27.21	42.35	30.44	72.79

Stability. Both quicklime and hydrated lime are reasonably stable compounds, although not nearly as stable as their limestone antecedent. Quicklime is completely stable at any temperature. It is vulnerable only to water; even moisture in the air produces a destabilizing effect by air slaking (see pp. 191 and 298). It will react chemically with acids, many other chemical compounds, and elements to form different calcium and magnesium compounds which are described in the ensuing sections, but it is almost impossible to decompose it into two or three component atoms (Ca^{++}, Mg^{++}, and O^{--}).

Hydrated lime is even more stable, since water will not cause a chemical change in its composition. Its perishability is jeopardized only by CO_2 absorption, for which it possesses a strong affinity in even dilute concentrations. It reacts similarly to quicklime with other chemical compounds and elements. However, unlike quicklime, hydrate will decompose and revert back to $CaO + H_2O$ when subjected to temperature of about 540°C.

Dolomitic types are slightly more stable than high calcium, for they will not absorb water or CO_2 so rapidly. Dead burned dolomite is completely stable under most conditions, except for slow reactivity with strong, concentrated acids.

Desiccant. Because it is so hygroscopic quicklime is an effective dessicant in absorbing moisture from the air or in removing water as liquid or vapor from chlorinated hydrocarbon solvents, alcohols, aldehydes, and ketones. (Its very faint solubility at elevated temper-

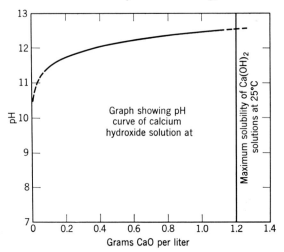

Fig. 7-7. pH values of $Ca(OH)_2$ solutions of varying concentrations at 25°C.

atures permits removal of the hydrate solid by sedimentation or filtration.)

Its moisture absorptive capacity is at least 24.3% of its own weight. Its rate of absorption varies greatly, depending upon its chemical reactivity and the vapor or water concentration. The rate of H_2O absorptive capacity of dolomitic quicklime is much less than high calcium.

pH. The pH of $Ca(OH)_2$ solutions at 25°C has been measured,[33] and its gradient is displayed in Fig. 7-7. This data is presented in tabular form as follows:

CaO g./l.	pH
0.064	11.27
0.065	11.28
0.122	11.54
0.164	11.66
0.271	11.89
0.462	12.10
0.680	12.29
0.710	12.31
0.975	12.44
1.027	12.47
1.160	12.53

This indicates that the addition of only a trace of CaO to pure distilled water will cause a rapid vertical rise in pH from neutrality at 7 to 11. From there the rise in pH is very gradual. A 50% saturated CaO solution gains very little in pH before reaching saturation. Temperature is a limiting factor on pH since rises in temperature that reduce solubility will decrease the pH slightly at fractional solubility concentrations. Similarly, pH values below 25°C are slightly higher at the lower temperature, probably as a saturated solution ascending to a pH of 13 at 0°C. Owing to much lower solubility, $Mg(OH)_2$ possesses a maximum pH of only 10.5; dolomitic hydrate is nearly identical to $Ca(OH)_2$.

Neutralization. Hydrated lime, either $Ca(OH)_2$, $Ca(OH)_2 \cdot Mg(OH)_2$, or $Ca(OH)_2 \cdot MgO$, ionizes readily in water into Ca^{++}, Mg^{++}, and OH^- ions and is the most widely employed base or alkali. Its hydroxyl ion, OH^-, combines readily with the acid hydrogen ion, H^+, in an aqueous solution with the hydronium ion (H_3O^+) to form water and calcium or calcium–magnesium salts with a substantial evolution of heat.

Unlike the other strong bases, sodium and potassium hydroxides, both $Ca(OH)_2$ and $Mg(OH)_2$ are diacid bases, as opposed to the former monoacid bases. Consequently, as displayed below, two molecules of a monobasic acid are required to neutralize one molecule of $Ca(OH)_2$ or $Mg(OH)_2$; with NaOH and KOH only one molecule of monobasic acid is necessary to effect complete neutralization.

$$Ca(OH)_2 + 2HCl = CaCl_2 + 2H_2O + 27,400 \text{ cal.}$$

$$Mg(OH)_2 + 2HNO_3 = Mg(NO_3)_2 + 2H_2O + 27,400 \text{ cal.}$$

$$NaOH + HCl = NaCl + H_2O + 13,700 \text{ cal.}$$

$$KOH + HNO_3 = KNO_3 + H_2O + 13,700 \text{ cal.}$$

Twice as much *heat of neutralization* is liberated by the diacid bases in the above neutralization reactions per gram-molecular weight in contrast to the monoacid bases. Other heat of neutralization values are:

	With H_2SO_4	With CO_2	With HCl
$Ca(OH)_2$	31,140	18,310	27,900
$Mg(OH)_2$	31,220	—	27,690

With dibasic acids, like sulfuric, one molecule each of hydrated lime and acid is required, whereas two molecules of the monoacid bases are essential for complete neutralization.

$$Ca(OH)_2 + H_2SO_4 = CaSO_4 + 2H_2O$$

$$2NaOH + H_2SO_4 = Na_2SO_4 + 2H_2O$$

Stoichiometric quantities of other acids and limes are contained in Fig. 7-8.

Since both $Ca(OH)_2$ and $Mg(OH)_2$ are very strong bases, they react with mild acids to form alkaline salts and with strong acids to form neutral or slightly alkaline salts.

In spite of possessing only very slight solubility as compared to the highly soluble bases, NaOH and KOH, both $Ca(OH)_2$ and $Mg(OH)_2$ are the strongest bases in that less weight is required to neutralize a given acid, as revealed in Fig. 11-2 on p. 368. $Mg(OH)_2$, derived from caustic magnesia, is the strongest base of all with highest alkali equivalent since the equivalent weight of its derivative MgO is less than CaO. That is why all types of dolomitic limes of equal purity possess greater basicity than corresponding high calcium limes. A graph showing the relative neutralizing power of all types of limes with five common acids is shown in Fig. 7-8.[34]

The faintly soluble lime hydrates neutralize acids with equal facility as unsaturated, saturated, and supersaturated solutions, but they are usually utilized in the latter form as suspensions. As the Ca^{++}, Mg^{++}, and OH^- ions react with the corresponding acid ions to form water and salts, the excess lime in suspension continues to dissolve and combines with the remaining free acid ions until all acid is neutralized or the resulting salt solution becomes supersaturated and crystallizes. Mechanical agitation, aeration, and elevated temperatures accelerate the neutralization reaction.

For further information on the use of lime for neutralization, see pp. 367–73.

Rate of Solution. Lime's efficacy as an alkali depends both upon its ultimate solubility and the rate at which it goes into solution or becomes available. Limes differ greatly with respect to the latter consideration.

First, particle size exerts some influence over rate of solution, particularly when quicklime is introduced into water. As the average diameter of the particle size diminishes, its dissolution increases, since

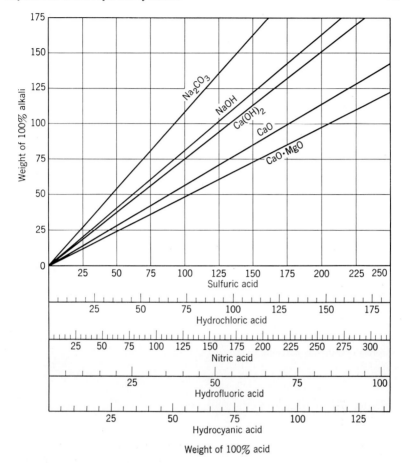

Fig. 7-8. Alkali neutralization equivalents for various acids. This graph may be used to determine the weights of alkalies required to neutralize a given weight of any of the acids indicated. Since the graph is based upon theoretically pure acids and alkalies, appropriate corrections should be made when applying the data. The weight of 100% acid = weight of dilute acid × the % concentration of acid present. A similar equation applies for alkalies.

its surface area is augmented. Adams equated this solubility differential in the following ratio: halving the diameter or doubling the surface area causes the rate of solution to ascend twice as rapidly, e.g., a quicklime of 100% —No. 65 mesh and +No. 100 mesh dissolves twice as fast as 100% —No. 35 and +No. 48 meshes.

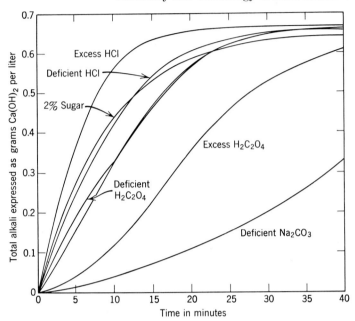

Fig. 7-9. Influence on high calcium limes rate of solution with various extraneous solvents.

Figure 7-9 shows how different solutes affect the rate of solution as follows:

1. The rate is directly proportional to the ultimate solubility of the lime, e.g., by approximately doubling the ultimate solubility (2.13 times) in a 2% sugar solution, the rate of solution is twice as rapid as in pure water.

2. If lime reacts with an acid that forms a soluble salt, like HCl, the rate of solution is increased, the gain in solubility being directly proportional to the concentration of the acid. The rate is further stimulated by excess acid, in contrast to a deficiency of acid; the latter only increases the rate slightly.

3. If lime reacts with an excess of acid that forms an insoluble salt, like oxalic acid, the rate of solution is retarded. The resulting insoluble oxalate salts encrust the lime particles impeding dissolution.

4. If lime reacts with a deficiency of acid that forms an insoluble salt, the rate of solution is slightly accelerated. An example is oxalic acid.

5. If lime reacts with alkaline salts that decrease its solubility, the rate of solution is similarly retarded by about the same proportion. This is due to the formation of an insoluble precipitate, $CaCO_3$, as the lime causticizes the solution, e.g., sodium carbonate.

Generally the above reactions are similar for both high calcium and dolomitic limes except that with acids that yield soluble salts the solution rate of dolomitic tends to increase more with excess acid but decreases more with deficient acid because of its MgO content.

Figure 7-10 demonstrates the difference in solution rates of typical high calcium and dolomitic limes of identical particle size in pure water.[35] At the outset, dissolution for both proceeds at a similar pace, but the ultimate rate of high calcium is more rapid because of its slightly greater solubility. In sugar solutions rates of both limes increase with the greater solubility induced by sugar, but the same differential in rate still exists in this medium.

Discordant and even paradoxical results have been obtained by some investigators. Statistically it can be contended that rate of solution increases with higher specific surface areas of hydroxide particles

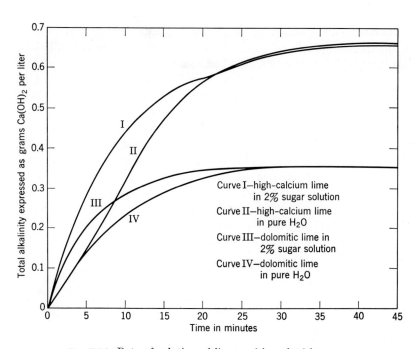

Curve I—high-calcium lime in 2% sugar solution
Curve II—high-calcium lime in pure H_2O
Curve III—dolomitic lime in 2% sugar solution
Curve IV—dolomitic lime in pure H_2O

Fig. 7-10. Rate of solution of limes, with and without sugar.

in a majority of cases, but there are disconcerting exceptions. Where rate of solution is slow and surface area high, it is postulated that diffusion of the solution is restricted by presence of minute capillaries and micropores, which may form stagnant, saturated solution pockets. This constricts the diffusion area, negating the advantage of high surface area, so that the total specific surface is unavailable to the dissolving medium. If this premise is correct, then the external surface area (rather than specific surface) would be a more reliable correlating factor in determining rate of solution.

This rate is determined with greatest accuracy by *electrical conductometric* methods. It is generally hypothesized that wetting and dispersion precedes dissolution in the first second of contact between the lime solids and water. This is, of course, associated with the *initial rate of solution* and is difficult to measure with accuracy; however, observable differences in the wetting times of different hydrates have been noted which again are believed related to the above surface area phenomenon. Staley and co-workers at M.I.T.[36] have recorded initial rates of solution of $Ca(OH)_2$ of different specific surface areas, ranging from 6.67 to 32.2 m.²/g. in 0.02 to 0.13 mg./1-mg.-sec.

Sterilization.[37] Because of lime's strong basicity and the high pH it imparts to aqueous solutions, its resulting residual caustic alkalinity has germicidal properties. In fact, some municipal potable waters are purified solely by detention periods in contact with excess free lime at a pH of over 9.5 (generally a pH of 10 to 11 is maintained). Below a pH of 9.5 only partial disinfection occurs.

Depending upon the degree and character of hardness in the water, the initial pH of the raw water, extent of pollution, and temperature, the magnitude of lime addition for sterilization is about 8 to 14 grains of CaO per gallon of water. A plant in London used 11.7 grains per gallon, or about 1 part of CaO per 5000 parts of water. Detention times will vary depending upon the extent of pollution (bacteria count) existing in the untreated water, but three to ten hours is a typical range of duration.

In a study of six municipal water-treatment plants in Ohio, the average B. coli index existing in the raw water was 2274/100 ml. After excess lime treatment (no chlorine was used), the average B. coli count was only 0.6/100 ml. At Columbus, Ohio the Scioto river raw water contains 300,000 B. coli/100 ml. After lime-soda water softening is effected, the bacteria of the colon and typhoid groups are reduced 99.67% in five hours detention. But if one grain per gallon of excess lime is added, 99.93% germicidal action is achieved, and with three grains excess, 100% bacteria reduction is attained. These waters meet purification standards for potable purposes. In addition to the B. coli and typhus bacteria, indications are that the high alkalinity of lime kills certain virus, such as hepatitus and infantile paralysis.

Carbon Dioxide. For all practical purposes *dry* CO_2 will not react with quicklime at ordinary atmospheric temperatures, although a few researchers claim to have observed slight traces of recarbonation. But if temperatures are increased, recarbonation will commence slowly at about 290°C. No significant adsorption of CO_2 will occur until about 400°C; starting at about 600°C and higher its affinity for CO_2 is very pronounced and rapid. However, complete recarbonation, even at elevated temperatures, does not occur. The reason for this is that this adsorption is a surface phenomena in which a shell of calcium carbonate is gradually formed around the CaO particle. As the shell of the $CaCO_3$ accumulates, a densification occurs making it more difficult for the CO_2 molecule to penetrate through the narrowing pores to unite with the interior CaO molecules. Percent of recarbonation varies depending upon size of particle, with greater CO_2 adsorption occurring in the smaller particles. With reactive quicklime dust recarbonation would be nearly 100%; with a 6 in. lump maximum adsorption might be only 25 to 30%. Heat-stimulated recarbonation occurs more readily with high calcium quicklime than with dolomitic. With the latter it is largely only the CaO component that recarbonates.

In the presence of *moisture*, vapor or liquid water, recarbonation occurs readily at atmospheric temperatures; in fact, water exerts a catalytic effect. Very little moisture is needed. Air slaking (p. 291), which precedes recarbonation, can occur slowly in arid climates of 15% relative humidity even though the volume of CO_2 in the atmosphere is only about 0.03%.

Essentially the same affinity for CO_2 occurs with *hydrates*. It is atmospheric adsorption of CO_2 that hardens lime mortars. But again CO_2 is impeded from penetrating into the interior of the mortar because of the carbonate crust that is formed on the mortar surface. Evidence of free lime hydrate in the interior of mortars that are hundreds of years old is common. Soda-lime is an established adsorbent for CO_2 vapor.

Artificially generated CO_2 can produce an exciting reaction with lime. Since limestone to quicklime to hydrate is chemically reversible, an Israeli chemist, Zalmanoff,[38] was intrigued with the possibility of making it a physically reversible reaction as well, to convert the lime hydrate paste into a carbonate substance as hard as the original stone. He claims to have achieved this in the laboratory and on a small pilot plant scale by recarbonation of both neat lime putties and lime-aggregate mixtures under certain optimum conditions of high CO_2 concentration, velocity, and pressure; elevated temperature; moisture content; and molding pressure. Strengths are obtained rapidly (in hours) and without autoclave pressure. The resulting carbonates possess strengths in compression of 5 to 10,000 psi, a magnitude of strength comparable to the original stone and far exceeding the 100 to 200 psi that can be obtained with atmospheric carbonation, only after a few years. Other researchers have claimed similar results. Utilization of this concept in manufacturing valuable commercial products has been thwarted thus far by inability to adapt this process economically to mass-production technique. Details on this process are proprietary, but it is likely that ultimately this process will be successfully commercialized.

Carbon Monoxide. There is some evidence that at about 500°C, just below the dissociation point of $Ca(OH)_2$, that carbon monoxide reacts with hydrated lime as follows: $Ca(OH)_2 + CO = CaCO_3 + H_2$.

Sulfur Compounds. Quicklime does not adsorb *sulfur dioxide* at atmospheric temperatures, but adsorption readily occurs at higher temperatures in the 300 to 400°C range, with rapid reaction occurring with temperatures over 400°C to form *calcium sulfite* as follows: $CaO + SO_2 = CaSO_3$.

As temperatures continue to rise, calcium sulfite partially decomposes to form other sulfur compounds, *Calcium sulfide* and *calcium sulfate*, and free lime combines with sulfur dioxide to form the same sulfates and sulfides, as follows:[39]

$$4CaSO_3 + heat = 3CaSO_4 + CaS$$

$$4CaO + 4SO_2 = 3CaSO_4 + CaS$$

A further temperature rise to about 1000°C decomposes most of the calcium sulfite. Continued heating at 1150°C causes the sulfide to be converted into the sulfate and elemental sulfur, as follows: $CaS + 2SO_2 = CaSO_4 + 2S$.

This latter reaction frequently occurs in lime calcination. The equilibrium conditions of the lime-sulfur system are complex, and there are no precise temperatures applicable to the above reactions. The lack of uniformity in temperatures probably stems from the varying surface reactivities of different limes and the variable concentrations and velocities of SO_2. By altering temperatures, empirically established for a given lime, *polysulfides* can also be formed.

Acid Gases. Acid gases are readily adsorbed and neutralized in lime hydrate suspensions. Acid anhydrides, like SO_2, NO_2, and P_2O_5, and hydrogen sulfide and carbon bisulfide react as follows:

$$2SO_2 + Ca(OH)_2 = Ca(HSO_3)_2 \text{ (calcium bisulfite)}$$

$$2NO_2 + Ca(OH)_2 = Ca(NO_3)_2 \text{ (calcium nitrate)}$$

$$\begin{cases} H_2S + Ca(OH)_2 = CaS \text{ (calcium sulfide)} + 2H_2O \text{ or—} \\ 2H_2S + Ca(OH)_2 = Ca(HS)_2 \text{ (calcium hydrosulfide)} + 2H_2O \end{cases}$$

Dry HCl gas does not react very readily with quicklime unless the temperature is increased to about 80°C, but it is very responsive chemically with $Ca(OH)_2$.

Peroxide. Hydrogen peroxide interacts with a lime slurry to precipitate crystals of calcium peroxide, containing eight molecules of water of crystallization. This compound loses its water of crystallization and becomes anhydrous when temperature is elevated to 130°C.

Halogens. Under dry conditions there is no reaction between quicklime and *chlorine*, but lime hydrates adsorb chlorine readily in either a dry or aqueous suspension form, creating calcium hypochlorite, an active bleaching agent, as follows:[40] $2Ca(OH)_2 + 2Cl_2 = Ca(OCl)_2 + CaCl_2 + H_2O$.

Fluorine readily displaces oxygen in quicklime, forming a fluoride, and reacts with avidity with hydrates.

Causticization. Lime reacts with all carbonate salts to form hydroxides by double decomposition, similar to the following equation (more on p. 355): $Li_2CO_3 + Ca(OH)_2 = 2LiOH + CaCO_3$.

Silica and Alumina. Lime reacts with available forms of silica and alumina although its reactions are quite complex, and there is not complete agreement on all aspects of these phase systems. Blank and Kennedy's[41] text on portland cement provides a most detailed, authoritative account of these involved reactions, as applied to cement. Since cement is beyond the scope of this book, lime's relationship with silica and alumina will mainly be discussed as it pertains to direct lime applications.

In the systems CaO–SiO_2–Al_2O_3, CaO–SiO_2, and CaO–Al_2O_3, lime combines with these compounds in varying degrees and ways at high temperatures (1200 to 2800°F) without the presence of moisture. These reactions are equally applicable with limestone if calcination temperatures are attained. Related information is contained on pp. 97, 273, and 375. In these systems, but only *faintly* with Al_2O_3, lime will combine slowly at ambient temperatures with compaction and under moisture-saturated conditions. These reactions are greatly accelerated by pressure of 10 to 75° atmospheric and/or elevated temperatures (200 to 500°F) which convert the moisture to steam. Pressure and/or heat can reduce the time of reaction from months to hours. Limestone will *not* react under these conditions.

Depending upon circumstances, the following four basic calcium silicate and aluminate compounds have been identified:

$$CaO \cdot SiO_2 \qquad 3CaO \cdot Al_2O_3$$
$$3CaO \cdot 2SiO_2 \qquad 5CaO \cdot 3Al_2O_3$$
$$2CaO \cdot SiO_2{}^* \qquad CaO \cdot Al_2O_3$$
$$3CaO \cdot SiO_2 \qquad 3CaO \cdot 5Al_2O_3$$

* Three forms of dicalcium silicate have been identified: alpha, beta, and gamma.

Two alumino-silicates are formed: $CaO \cdot Al_2O_3 \cdot 2SiO_2$ and
$$2CaO \cdot Al_2O_3 \cdot SiO_2$$

The di- and tricalcium silicates and all four calcium aluminates possess pronounced hydraulic setting properties with water, but the strongest cementing components are the tricalcium silicates and aluminates, of which portland cement is largely composed (p. 97).

Kalousek and co-workers[42] have synthesized hydrous dicalcium silicates of the alpha, beta (several varieties), and gamma types experimentally by autoclaving mixtures of high calcium lime slurry and ground quartz or silicic acid at various pressures, temperatures (130 to 350°C), different durations (one to thirty days), various water contents, and lime-silica ratios. Modifications of these variables in this hydrothermal process produced diverse forms of synthetic hydrous calcium silicates, which were identified by a combination of X-ray, differential thermal analysis, and micropetrography techniques. Often two or more solid phases of these different calcium-silicate reaction products were produced in the same specimen. Lime–silica ratios employed were 1:1.5–3 ($CaO:SiO_2$), and the water contents ranged from 0.3 to 1.25 mol. of H_2O per mol. of compound.

The pronounced hardening effect that occurs when hydrated lime is mixed with a graded silica sand, compacted, subjected to live steam, temperatures between 150 and 350°C under varying pressure in an autoclave, is generally explained by the formation of hydrocalcium silicate, a cement-like reaction product. Initially this material is a gelatinous substance when wet, but on drying it binds the coarse and medium sand grains together into a hard mass. The reaction of lime occurs mainly with sand fines, much of which are available forms of silica. The calcium silicate mineral tobermorite has been frequently identified and is regarded as one of the most effective cementing gels.

Eades and Grim[43, 44] with X-ray diffraction and differential thermal analysis studies have detected complex calcium silicates that are formed by the reaction of lime and available silica contained in clay minerals (kaolinite, montmorillonite, illite, chlorite, etc.), compacted under optimum moisture conditions, and exposed either to normal climatic and temperature conditions or laboratory curing. Development of such silicate reaction products requires at least one to three months of warm temperature, 70 to 80°F. The silicate formation is slow and gradual, occurring over several years. This hardening action, resulting from the formation of calcium silicates, can be greatly accelerated by increased temperature. In the laboratory Eades obtained about the same compressive strength for lime–soil mixtures in only seven days with accelerated curing as is obtained in the field in ninety days of normal summer temperature. The specimens were subjected to 140°F in an oven for three days, cooled to room temperature, and then immersed in water for four days.

Eades reports the reaction of lime and clay minerals or soils to consist first of base exchange with the Ca^{++} cations displacing H^+, Na^+, or K^+ ions in the clay. Much of the free lime carbonates by absorbing CO_2. However, along with carbonation other free lime attacks the edges of the clay mineral actually destroying or eroding these microparticles with formation of non-crystalline gelatinous calcium silicates that on drying behave like a crystalline cement in binding these particles together. The rate of reaction varies with different clay minerals, but generally kaolinite is the most reactive, followed closely by some types of montmorillonite (Aberdeen and Grundite). The amount of lime also varies from 2 to 12%, depending largely upon two variables: base exchange capacities of different clay minerals and their rate and extent of lime–silica reactivity. Attainment of a high pH of over 12.0 (Eades contends it to be 12.4) is essential to this reaction. Below this level there will be a change in the physical characteristics of the soil, but no hardening of consequence will occur. Gains in compressive strength for varying soils and increments of lime are illustrated in Fig. 7-11. For related in-

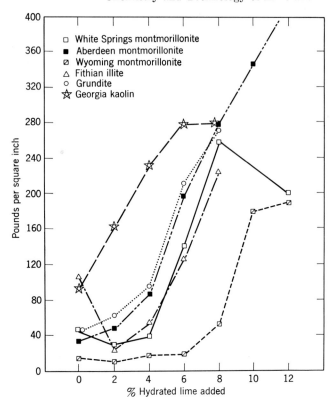

Fig. 7-11. Varying compressive strengths obtained with different percentages of lime and clay soils.

formation on this property of lime as applied to soil stabilization, see p. 412.

In effect, such soils containing clay behave like a pozzolan with lime. Similar calcium silicate reactions occur when lime is intimately mixed with other finely divided natural or sintered (synthetic) siliceous materials and wastes. These materials are:

Natural	*Artificial*
Volcanic ash	Pulverized blast-furnace slag
Tripoli	Fly ash (utility plant waste)
Diatomaceous earth	Ground clay brick or tile
Silica flour (from sand)	Certain pulverized gangues from ores
Pumicite dust	Stack dust from expanded shale (lightweight aggregate plants)

The degree of reactivity of the above pozzolans (even among the same types) with lime vary tremendously, depending upon the chemical analysis, particularly the amount of available silica present; fineness of pozzolan; and reactivity and purity of lime. Evaluation of relative pozzolanic efficacy can be determined by test (p. 497). As a result, the lime-pozzolan ratios may extend over a wide scope, from 1:1.5–10, depending also upon the application and the degree of cementing required.

There is no reaction between lime and ordinary sand, except at high temperature and pressure, since otherwise the silica present is unavailable. The reaction of MgO and silica is much more obscure, however it has been definitely established that *magnesium silicates* are formed. Such silicates are widely formed in nature.

Iron. Lime reacts with iron at high temperatures of at least 1600°F to form calcium and dicalcium ferrite ($CaO \cdot Fe_2O_3$ and $2CaO \cdot Fe_2O_3$). Mixtures of 25 to 30% CaO with 70 to 75% Fe_2O_3 melt at 1220°C. If some of the Fe_2O_3 is reduced to FeO, the melting point is depressed. For related information see pp. 154 and 269.

Effect on Metals. When in contact with metal equipment, lime does not affect steel or cast iron to the slightest extent. In fact, by coating these metals with a lime whitewash, it acts as a conservation agent by protecting the metals from oxidation. However, any form of lime or strong alkali will disastrously attack and destroy *aluminum,* except special alkaline resistant alloys of this metal. *Lead* and *brass* are also readily attacked, and under some circumstances it will literally dissolve lead.

Carbon. At extremely high temperatures of 3000°F+, carbon will displace oxygen in quicklime by fusion, forming calcium carbide and carbon monoxide (p. 356).

Phosphorus. Elemental phosphorus will react with lime at incandescent heat to form a mixture of phosphate and phosphide.

Calcium Dioxide. As early as 1818 several scientists claimed to have made a fairly stable compound, calcium dioxide, octohydrate ($CaO_2 \cdot 8H_2O$) by interaction of an excess of hydrogen peroxide and milk-of-lime. Nogareda, a Spaniard, confirmed the above compound and claims to have made an anhydrous calcium dioxide that was partially decomposed into calcium hydroxide.[46, 47]

The anhydrous product is a microcrystalline powder whereas the hydrated dioxide contains well-defined crystals. Both types destabilize in the presence of water, liberating one atom of oxygen. Reportedly these two dioxides are stable in a dry atmosphere up to temperatures of 200°C. There is disagreement on the decomposition point when these compounds revert to a $CaO \cdot Ca(OH)_2$ mixture.

These products are just laboratory curiosities, and little is known about their properties. It is the only authentic method reported for *oxidizing* lime and might be fruitful for further investigation.

Several early investigators reported the existence of an unstable *dihydrate of lime* [$Ca(OH)_2 \cdot H_2O$], claiming as evidence that the solubility curve at 60°C had a break or apparent transitive point when a saturated lime solution was reprecipitated by gradually elevating the temperature. Haslam and his co-workers[20] attempted to isolate this alleged compound by every conceivable method and concluded that no such compound exists. They contend that the break in the solubility curve was the result of a previous laboratory error. Consequently, in spite of the above and other assertions that this and other forms of hydrate exist, the writer believes that only one form prevails, the monohydrate. The sole exception to this would be when a saturated solution of $Ca(OH)_2$ freezes, the ice forming creates a supersaturated condition. Later, on melting, *hemihydrate* crystals [$Ca(OH)_2 \cdot \frac{1}{2}H_2O$] are deposited. However this form is very unstable and quickly reverts to a normal hydrate.

Magnesia Compounds. The existence of a *magnesium dioxide* (MgO_2) has been reported, but it is even less stable than its CaO_2 counterpart.[48]

In studying the vapor pressure of $Mg(OH)_2$, Johnston discovered some "bumps" in his vapor pressure gradient at 80 and 130°C and conjectured that two unorthodox, fugitive hydrates occur in addition to the normal $Mg(OH)_2$.[19] These are: $2Mg(OH)_2 \cdot H_2O$, which is stable up to 80°C, and $4Mg(OH)_2 \cdot H_2O$, which is fairly stable between 80 and 130°C. Only $Mg(OH)_2$ can exist above 130°C. These forms, which are only of academic interest, have been confirmed by subsequent investigators.

Some other combined forms of $Mg(OH)_2$ exist in nature, such as:

Mineral	Chemical Composition
Artinite	$MgCO_3 \cdot Mg(OH)_2 \cdot 3H_2O$
Hydromagnesite	$3MgCO_3 \cdot Mg(OH)_2 \cdot 3H_2O$
Landsforite[a]	$3MgCO_3 \cdot Mg(OH)_2 \cdot 21H_2O$

[a] This is very unstable.

For other fundamental data on lime, refer to Chapters 6 and 9.

REFERENCES

1. ASTM *Standards,* Part 9 (1964), p. 46.
2. W. Rudolfs, "Lime Handling and Application in Treatment Processes," *Nat'l. Lime Ass'n Bull.* (1949), p. 13.
3. Norton, *J. Am. Ceram. Soc.* **8** (1925), p. 799.
4. R. Sosman et al., *J. Wash. Acad. Sc.* **5** (1915), p. 563.
5. Schumacker, *J. Am. Chem. Soc.* **48**, (1926), p. 396.
6. Born, *Z. Elektrochem.* **31** (1925), p. 309.
7. R. Doman, J. Barr, R. McNally, and A. Alper, *J. Am. Ceram. Soc.* **46** (Jul. 1963), p. 313.
8. F. Smyth and L. Adams, *J. Am. Chem. Soc.,* **45** (1923), p. 785.
9. J. Murray, "Specific Heat Data for Evaluation of Lime Kiln Performance," *Rock Prod.* (Aug. 1947), p. 148.
10. W. Roth and Bertram, *Z. Elektrochem.* **35** (1929), p. 297.
11. A. Mitchell, *J. Chem. Soc.* **123** (1923), p. 1055.
12. L. Pankratz and K. Kelley, "Thermodynamic Data for MgO (Periclase)," *U.S. Bu. Mines R.I. 6295* (1963).
13. S. Brunauer, D. Kantro, and C. Weise, "Surface Energies of CaO and Ca(OH)₂," *Port. Cement Ass'n Bull.* (1955).
14. G. Jura and C. Garland, *J. Am. Chem. Soc.* **74** (Dec. 1952), p. 6033.
15. R. Hedin, "Processes of Diffusion, Solution, and Crystallization in System Ca(OH)₂ —H₂O," *Swed. Cement & Conc. Res. Inst. Bull. 33,* 92 pp. (1962).
16. *Int. Crit. Tables,* **5**, pp. 98–99.
17. *Int. Crit. Tables,* **5**, pp. 195–196.
18. W. Hatton et al, *J. Am. Chem. Soc.* **81** (1959), p. 5028.
19. J. Johnston, *Z. Phys. Chem.* **62** (1908), p. 336.
20. R. Haslam. G. Calingaert, and C. Taylor, *J. Am. Chem. Soc.* **46** (1924), p. 308.
21. R. Haslam, W. Whitman, and J. Cochrane, *Research Rept. to Nat'l. Lime Ass'n.* (1924).
22. H. Bassett, Jr., *J. Chem. Soc.* (1934), p. 1270.
23. J. Herold, *Z. Elektrochem.* **II** (1905), p. 417.
24. A. Siedell, *Solubilities of Inorganic & Metal Organic Compounds,* Van Nostrand (Princeton, N. J.), pp. 309–319.
25. J. Johnston and C. Grove, *J. Am. Chem. Soc.* **53** (1931), pp. 3976–3991.
26. G. Ringquist, *Swed. Cement & Conc. Res. Inst. Bull. 19,* 56 pp. (1952).
27. N. Knibbs, *Lime & Magnesia,* E. Benn (London, 1924), p. 71.
28. H. Remy, A. Kuhlmann, *Z. Anal. Chem.* **65** (1924), pp. 1–24, 161–181.
29. W. Herz and G. Muhs, *Z. Anorg. Chem.* **38** (1904), p. 140.
30. H. Fredholm, *Z. Anorg. Chem.* **217** (1934), pp. 203–213.
31. Maigret, Bull. Soc. Chim. (3) **33**, 631.
32. "Chemical Lime Fasts," *Nat'l. Lime Ass'n. Bull.* (1949), p. 30.
33. F. Lea and G. Bessey, *J. Chem. Soc.* (1937), pp. 1612–1615.
34. "Chemical Lime Facts," *op. cit.,* p. 41.
35. F. Adams, *Ind. Eng. Chem.* **19** (1927), pp. 589–591.
36. H. Staley, *M.I.T. Research Rept. to Nat'l. Lime Ass'n.* (1946).
37. M. Riehl, *Water Supply & Treatment,* 9th Ed. (and earlier editions by C. Hoover), Nat'l. Lime Ass'n. (1962), pp. 66–68.

38. N. Zalmanoff, "Carbonation of Lime Putties to Produce High Grade Building Units," *Rock Prod.* (Aug. 1956) p. 182, and (Sept. 1956) p. 84.
39. Forster and Kubel, *Z. Anorg. Chem.* **139** (1924), p. 261.
40. T. Miller, "Improving $Ca(OH)_2$ for Better Hypochlorite Production," *Azbe Award Bull. 4,* Nat'l. Lime Ass'n. (1964).
41. R. Blanks and H. Kennedy, *The Technology of Cement & Concrete* **1,** John Wiley (New York, 1955).
42. G. Kalousek, J. Logiudice, and V. Dodson, *J. Am. Ceram. Soc.* **37** (Jan. 1954), p. 7.
43. J. Eades and R. Grim, "Reaction of Hydrated Lime with Pure Clay Minerals," *Hwy. Res. Bd. Bull.* (1960), 262.
44. J. Eades and R. Grim, *Hwy. Res. Bd. Bull. 335* (1962), pp. 31–39.
45. Knibbs, *op. cit.,* p. 49.
46. C. Nogareda, *Ann. Soc. Espan. fis. Chim.* **28** (1930), p. 461.
47. C. Nogareda *Ann. Soc. Espan. fis. Chim.* **29** (1931), p. 131.
48. Knibbs, *op. cit.,* p. 75.

CHAPTER EIGHT

Lime Manufacture

From a preceding chapter on the Theory of Lime Calcination it should be apparent that rigid industry-wide standardization of lime's chemical and physical properties is untenable since there are so many pronounced differences in limestones, even when the chemical analyses are approximately identical. As a result, most lime material specifications are necessarily quite general and liberal in their provisions, as Chapter 11 portrays. Many of these anomalous characteristics are still scientifically inexplicable. As a natural consequence, few plants will manufacture lime of *exactly* the same properties. Each plant must determine its own optimum manufacturing conditions and procedures by numerous compensations for limitations and/or full exploitation of natural advantages. Frequently optimum quality and economy are incompatible, so that compromises are judicious between these two extremes in the manufacture.

In view of the highly individualized nature of lime burning, which involves countless kiln design modifications, diverse fuels, varying degrees of automation, and totally different operating and labor conditions, its manufacture will be treated somewhat generally, with emphasis on *current* practices, performance, theories, operating problems, and possible solutions . . . rather than a "how to do it" detailed explanation which might only be applicable to a small percent of actual situations. Recommended supplemental reading on a "case history" type of plant that includes specifications and brand names and models of equipment would be past issues of *Pit and Quarry*[1] and *Rock Products*,[2, 3] trade journals many of whose articles are cited in the selected bibliography at the end of this chapter. For those who read German, *Zement, Kalk, Gips* (a trade journal) is a fruitful source. The English journal *Cement, Limes, and Gravel* is also suggested. For detailed information on equipment the annual *Pit & Quarry Handbook*[4] is the most prolific source of information. For specific

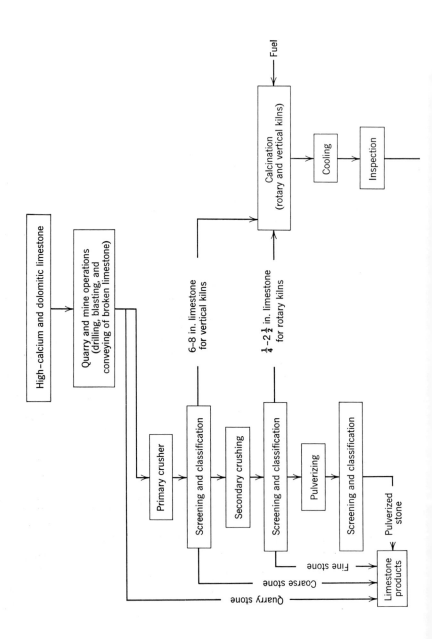

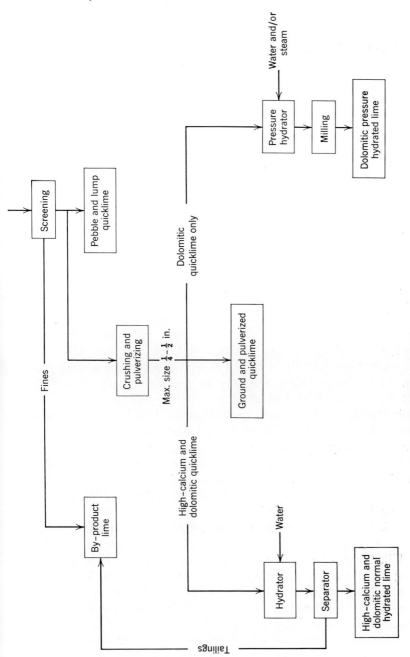

Fig. 8-1. Simplified flow sheet of integrated lime operation.

plant design and construction details, consulting engineers experienced in lime are recommended.

A simple flow diagram of a typical integrated rotary and vertical kiln lime operation is exhibited in Fig. 8-1, including the stone processing through to the hydration of the quicklime. In the United States, however (probably to a greater extent than other countries), there are quite a number of successful plants that purchase their stone for kiln feed instead of quarrying and processing their own (see p. 101). Such plants would be regarded as unintegrated.

KILNS

The "mainspring" of a lime plant, of course, is the kiln in which the limestone is calcined. The design and operation of kilns have been metamorphosed from the days of the ancient Egyptians into a modern chemical-process industry. Most of the advances have occurred since 1900, with greatest progress since 1935. Before this period there was little difference in the kilns of the nineteenth century and those of antiquity. In fact, even in the 1960's in some of the less prosperous of the civilized nations, primitive lime production still flourishes, resembling almost a "black art." A few such kilns are still operated sporadically in rural Pennsylvania for agricultural purposes. Such obsolete kilns will only be reviewed briefly since they are of scant commercial significance.

Probably the most complete treatise on the kiln types of the past is contained in Searle's *Limestone and Its Products,* published in 1935.[5] Of the fifty types described and illustrated by Searle, probably only four or five types were operated in the United States in 1964; most of these kilns have now been obsoleted in many countries, but nevertheless, a student of thermodynamics might find the theories of these old kilns of academic interest.[6, 7]

The first such kilns were *field* or *pot kilns,* intermittent types, that were usually crudely constructed of stone, often on the side of a hill. They were of low height and had steel plates at the top serving as a windbreak. Stone was placed through the top of the kiln with the largest stone introduced first forming an arch. Layers of increasingly smaller stone were then added; a wood fire was ignited in the hearth below, later supplemented with coal or more wood. After burning for three to five days, the fire is extinguished, and the lime is drawn after cooling by hand poking to the hearth below, with usually about 25% overburned, the same percent underburned, and the balance a passable grade of quicklime. Heat losses, of course, are stupendous since the kiln cools down during each discharge, and there is no heat recovery from the waste gases. The extensive hand labor involved is prohibitive in the United States.

The next major improvement was the *vertical mixed-feed kiln* of greater height and capacity. These were at first crudely constructed with field stone (Fig. 8-2), but unlike the field kilns, since these kilns were largely continuous, they were usually lined with refractory brick. Alternate layers of stone and fuel (coal, coke, or wood) ad infinitum were charged into the top of the vertical stack. As a layer of stone was calcined, it was drawn at the bottom and additional stone and fuel were added at the top. Such kilns improved fuel efficiency and capacity and were inexpensive to construct, but the ash contaminated the lime since the fuel and lime were intermingled in the kiln and discharged together. This type of mixed-feed kiln should not be confused with modern versions that will be described in more detail later in the chapter. There were many modifications of these two old basic kilns.

Since 1900 there have been almost countless varieties of patented kilns invented and promoted. The U. S. patent office and those from other countries have granted literally more than a thousand patents, most of which were never used or were soon proven unfeasible. Such

Fig. 8-2. Oldest type of mixed feed kiln.

kilns are not worthy of mention. Modern kilns can be classified as follows (and are referred to throughout by their number designations):

1. Vertical
   a. Standard types
   b. Producer gas
   c. Large capacity, gas-fired with center burners (Azbe and others)*
   d. Mixed-feed
2. Rotary
   a. Conventional types
   b. Conventional types equipped with preheater, cooler, and heat exchangers
   c. Grate-kiln system*
3. Miscellaneous
   a. Fluo-solids*
   b. Rotary hearth with traveling grate (Calcimatic)*
   c. Ellernan*
   d. Inclined vibratory vertical*
   e. Horizontal ring (Hoffman)—now obsolete

Vertical Kilns

Of these kilns the vertical type, continuous or semi-continuous, is by far the most widely employed in the world. There are hundreds of modifications in its basic design; in fact, seldom is an identical kiln installed, even in the same plant. Usually the manufacturer or engineer embellishes the basic design in an attempt to adapt the kiln to a particular quality or gradation of stone, type of fuel, consumer requirement, labor supply, desired capacity, etc. Frequently, these modified designs produce disappointing or adverse results, but undaunted, the lime manufacturer will continue his quest for improved efficiency and quality control, always hoping to build a better mouse trap. Competitive pressures simply do not permit complacency with existing kiln efficiencies.

Figure 8-3 shows a plant drawing of a conventional vertical kiln and quarry layout.

As displayed in Fig. 8-4, all modern vertical kilns are divided into four distinct zones by imaginary horizontal planes. From top to bottom in sequence are: (1) *stone storage,* a vertical or often

---

* Kiln is partially or entirely covered by patents to a greater extent than those without an asterisk.

Fig. 8-3. Sketch of conventional, integrated vertical kiln plant layout, with quarry in background.

a modified hopper-shaped zone, (2) *preheating* zone that is designed to heat the stone near dissociation temperatures, (3) *calcining zone* where combustion occurs, and (4) *cooling and discharge* that is usually shaped like an inverted truncated cone, at the bottom of which the lime is discharged.

The proportioning and contouring of these four zones are what constitute the "art" of vertical kiln design.

Often other subsidiary zones are referred to, such as precalcination, which is located at the lower quarter of the preheating zone or the finishing, which is located at the extreme lower part of the calcining zone.

Figure 8-5 illustrates the traditional *standard* (1*a*.) vertical kiln that was so common in the United States up to 1940 and which has been on the descendancy because of its considerable hand labor (automation is difficult), inadequate quality control, low capacity per cubic foot of kiln area, and mediocre-to-poor thermal efficiency. One of the least efficient was the *direct hand-fired* vertical, usually coal-shoveled by hand at $\frac{1}{2}$, 1, or 2 hour intervals. On each firing generally considerable heat is lost through the stack due to the emission of thick smoke containing volatile hydrocarbons of high calorific value. This was caused by volatilization of the coal in sudden contact with the hot flame temperature and incandescent lime. The fuel simply temporarily exceeded the available air supply, creating an imbalance and innundating the calcining zone with more heat

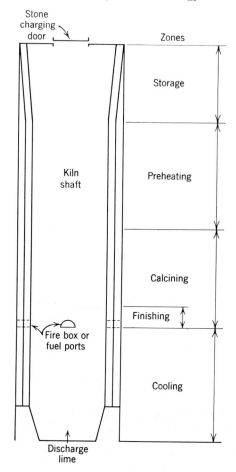

Fig. 8-4. Zonal sections of a vertical kiln.

than could be assimilated. Often secondary combustion occurred as the hydrocarbon-laden smoke billowed from the stacks—a deplorable waste of heat calories.

Conversely, the design of many of these kilns demanded slow firing before the hot lime could be safely discharged. This was achieved by introducing large excesses of air into the kiln. The resulting wide fluctuations in kiln temperature not only wasted heat but yielded an uneven distribution of heat and faulty transference that vitiated the quality of the lime by causing excessive amounts of over- and underburned lime.

The *indirect gas-fired kiln* (1b) was a distinct improvement in most instances (Fig. 8-6). Solid fuels are supplied to separate furnaces, forming gaseous fuels for combustion. These fireboxes may be installed into the kiln structure or as adjuncts, separated from the kiln except for ducts and tuyeres. In some plants a large independent furnace is used to supply fuel for a battery of kilns. These are also known as *producer gas* kilns. Generally a more even heat results from the injection of the hot gases, either neat or diluted with varying amounts of primary air. Hot secondary air is propelled through the cooling zone, providing more oxygen for combustion and the necessary draft. Stone is fed into the top of the kiln, similar to the hand-fired vertical, but unlike the latter the stone does not make physical contact with the solid fuel—only with the hot combustion gases. This obviates possible degradation of quality from ash. Correlating the velocity, flow, and concentration of the gas from the producer into the most efficient sections of the kiln with such other variables as the size and gradation of the kiln feed, the volume of secondary air, draft, etc. requires complex coordination.

Tapered boiler-type steel stacks that greatly narrow the opening are not employed as widely as in former years, and only rarely in the United States. The increasing use of forced and induced draft along with more generous kiln heights eliminates the necessity of

Fig. 8-5. A battery of standard, low-capacity vertical kilns that are fired with coal. Such kilns are now passe in the United States.

Fig. 8-6. Close-up sketch of vertical kiln, showing external firebox, refractory lining, and kiln details.

such stacks. Completely enclosed kilns vent their exhaust gases through dampers in the kiln ceiling or through flues constructed on the kiln periphery. Such closed tops may be designed to create a hydrostatic pressure in the kiln (Schmatolla principle). The tops are equipped with hatches through which the stone is charged. The outer shell of all shaft kilns is usually heavy steel boiler plate, which encases insulation and refractory brick of varying thickness.

Cooling Zone and Discharge. Air for combustion is supplied through the cooler up to the calcining zone. However, this air performs a dual purpose. When the cool air contacts the hot lime in the cooler, it absorbs heat, thereby cooling the lime for discharge and reintroducing the recuperative heat into the calcining zone as secondary combustion air. In this manner some of the heat used to preheat the stone is indirectly conserved. The amount of heat recovered from the cooler, of course, varies greatly depending on kiln design, operating skill, etc., but it can be considerable, as much as 900,000 Btu/ton of lime. With the most efficient coolers it is possible to handle with bare hands lumps of quicklime directly from the cooler discharge opening (as low as 90°F). This alone is indicative of low heat loss.

Proportioning and contouring the cooling zone and designing the size, number, and location of the discharge openings at the base of the kiln determines whether there will be adequate (or too much) air for cooling and as a draft for combustion. Theoretically, the minimum amount of air required is about twice the weight of the lime produced. This assumption is predicated on the fact that 10 lb. of air are necessary for the combustion of 1 lb. of solid fuel. Then if 1 lb. of coal, or its equivalent in other fuels, is required to produce 5 lb. of lime, the air requirement is 10 lb. In practice, however, additional air is needed up to about three times as much to prevent an uneven distribution of heat or excessive temperature. Some vertical kilns are so designed that they operate strictly with natural draft. In others forced draft propelled by fans is used entirely or as a supplement to natural draft in order to obtain the desired balanced draft and higher capacity. Too much air or draft is also highly undesirable, since it will temper the flame temperature unduly and contribute to an excessive heat loss with hot oxygen in the exit gases ascending the stack.

Most cooling zones are tapered at the bottom in varying degrees. One common problem with this design is that the lime tends to slide down the inclined sides to the discharge opening at a faster rate than in the center. Two or more openings at the bottom assists in a more even descent. Reduced inclination of the inverted cone to no less than a 70° angle is also a plus factor. A reversed slight outward inclination of the sides that widens the bottom encourages more rapid flow through the center than at the cooler's periphery. Double discharge mechanisms have been developed that induce an even flow. Small hopper chutes at the base are also beneficial in automatic discharge systems in abetting an even, steady flow. Manual discharge systems with doors, horizontal sliding plates, and other devices are rapidly being replaced with automatic discharge methods (Fig. 8-7).

Means of transporting lime from the cooler's discharge opening for shipment, hydrating, or classifying has progressed over the years from the wheelbarrow, often accompanied by hand sorting, to beneficiation of the lumps spread on the floor, to special cars on tracks that run underneath the discharge hopper, to modern conveyor belts synchronized with automatic discharge mechanisms. The latter method, of course, is the most efficient.

Fig. 8-7. Vertical kiln discharge mechanism, illustrating draw gates and draw hopper with lime conveyor underneath.

Calcining Zone. There are two basic methods of operating shaft kilns. These are dubbed such unerudite names as the (1) "hanging" and (2) "follow" methods. The first practice is dependent on the tendency of certain gradations of stone to arch or "hang" as well as for some of the basic lime, heated to incandescence, to adhere to the acid refractory linings through a pseudofluxing effect. Designing the interior walls of the kiln in the calcining zone with a slight constricted or elliptical shape accentuates this arching tendency. When the fireman determines that calcination is terminated, he loosens the semi-coalesced lime mass by prodding with long iron bars through the firebox or special poke holes. The lower strata of lime slides down to the cooler. Some utilize this technique in drawing from the cooler, but obviously considerable heat loss results and it is considered a malpractice.

The "follow" method is predicated on the stone and lime moving steadily and evenly downward through the different zones without "hanging," and discharging is performed either continuously or at short intervals. Interior wall designs are critically important, usually very cylindrical but in some cases with a slight oval-shaped calcining zone to minimize arching.

Essential to the operation of any vertical kiln is that ample voids are present in the stone charge for unimpeded circulation of fuel gases and draft around and through the mass of stone and lime in order to enhance uniform heat transference.[8] This is one reason why small stone of less than 2 in. or broad gradations cannot be calcined successfully in vertical kilns since there is insufficient void size and space for proper circulation. The larger the stone charge and the more restricted the gradation, the greater are the void sizes. An ideal percent of voids in the calcining zone is 45%. Balanced against large stone and high voids is the incontrovertible fact that small stone calcines much faster. The heat transfer rate is calculated to be about 200% faster when stone size is halved; some contend even more.

The proportioning of the calcining zone will also depend on the type of fuel and its rate and extent of heat generation above the dissociation point of limestone (high-level heat). Higher temperatures require shorter calcining zones than lower or more moderate temperatures, and certain fuels generate higher temperatures than others. While generally capacity and thermal efficiency reach their zenith at the highest rate of heat transmission from fuel to stone, i.e., highest temperature, such heat is usually completely impractical. The advantage of maximum or high temperature is more than counterbalanced by the costly attrition on refractory linings and the adverse effect on lime reactivity caused by hard- or overburning. (See p. 143.)

Channeling and stratification of combustion gases, which induces uneven distribution of heat, is minimized by closer gradation of kiln

Fig. 8-8. Poke holes in exterior shell of vertical kiln.

feed and use of more cubically shaped stone as opposed to laminar pieces.[9] With the former stone its more circular or rectangular void shape is more conducive to the uniform circulation of gases.

Regardless of whether the "hanging" or "follow" technique is employed, an adequate number of strategically located *poke holes* with tightly fitting doors are needed at the calcining zone so that all of the inside walls of the hot zone are accessible for trimming (poking), as viewed in Fig. 8-8. Usually they are located in easy proximity to constricted areas of the kiln lining or any area where undesirable hanging or obstruction is apt to occur. Poke holes are convenient at the top and bottom of the calcining zone. Poke holes can also be employed as a supplementary source of air in regulating combustion.

Small *inspection holes* may also be placed in the external shell of the kiln as a means for the operator to observe progress of calcination and degree of luminosity of flame. A dual purpose of such holes would be use of pyrometers and thermocouples for checking kiln temperatures at adjacent kiln sections.

With externally fired kilns from coal fire boxes (producer gas), natural gas, or liquid fuels, a series of side ports introduces the fuel gases and primary air into the calcining zone in diverse ways. Some kilns also use air flues, controlled by dampers, as supplementary air for combustion, introducing it just below the calcining zone. This air is partially heated by radiation from the kiln linings in which the air flues are encased.

The calcining zone has been proportioned and shaped in every conceivable manner by enterprising engineers. Circular, oval, square, and rectangular vertical shafts as well as various degrees of constricted and bulging contours have been designed. The theory is that contoured walls are conducive to the stone rolling over or tumbling as it descends the shaft, exposing all sides to equal heat, whereas the vertical sides tend to induce the stone to sink down through the kiln with little change in position. Another theory in support of constricted contours is that they compress the calcining zone and promote higher temperature and more even heat distribution that accelerates calcination and maximizes throughput. While this may be true, the attrition on refractory linings increases and periodically severe hanging problems arise. Generally most knowledgeable engineers prefer straight, cylindrical, rectangular, or possibly faintly contoured kiln linings. There is no conclusive preference.

The hot exhaust gases, including much CO_2 from the calcining zone, migrate upward to the preheating zone, where the incoming stone is heated. These gases contain more heat than is necessary for preheating, so the excess is often recaptured and recirculated back to the calcining zone as secondary combustion gases. In addition to fuel economy, frequently this waste CO_2 gas is desired for dilution of the rich fuel gases to temper the flame temperature in order to avert overburning. It is inevitable that some of this migratory heat will escape up the stack or flues or through the stone storage zone. Other principal heat losses are through the cooler, along with the still warm lime, and by radiation and convection.

Preheating and Storage Zones. In many vertical kiln plants the preheating zone is synonymous to the lower half of the storage zone, since when the kiln linings are cylindrical, the sections are indistinguishable. Walls of storage zones are also tapered both inward and outward from the top in varying degrees, forming in effect a normal and inverted hopper-shaped impression. There are proponents of many designs and proportions. The prime consideration is that the kiln feed descends uniformly down the shaft and that there is no segregation of stone sizes, if possible. With some stone gradations mild contouring may abet this objective. If the kiln or storage zone lacks height, inadequacy in the draft may be rectified by superim-

Fig. 8-9. Charging stone into vertical kiln from cable car with retractable sides.

posing a chimney over the storage zone, but extending the height of the kiln is usually preferable.

Although some vertical kilns are open at the top, there are serious disadvantages of winds interfering with the draft, even creating downdrafts, and from rain saturating the kiln feed. As a result, all modern verticals are covered either with specially designed retractable doors or hatches, or chimneys, or flues to vent exhaust gases, or roofs with corrugated metal side walls attached to gantries.

Inclined conveyor belts are largely employed in *charging stone* to modern kilns and have substantially replaced the side dump, bottom dump, and removable container types of cars that were propelled by an aerial cable on inclined tracks to the top of the kilns. The latter type was the universal trademark of U. S. vertical plants in the 1910–1940 era (Fig. 8-9). Bucket elevators and skip hoists are still employed. Page 223 describes and Fig. 8-10 illustrates a highly mechanized German charging method. Less segregation of stone occurs in modern conveyor charging processes.

The special kinds of advanced shaft kilns (Ic.) may be *gas fired* from natural gas, producer gas, fuel oil, or even wood. In the United States natural gas predominates. The introduction of *center burners*[8, 9] on two or four sides of the kiln into the calcining zone greatly improved kiln performance, fuel efficiency, and lime quality (Figs. 8-11 and 8-12). In many of the other vertical kilns combustion gases and flames are often poorly distributed in the burning zone, tending to gravitate to the walls of the kiln, thus causing a heat imbalance in the center. Lack of uniformity in the lime results with some burned properly and some underburned, or soft-burned, with the remainder over- or hard-burned. Center burners provide a much more even temperature to all sections of the calcining zone. The gas injected into the center ignites and diffuses readily toward the periphery of the kiln. Usually these kilns are also equipped with sideburners.

Figure 8-13 illustrates a modern, high-capacity, *Azbe type of center burner* vertical kiln. By pioneering the center burner, Azbe further upgraded the shaft kiln efficiency established by Schmatolla, whom Eckel cited as having established the most efficient vertical kiln in the United States in 1920. Lower fuel consumption resulted, but probably of greatest importance the capacity of the kiln, measured in square feet of horizontal kiln cross-sectional area, was greatly increased from $\frac{1}{4}-\frac{1}{2}$ tons to 1–2 tons/day/ft.^{2} Another yardstick of capacity and efficiency is tons of lime per cubic foot of total interior kiln volume. A similar proportionate increase in capacity was registered on this basis. This was achieved largely by accelerating the

Fig. 8-10. Monrail propelled, overhead limestone and coke charging equipment for modern German mixed-feed kilns. Retracting bottom of feeder is shown open after discharge.

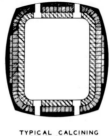

TYPICAL CALCINING
ZONE SECTION

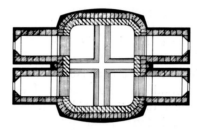

SECTION THROUGH BURNERS

Fig. 8-11. Below is kiln section
showing four center burners; above
shows modified rectangular calcining section above center burners.
(Courtesy, Azbe Corporation.)

burning process so that lime was produced in six to twelve hours
(between charging and discharging) instead of two or more days
in the more sluggish standard kilns. Stated another way, the same
size of cylindrical shell that produced 15 tons/day in standard shaft
kilns is able to produce 40 to 60 tons/day with such modern refinements as the center burner, improved preheating, cooling, charging,
discharging, and draft. As a result, these kilns cannot be considered
anymore as miniscule production units. Capacities of various kilns,
depending on size, may range from 50 to 250 plus tons/day, with
fuel consumption reported in the 4 to 6 million Btu/ton bracket.
Further modifications[11, 12] have developed much higher capacities of
up to 585 tons/day; however, the quality of the lime at such rates
of production is considered erratic with high core content (underburning and uneven burning resulting intermittently).

Exterior dimensions of standard and all gas-fired verticals range
widely from only 6.5 × 16 ft. to 25 × 100 ft., with about 12 × 50

ft. as average, diameters and heights respectively. Most recent verticals are of the larger sizes. Interior dimensions that allow for the kiln casing, insulation, and refractory brick average in diameter 2.5 to 3.5 ft. less than the above exterior dimensions, not allowing for elliptical or concave interior kiln sectional zones.

To achieve these high capacities and efficiencies with modern gas-fired vertical kilns, meticulous kiln design and combustion control is essential. Heat transference is not only affected by radiation and convection but by surface combustion to the stone. This latter heat generative reaction occurs as the flame spreads in direct contact with surfaces of the incandescent lumps of lime. For maximum combustion efficiency both air and gas should be preheated before uniting in the calcining zone. When hot (near 1650°F) their turbulence increases the spread of flame temperature more rapidly and uniformly. As excess air is increased, temperatures are lowered, the rate of calcination is retarded, and thermal units are wasted. As a consequence, critical kiln design considerations are the number and positions of the air and gas inlets, controls for adjusting velocity of air and gas, and position of the exhaust gas offtake. With some modern kilns the trend is for smaller calcining areas, larger cooling zones, and use of concentrated gas, undiluted with primary air.

To assure maximum voids in the center of the kiln, some successful plants with center burners favor charging kiln feed around the periphery of the kiln rather than the center. With side ports the tendency is to charge stone into the middle, forming a cone, so that the largest stone rolls to the sides yielding the greatest voids at the periphery. The theory behind this practice appears somewhat redundant, since it is based on the assumption that gases from the kiln periphery penetrate more readily into the interior through the larger voids and that the acknowledged higher wall temperatures resulting from side burners are not as likely to overburn the largest stone adjacent to the sides while the lower temperatures in the center are calcining the smaller stone. Striving for uniform heat and kiln feed distribution should be more rewarding than attempting to control an irregularity. Many of these practices and theories are controversial among experienced lime burners.

In spite of producing an inferior quality of lime on an average, *mixed-feed kilns* (Id) continue to flourish with many new refinements and improvements.[13, 14] The continuance of mixed feed is due to one attribute—low fuel consumption. On an average, these kilns have the lowest fuel consumption of any kiln type. Thus, when quality is relatively unimportant and when fuel is costly, it is the logical

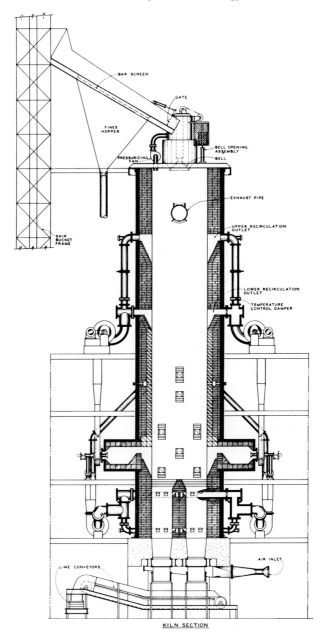

Fig. 8-12. Schematic drawings of two sides of Azbe high-capacity vertical kiln with center burners—150 to 250 ton/day capacity. (Courtesy, Azbe Corporation.)

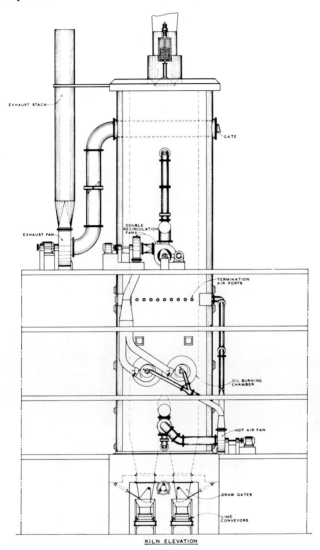

EXHAUST STACK

GATE

DOUBLE
RECIRCULATION
FANS

EXHAUST FAN

TERMINATION
AIR PORTS

OIL BURNING
CHAMBER

HOT AIR FAN

DRAW GATES

LIME
CONVEYORS

KILN ELEVATION

Fig. 8-12. (*Continued*)

kiln to operate. Many large captive lime plants in the United States alkali and beet sugar industries also prefer these kilns because of the concentrated CO_2 gas that emits from the stacks when operated at optimum efficiency and that is recovered for use in these chemical-process industries.

Fig. 8-13. Large-capacity Azbe vertical kiln, featuring center burners and series of side burners at multilevels. (Courtesy, Azbe Corporation.)

For maximum thermal efficiency a dense, metallurgical grade of coke of low reactivity or anthracite coal, both of which are low in ash and volatiles, are admixed intimately with the kiln feed at a predetermined proportion. The fuel and stone should be of comparable size gradation. This method is far superior to traditional mixed feed charging of alternate beds of fuel and stone, since it obviates excessive localized temperatures with resultant overburning and promotes more even heat distribution throughout the whole kiln cross-section.

Furthermore, there is less chance for the product of combustion, CO_2, to be reduced subsequently to carbon monoxide (CO) by contact with heavy concentrations of the hot fuel and, as a result, dissipated in the exhaust gases by the reaction: $CO_2 + heat + C \rightarrow 2CO$. Temperature control is delicate since high rates of calcination that demand elevated temperatures can cause fuel losses through this undesirable CO reaction. Yet slow rates of calcination at depressed temperatures can be equally wasteful through incomplete oxidation of the carbon in the coke. Not only are calories lost, but the carbon discolors and contaminates the lime on discharge. If only 1% of CO is present in the exit gases, about 6% actual fuel value is lost, totaling about 240,000 Btu/ton of lime. To forestall this eventuality modern plants exercise close control over the proportion and rate of fuel feed and conduct continuous sampling and testing of exhaust gases for CO, CO_2, H_2, and hydrocarbons as well as spot checks with the Orsat apparatus for O_2 content.

Coke or coal fines should not be charged since they tend to filter down through the voids disturbing the homogeneity of the stone-fuel proportion. In addition, these fines contain a higher ash content. Bituminous coal is rarely ever employed in these kilns.

Because of high unit cost of fuel, these kilns have been largely employed in post-World War II West Germany. The German Seeger mixed feed kilns have been steadily improved until one large German company reports fuel consumption with forced draft of about 3.0 million Btu/ton, or 85% of theoretical thermal efficiency, based on low heat value.[15] This spectacular performance is achieved with $2\frac{1}{4} \times 3\frac{1}{2}$ in. kiln feed in large 15×95 ft. (inside diameter), highly automated kilns having an interior sectional area of 171 ft.[2] and a capacity of 270 tons/day. Figure 8-14 displays such an ultramodern mixed feed plant. This type of plant obtains the greatest fuel efficiency with the smallest and most uniform size of kiln feed. It features fully automated overhead equipment on a monorail for charging the kilns. Figure 8-10 illustrates the retractable bottom of the feeder open after a load of stone and coke is charged into the kiln. Discharging from the cooler is similarly automated with special two-ring plates and a rotating tripper to a lime bin and conveyor belt. The limestone and coke are charged intimately mixed with the stone and coke of comparable size ($2-3\frac{1}{2}$ in. or 3–5 in.)

Most authorities contend that minimum heights of mixed feed kilns should be at least 50 ft. for proper draft. If shorter kilns must be used because of break-up of soft stone, superimposed chimneys can compensate for lack of height and provide adequate natural draft. The diameter of the kiln, not the height, is the major determinant of capacity. If a strong natural draft propels excessive air into the

Fig. 8-14. Modern high-capacity German mixed-feed kilns of 270 tons/day capacity each.

kiln, additional kiln height will prevent undue heat losses from the exhaust gases.

The *size of stone* charged into verticals is much larger than any other type of kiln (Fig. 8-15). The feed extends from as high as 7–8 × 12 in. to a minimum of 2 × 3–4 in. A common size distribution is 4 × 8 in. Stone that decrepitates is futile to use in vertical kilns. There have been a few sporadic reports of successful calcination with smaller sizes, such as 1 × 2 in., but most reports lack conclusive confirmation. In contrast, there have been numerous authentic reports of consummate failures. Probably the most compelling research-engineering objective of shaft kiln operators has been to develop a small stone vertical kiln. But with the sustained effort expended and the generally fruitless results obtained, it is the author's opinion that only a few types of stone *might* lend themselves to successful calcination in small stone shaft kilns.

This is the main disadvantage of vertical kilns per se. In a large shaft kiln operation a vast amount of spalls (undersized stone) is necessarily accumulated in proportion to the kiln feed, usually sizes less than 3 in. Depending upon the fracturing qualities of the rock, spalls may comprise from 30 to 70% of the total tonnage of stone produced. Often this tonnage is too vast to market profitably, even on a by-product basis, at least consistently. The reduction of spalls that would ensue if stone in the ½ to 3 in. range could be calcined would be a great boon to these manufacturers. This was the major initial factor in the rapid growth of rotary kilns and has led to many plants installing both types of kilns as a solution to the persistent spalls problem in the United States.

In spite of great strides in improving quality, on an average the vertical cannot equal the modern rotary kiln in chemical reactivity. With large-sized stone more segregation is apt to occur. This contributes to over- or, more often, underburning and less uniformity.

Fig. 8-15. Typical size of kiln feed (stone) for vertical kiln, exemplifying a satisfactory percent of voids between stone.

Some of the lumps may be perfectly burned and equal or superior to the rotary lime, but not if a composite sample collected over several days is compared. The larger stone is much slower to calcine because of greater thickness for heat to penetrate as dissociation proceeds (p. 150). The rate of dissociation can be accelerated with elevated temperatures, but then overburning of the surface results, yielding even poorer reactivity (p. 138). Inherently, the nature of the operation does not lend itself to the same degree of instrumentation and quality control that prevails with the rotary. Restricted sizing, such as 4×5 in., would alleviate this disadvantage, but the meticulous classifying would be prohibitive economically by augmenting the spall pile.

Yet verticals have a decided advantage[10] over rotaries in fuel consumption, although latest rotary developments are narrowing this gap, as will be discussed in the next section. Greater flexibility due to smaller producing units and considerably lower capital investment are its greatest virtues. With maximum automation, which is commencing to approach rotaries in efficiency, this investment attribute is minimized.

Rotary Kiln

The United States is easily the world's leader in rotary kiln production with nearly 85% of the open-market lime capacity provided by this kiln in 1964. Captive lime production is much less, under 50%. However, elsewhere in the world rotaries are scarce, with the various vertical types comprising over 90% of the total productive capacity. England has only one rotary plant; West Germany has one. There are a few rotary plants in Finland, Sweden, Denmark, and Belgium. Insofar as the more meager information indicates, the iron curtain countries are exclusively vertical (primitive kilns excepted). There is one huge rotary plant in the Union of South Africa. Adding momentum to its growth in the United States is increasing demand for this lime by many chemical consumers, but usually only when its cost is at parity with the vertical kiln lime. Figure 8-16 is an over-all view of a modern rotary kiln plant. The operating principle and lay-out of the rotary kiln is depicted in a diagrammatic sketch (Fig. 8-17).

Rotary kilns vary greatly in diameter and length from 5×60 ft. to 13×450 ft. Average sizes have advanced; kilns less than 150 ft. in length are rarely installed in the post-World War II era. In the larger kilns the ratio of length to diameter generally extends

Fig. 8-16. Over-all view of modern rotary kiln lime plant.

between 30–40:1. The rotary is installed at an incline of 3 to 5° on four to six foundation piers and revolves on trunnions at each pier (Fig. 8-18). Rotation speed is adjustable with variable speed drives and operates generally in the range of 30 to 50 sec./rev. Kilns are lined with 6 to 9 in. of refractory brick. Stone is charged into the kiln at the elevated end and discharged at the lower end, moving countercurrent to the flow of combustion gases, derived from fuel injected at the lower end. Unlike verticals, these kilns are only charged with a maximum of 10% stone, so that at least 90% of the interior space of the kiln is confined to the flame and hot gases. Figure 8-19 presents a sketch of kiln feed in a rotary kiln.

The original conventional rotary (2a) of short length had deplorable thermal efficiencies, providing no heat recovery and maintaining excessive heat radiation losses. Enlarging kiln length reduced heat losses, but the major thermal improvement stems from a combination of different types of auxiliary equipment (type 2b) that are either integral or exterior appurtenances to the kiln.[17] These are the preheater, heat exchangers, and cooler. In addition, to appreciable reduction in heat losses, significant boosts in capacity are achieved. Consequently, modern rotary plants are invariably augmented with

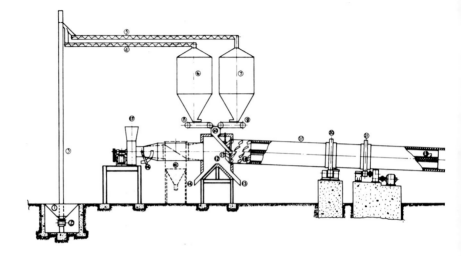

Fig. 8-17. Schematic drawing of modern rotary kiln with complete

this equipment. As a result, in the most efficient plants fuel consumption has been practically halved as compared to former years. Optimum fuel efficiencies are now as low as 5.5 to 6.5 million Btu/ton of lime, whereas standard kilns without this equipment were usually 10 to 14 million Btu/ton, with the lowest about 8.5 million.

These types of supplemental, thermal, recuperative equipment are summarized as follows:

1. *Coolers*
   a. Several modifications of *contact* coolers, also called "deheaters," are in operation, and some provide excellent heat recuperation and cooling functions for lime. The so-called Niems cooler[20] has strong adherents.
   b. *Rotary* coolers of different designs are effective in cooling the lime but generally yield mediocre results in thermal recuperation. They are widely employed in rotary portland cement plants.
   c. *Grate* coolers, commonly employed in cement plants, require huge masses of air, more than the kiln can accommodate, so that heat recovery is quite incomplete, although cooling is highly efficient.

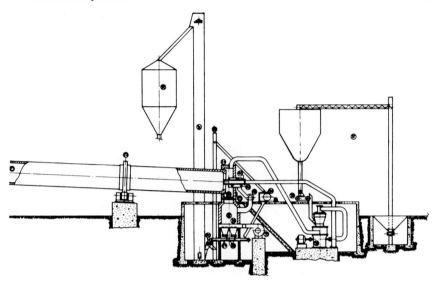

auxiliary heat recuperative equipment. (Courtesy, Azbe Corporation.)

  d. Multicooler tubes (planetary or satellite type) that are installed
  concentricly around the periphery of the discharge end of the
  kiln receive the lime through ports in the kiln (Fig. 8-20).
  Secondary combustion air, flowing countercurrent to the discharge, is recycled through the cylinders into the kiln. This
  differs from coolers 1*a, b,* and *c* in that it comprises a *single*
  unit with the kiln; the others are *separate* units connected as
  appurtenances to the kiln.

  2. *Heat Exchanger Cross-sections*[17, 19] are designed to increase the
interior surface area of the kiln so as to effect a more complete
heat transfer of the hot exhaust gases to the kiln feed, reduce radiation
losses, and increase through-put. They are composed of either refractory or special heat-resistant metal alloys that sectionalize the
kiln into quadrants, the number of such compartments depending
on kiln dimensions and experience. The refractory quadrants are generally installed at the hottest end of the kiln, such as the lower
end above the calcining zone. The metal quadrants are placed at
the cooler inlet of the kiln. They do not impede the flow of lime
or stone through the kiln but absorb considerable (otherwise exhaust)
heat, adding to thermal efficiencies and augmenting capacities by
25 to 30%. Their thermal efficiency can be measured by the decrement

Fig. 8-18. Close-up of rotary kiln, showing trunnions.

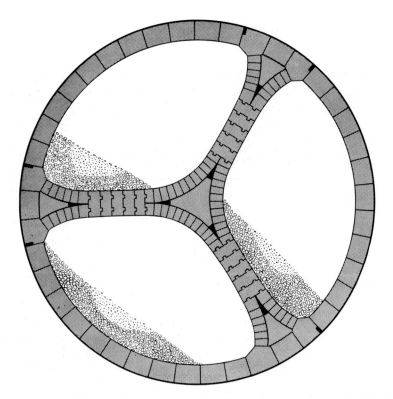

Fig. 8-19. Cross-section of rotary kiln illustrating appearance of kiln feed on trefoil segmentation.

in temperature of the exit gases. Figure **8-21** shows a quadrant as well as the refractory lining of a rotary kiln. Also see Figure **8-19**.

3. *Preheaters*

   a. The *stacked-tower* preheater is located before the inlet end of the kiln as a separate appurtenance and acts as the kiln feed storage hopper. Considerable heat is absorbed from the hot exhaust gases passing through this stone storage. From there gases diffuse to adjoining dust collectors before venting up the stack.

   b. The *grate-kiln* system, which is described later in this chapter.

The *size of kiln feed* is almost invariably smaller than the vertical kiln. In fact, large stone cannot be efficiently burned in rotaries. Generally 2 in. is the maximum size that is calcined, although a few plants have extended this range to $2\frac{1}{2}$ in. Generally $\frac{1}{4}$ in. is the most diminutive that is calcined. Sizes smaller than this contribute to ring formation in kilns, lack of uniformity in combustion, and a more serious dust-control problem. To minimize fines most plants will

Fig. 8-20. Satellite cooler units around the periphery of the rotary kiln's firing hood.

Fig. 8-21. Interior refractory lining of rotary kiln, showing quadrant section.

screen the kiln feed just before it enters the kiln or preheater. However, size distribution is also important; the more restricted the gradation, the more uniform the lime quality. Plants with multiple rotary kilns classify their stone to $\frac{1}{4}$–$\frac{1}{2}$ in. for one kiln, $\frac{1}{2}$–1 in. for the next, 1–1$\frac{3}{4}$ in., to 1$\frac{1}{2}$–2$\frac{1}{2}$ in., etc. However, superior quality and uniformity is obtained by even closer sizing, such as 1 to 1$\frac{1}{2}$ in., but this advantage must be balanced against the increased cost of stone classification.

As the stone slowly gravitates through the inclined kiln, a gentle tumbling action for all particles rather than a sliding or channeling effect is propitious for optimum heat transfer, quality, and output. To induce this movement dams and obstructions, made usually from refractories, are constructed into the linings. The same approximate effect is achieved also by altering the internal diameter of sections of the kiln by about 6 in. Refractory quadrants and metal cross-sections that apportion the kiln's interior tend to abet this function, although they are primarily installed to secure improved heat transference. If the tumbling action becomes too turbulent, increased degradation is the consequence, with an accumulation of excessive stone and lime fines and dust that impairs kiln performance and imposes a heavier burden on the dust collection system.

The Bureau of Mines[22] developed the following formula on calculating the speed with which stone may flow through a rotary kiln:

$$t = \frac{1.77 \sqrt{\theta} L}{SDN} \text{ Factor}$$

t = time of stone in the kiln (minutes)
L = length of kiln (feet)
D = interior diameter of kiln (feet)
S = slope of kiln (degrees)
N = revolutions/minute of kiln
θ = angle of repose of the stone
Factor = arbitrary allowances for internal obstructions and dams or changes in the kiln diameter, which accelerates the movement of the kiln feed and expands production. If no irregularities are present, the factor is 1; with obstructions it may be 2, 3, or 4, depending on the number present.

Solid, liquid, and gaseous fuels are employed, but pulverized bituminous coal is the most widely used and is blown into the kiln at the discharge end with primary air. Lump coal cannot be employed, so every coal-consuming rotary plant maintains its own pulverization equipment contiguous to the kilns. Majority experience among lime plants indicates superior thermal efficiencies are obtained with coal as opposed to gas, whether the latter is derived from coal, oil, or natural gas (see p. 253). The exact reason for this anomaly has never been adequately explained, other than heat transfer to stone is more efficient, and coal has higher inherent heat value. As a result, a few plants are known to purchase coal rather than natural gas, even though the latter offers a slightly lower cost on an equivalent Btu basis. (This difference in relative fuel efficiencies is inapplicable to vertical kilns; in fact, the results tend to be the reverse with gaseous fuels, offering superior average efficiencies; the sole exception would be coke-fired mixed feed kilns.)

Pulverized coal offers one disadvantage in that it accentuates the problem of *kiln ring* formations much more than gaseous fuels. A traditional, chronic problem with many rotary kilns is the formation of "rings" that gradually accumulate on the interior kiln linings. Unignited, pulverized coal tends to adhere to these rings along with accumulations of stone, and lime dust, and coal ash. Another cause of this build-up is uneven kiln temperature. If concentration of fuel gas is too high near the kiln lining, some of the small lime particles tend to fuse and adhere to the lining. Elimination of fines from kiln feed and use of larger sized stone of a minimum of $\frac{1}{2}$ to 1 in. is also a plus factor. Use of low ash coal and finer and more uniform comminution of the coal tends to alleviate this problem. Most kilns that burn gas or oil report either minor or no trouble with rings, particularly the latter. When rings become too thick, over-all kiln performance is seriously impaired. So the kilns are cooled slowly and shut down. Then the rings are most commonly removed by shooting with specialized industrial guns. Considerable trial-and-error experimentation in operating practices has enabled some coal-burning plants to eliminate injurious rings, at least avoiding shut-downs between the necessary periodic kiln relining operations. Rings increase the attrition on refractory linings and curtail capacity.

Maximum combustion efficiency in rotaries[16, 19] is fostered by such largely empirical relationships as kiln diameter, heat release rates, flame lengths and shape (invariably high luminosity is the object), flow pattern of secondary air entering the kiln, amount of kiln feed loading, size distribution of stone, dust loading of combustion air, location and direction of the primary air-fuel stream in relation to kiln axis, and draft, velocity, and proportion of air-fuel mixture.

Most rotaries are fired by a single large burner extending through the firing hood, at least up to and usually slightly past the kiln opening. An empirical observation is that when the burner penetrates beyond the kiln opening, at least to the point where preheated air from the cooler is introduced, the resulting flame is smoother and more stable. On an average, rotaries operate at higher temperatures than vertical kilns, but there is no standard temperature employed; it ranges between 2300 and 2650°F in the hottest zone. Fuel gases flow through the kiln with considerable turbulence, but stratification of gases can occur, which retards efficiency. This usually stems from insufficient mixing of fuel gas and air on entrance into the calcining zone, resulting in dissipation of unoxidized combustibles. Affecting the turbulence and flow of gases are interior kiln constrictions, such as quadrants, dams, and scale "rings." Internal kiln reactions are often unpredictable and unexpected. That is why complete instrumentation is essential to appraise performance so that a few of the many variables can be adjusted to produce the desired result.

The most modern innovation in rotary kiln design (Fig. 8-22) is the *grate-kiln system*, which previously had been perfected for portland cement manufacture.[22, 24] This consists of a much shorter rotary kiln with a much larger proportionate diameter than other rotaries possess for comparable lime capacity. It features a *traveling grate* that serves the dual purpose of preheater and partial calciner that is rectangularly housed at the entrance to the elevated charging end of the kiln. A much broader size gradation of kiln feed is accommodated (No. 4 mesh to 2 in.). The coarse stone is fed onto the traveling grate; then the fine stone is introduced on top of the coarse stone, providing a bed of 6 to 10 in. deep. The fine stone is calcined from hot, surplus exhaust gases transferred from the short kiln; the coarse stone is preheated. The fine precalcined lime is then first discharged to the rotary kiln, where it is submerged underneath the preheated coarse stone, reportedly protected from overburning. After the coarse stone is calcined, the lime is discharged into a revolving, counterflow recuperative cooler. Waste heat, along with primary heat, is reintroduced into the discharge end of the kiln.

High thermal efficiencies are reported of 5 to 5.5 million Btu/ton of lime. A compact plant of very high capacity results from the grate-kiln combination. Compared to other rotaries this kiln is relatively very short and wide. Equal or more capacity is obtained from this kiln with one-third the length of other rotaries and the same diameter. Limited experience indicates that lime of satisfactory

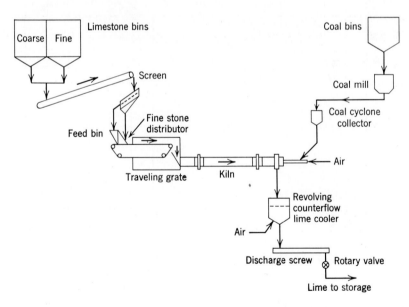

Fig. 8-22. Flow diagram of traveling grate rotary kiln (AC—Lepol).

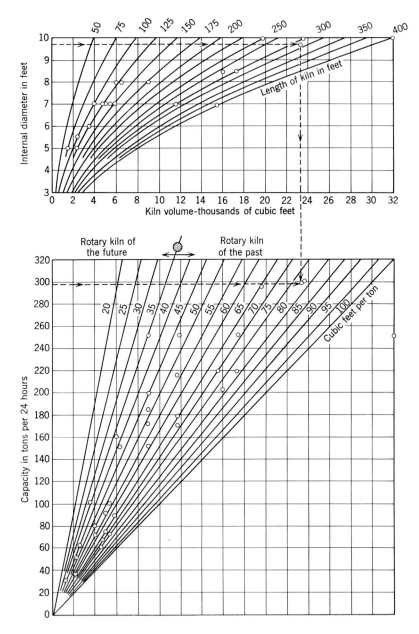

Fig. 8-23. Relation of rotary kiln volume to capacity.

quality is produced, but these plants are more difficult to coordinate at the outset than most other kiln types, requiring a much longer "shake-down" period. As more experience is gained, this disadvantage may diminish eventually.

Rotary capacities differ, of course, depending upon dimensions, but variance occurs even with kilns of the same size and auxiliary equipment. Again, the varying physical properties of the stone are manifest as well as differences in operating skills. For the largest kilns operating rates as high as 450 to 550 tons/day have been established. Generally, a modern kiln 9 ft. in diameter and 200 ft. in length, fully equipped, will produce 200 to 250 tons of lime per day. The grate-kiln, however, has the greatest capacity on the basis of cubic feet of kiln area/ton of lime. In Fig. 8-23 Azbe found a wide range in capacities in relation to kiln volume, exclusive of the grate kiln, on performances that he has investigated; 30 to 40 ft.^{3} of kiln area/ton of lime production appears to be optimum; the poorest capacity performance is 75 to 85 ft.^{3} of kiln area. One last observation: those kilns with the highest proportionate capacities usually also have the lowest unit fuel consumption—a two-pronged reason for striving for improved thermal efficiency.

Miscellaneous Kilns

A post-World War II development for calcining very small stone, approximately No. 8 to 65 mesh is the *Fluo-Solids* kiln.[25, 26] Externally this kiln appears to be a large covered-shaft kiln, but this is its only similarity; it operates on an entirely different principle. This patented kiln, which is also called a "reactor," calcines the fine kiln feed particles while they are maintained in dense suspension through emission of low-pressure air and combustion gases. The light, fine material is actually air-floated or "fluidized" in the preheating and calcining zones. The kiln is divided into three or five actual compartments, as Fig. 8-24 illustrates schematically. The compartments comprise preheating, calcining, and cooling (some kilns have two or three preheating sections). Stone storage is not part of the kiln. The limestone particles flow from one section to the other through vertical outlet conduits, as the particles rise above the outlets' height. Kilns can be operated at lower temperatures because of the fine stone size, 1650 to 1800°F. Fuel consumption is relatively low. With fuel oil of 14,000 Btu/lb., fuel-lime ratios are between 1:5 and 1:6, or about 5 million Btu/ton of lime (possibly slightly less). Most existing kilns have a capacity of 100 to 125 tons/day. A 200-ton

Fig. 8-24. Schematic drawing of Fluo-Solids kiln, showing interior of kiln.

unit went into operation in 1964. The quicklime produced is the most highly reactive of any commercial lime. It is very soft-burned and virtually coreless. Exacting instrumentation assures precise quality control.

The main deterrent to more extensive use is the cost of obtaining the meticulously, finely classified kiln feed, which would require prohibitive grinding and washing with most hard limestones. However, with soft limestone or heavy crystalline stone that thermally decrepitates, it may provide the most practical and economical (or only) solution. It probably operates at the lowest temperature of all kiln types. Similar to the grate-kiln, it is difficult to coordinate its operation at the outset. Being as highly automated as the rotary kiln, it is only 20 to 25% less than the rotary in capital investment per ton of capacity.

An entirely new concept of calcination, radically different from any other lime kiln, is the patented *Calcimatic kiln*,[27, 28] as illustrated in Fig. 8-25. It features a large diameter, *circular refractory hearth* that can be operated at various speeds of 35 to 200 min./rev. Controllability of this kiln is unique since the hearth is divided into heating zones, and through instrumentation precise temperatures can be maintained uniformly at these different zones. This, coupled with the adjustable speed of the hearth, enables a plant to control the time of calcination for diverse stone sizes or for varying qualities of lime ranging from hard- to soft-burned, including reportedly coreless lime. Thus, it appears that this kiln is sufficiently flexible to be readjusted each day to produce a different specified quality, various sizes of lime, and at divergent rates of production, contingent upon demand. The first commercial kiln uses stone ranging from $\frac{1}{4}$ to

Fig. 8-25. Calcimatic kiln, featuring traveling hearth.

4 in. Its circular hearth has an external diameter of 55.5 ft., only 8 ft. in height, and 16 ft. in width. The rotating hearth, supported on two concentric tiers of rollers, is independent of the stationary kiln. Heat is supplied by multiple burners. Stone is fed from a pre-heater chamber above the kiln through special metal cylinders onto the hearth in an even bed 1 to 4 in. thick, as desired. A continuous loop drag conveyor removes all lime at the discharge point as it completes a revolution.

This size kiln is claimed to produce 100 to 125 tons/day, with fuel consumption of 5 million Btu/ton of lime (or slightly less). Automation is complete. Consequently, it appears that this kiln provides the greatest flexibility in operation of any kiln type, and it is predicted that its application will be steadily extended to other plants. On an average, cost of this kiln is less than the rotary and and probably slightly less than the Fluo-Solids kilns.

Two other patented postwar kilns of novel design are the *Ellernan*[32] (Fig. 8-26) and the *Corson vibratory, inclined.*[33] Both types probably represent the lowest average capital investment cost of any modern kiln for a given capacity; the kiln and auxiliary equipment is relatively simple and compact. The Ellernan consists of a rectangular calcining chamber with horizontal tunnel beams. Heat is generated in a separate adjacent firebox with oil or gas, and it enters the kiln with the combustion gases and the flame rising under induced draft. Stone spalls (from the standpoint of shaft kilns) of $\frac{1}{4}$ to $\frac{3}{4}$ in. or $\frac{1}{2}$ to $1\frac{1}{2}$ in. in size are calcined. Low-production units of 6 to 15 tons/day have been operated.

Since it is the newest of the postwar kilns, information on the patented Corson kiln operation is still proprietary and incomplete, as it is still in the commercial development stage. Its capacity and capital cost is greater than the Ellernan. The initial popularity and operation of the Ellernan kiln has waned due to reportedly erratic results with different kinds of stone that yield lime of high core content.

One other kiln will be described, even though it has entirely lost acceptance, since it is so radically different from all other types. This is the *Hoffman, or "ring,"* kiln[5] of German origin which was widely used throughout Europe since its invention in 1865, although increasingly less since 1925 and rarely employed since World War II. This kiln has never been employed in the United States on a commercial scale.

There are many modifications in design, but essentially it is a horizontal process consisting of an annular tunnel (circular, elliptical,

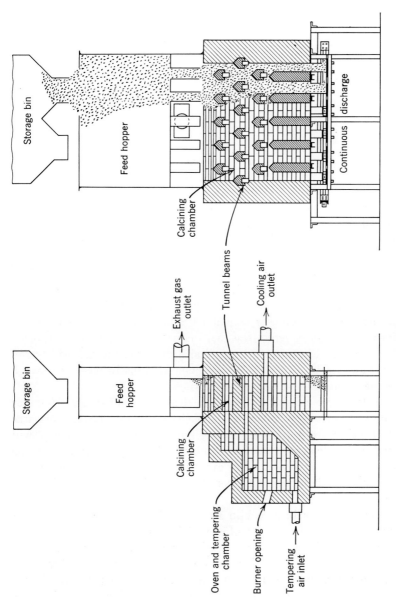

Storage bin

Feed hopper

Calcining chamber

Continuous discharge

Tunnel beams

Cooling air outlet

Exhaust gas outlet

Storage bin

Feed hopper

Calcining chamber

Oven and tempering chamber

Burner opening

Tempering air inlet

Fig. 8-26. Sketch of Ellerman kiln.

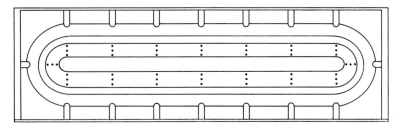

Fig. 8-27. Drawing of Hoffman (tunnel-type) kiln layout.

or rectangular) in which hot gases from the direct firing of solid fuel travels slowly horizontally through the tunnel preheating and calcining the stone charge (see Fig. 8-27). Considerable cool air is induced by fans into the tunnel and is heated by the hot lime, thus serving both to cool and preheat the stone. The tunnel of 8 to 10 ft. in height is divided in twelve or twenty imaginary or partially constructed chambers about 12 ft. in length into which stone is carefully charged by hand through feed holes in the arch of the tunnel, forming pillar-like masses 3 to 4 ft. apart. Small sized coal is added in layers as the stone pillar is assembled, and through various networks of flues the hot gases ignite the coal in succeeding chambers. Depending on the skill of the operator and size of stone, the fire will travel slowly at the rate of 2 to 4 ft. per hour, eventually passing through all of the chambers and encircling the tunnel.

In a twenty-chamber kiln under ideal conditions: one chamber will be empty; one chamber will be charging; seven chambers will be preheating from the circulating hot gases; four chambers will be under fire (calcining); six chambers will be cooling and at the same time heating the air for eventual combustion of the fuel in succeeding chambers; one chamber will be discharging from side openings in the tunnel.

At one time this bizarre kiln was considered to have the greatest thermal efficiency of all kilns (45 to 55%), since there is considerable heat recovery inherent in its operating principle. There was also less contamination of lime quality from ash than in mixed-feed vertical kilns. However, much more hand labor is required per ton of lime produced than in any other type of kiln, especially for the cumbersome charging and discharging methods that are necessary—prohibitively costly for the United States and most countries in the postwar era. Essential to the success of this kiln is extreme (strictly empirical) operating skill, more than is necessary for other kilns. Daily production capacities ranged from 25 to 80 tons/day. Capital investment costs were high.

Other kilns are simply slight modifications of the kilns previously described, e.g., the Spencer and Priest kilns are English producer gas vertical kilns with special proprietary features. A new patented Hungarian kiln (producer gas variety) developed in 1955–1960, appears to have provoked considerable interest in Europe.[34] Reportedly, Czechslovakia has constructed four new plants modeled after the Hungarian kiln and plant layout. Italy and Yugoslavia have also built similar plants and have ordered others. The kiln allegedly features the use of low-grade, low-cost, brown bituminous coal or lignite in the patented gas producer; large kilns of 50 to 100 tons daily capacity; fuel consumption of 4 to 4.5 million Btu/ton of lime; rather small stone of 2×3 in. (and larger); considerable automation. Licensing of the kiln and plant is through the Hungarian government.

Two other European kiln developments of possible future impact are the Fellner and Ziegler German rotary kiln,[36] the proponents of which claim fuel consumption of 4 million Btu/ton of lime, and a large Austrian, high-capacity vertical kiln, which features dual and triplicate calcining and/or preheating zones within the single large vertical shell and claims of very low thermal efficiency of 3 to 3.5 million Btu/ton of lime (Wopfinger Stein & Kalkwerke).[37] More confirmation is needed on these two *apparently* spectacular developments. A bizarre design of a new American vertical kiln involves a *doughnut-shaped* cross-section, with an air-cooled vertical shaft in the center.[8] Kiln dimensions are 57.5 ft. high and 11 ft. interior diameter; circular calcining area is only 3 ft. in width or 75 ft.^{2} cross-sectional calcining area (center is 5 ft.); capacity is 100 tons/day. A new patented German kiln (Heiligstaedt) is evoking interest.[39] A combination rotary and vertical kiln comprises Azbe's patented *terminal calciner*. Calcination is completed in a small vertical at discharge end of the rotary.[18, 40]

Chemical Analysis. The chemical analysis of the lime depends predominately on the stone analysis from which it is derived and to a lesser but important extent on the purity of the fuel and whether these impurities are absorbed. The influence of stone on calcination and quality of lime was discussed in Chapter 6.

The following is regarded as a typical range in chemical analyses[41] of both high calcium and dolomitic commercial limes in the United States.

Component	High calcium, range, %	Dolomitic, range, %
CaO	93.00–98.00	55.00–57.50
MgO	0.30–2.50	37.00–41.00
SiO_2	0.20–2.00	0.10–1.75
Fe_2O_3	0.10–0.40	0.05–0.40
Al_2O_3	0.10–0.60	0.05–0.50
CO_2	0.40–2.00	0.40–2.00
SO_3	0.01–0.10	0.01–0.10
P	trace–0.05	trace–0.05

See p. 156 on trace elements found in limestone that occur in their derivative limes.

The values in the above range do not necessarily represent minimum and maximum percentages.

Refractory Linings

All kilns, regardless of type, are refractory lined. This is another area in which manufacturers are constantly experimenting with different kinds, shapes, and thicknesses of refractories, always hoping to discover a refractory combination of the lowest installation and maintenance cost and longest life. The results of these probings are too divergent to form a consolidated opinion. Literally most types of refractories have been tried—ordinary and special fireclay brick; silica, high-alumina, and basic brick; as fabricated units, castable, or rammed types; in thicknesses varying from 6 in. to 1.5 ft. (See p. 256 for details on a German lining of 2 ft. 9.5 in.)

One inevitable result from these studies is that no one has discovered a refractory lining that will endure indefinitely. Eventually attrition and spalling of the lining occurs due to abrasion, excessive temperature, or sudden changes in temperature, and before disintegration progresses too far, the kiln must be cooled, shut down, and repaired or partially relined. To minimize damage, the kiln is first slow-fired and cooled gradually; similarly, it is reheated gradually after repair. Thus, it is rare for a kiln to operate continuously for a full year. So in calculating total lime capacity, allowance must be made for "down-times." Assuming repairs to linings are necessary about twice a year, experience indicates that the annual down-time will average about thirty days. So capacity should be calculated on 335 days. If kilns are forced to produce more than their rated capacity, attrition on linings is, of course, heightened. In former years shutdowns were more frequent or longer, aggregating more down-time of fifty to sixty-five days per year. Development of superior kiln instrumentation, smoother flow of kiln feed, and improved refractory quality and installation have combined to extend refractory life.

To minimize damage to refractories, cooling of kilns is avoided, if at all possible. If demand for lime subsides when storage silos are filled to capacity, production obviously must be curtailed in some kilns. Under such circumstances *slow-firing* is practiced to maintain heat in the kilns with the expectation that production will be resumed in a few days. The cost of such wasted fuel is generally less than relining maintenance and start-up, if the kiln is permitted to cool. This would not be true with prolonged slow-firing. Such decisions are often subject to speculation and are calculated risks. Slow-firing is also practiced by those who burn natural gas and who have no stand-by fuel in the event of gas curtailments by utility companies during cold weather. Utility companies are cognizant of the potentially costly damage to kilns that results from cooling and will usually provide enough gas to maintain some temperature (slow-firing).

Different kilns perform better with certain refractories than others; also diverse temperatures prevailing in various kiln zones usually dictate specific refractories. A few general, concensus observations[42] on refractory preferences are enumerated as follows:

1. Superduty (very dense, hard) fireclay brick is used at feed end of rotary kilns and storage and preheating zones of vertical kilns because of its high abrasive resistance.

2. Basic brick and periclase are generally more preferential for rotary than vertical kilns.

3. For vertical kilns 60% high-alumina brick is favored.

4. Special high-alumina (80% Al_2O_3) refractory, derived from fused alumina, are preferred in the burner arches of center-fired vertical kilns and in the hottest zones of the rotary kilns. The same applies to corundum, a 90% alumina composition.

5. Forsterite-type magnesium silicate refractory enjoys acceptance in the burning zones of dead-burned dolomite rotary kilns.

6. High-heat-duty fire clay of moderate strength is probably most widely employed in vertical kilns per se.

7. A popular refractory in the calcining zone of vertical kilns is silica brick.

Yet there is not too much unanimity of opinion among the producers on refractories. A combination of refractories that performs well for one may fail in another apparently similar type kiln. An empirical analysis is most rewarding for each individual plant.

Flexibility

The loss of production due to periodic kiln relining exposes the lack of flexibility in single kiln plants as compared to multikiln plants. The former must maintain proportionately much larger lime storage facilities. Meticulous preparation is necessitated for each shutdown in order to amass adequate inventories for customers. A multikiln plant, on the other hand, of 10 to 20 smaller kilns simply assumes that about 10% of its kilns will be "down," in event of capacity production, and plans a systematic rotational maintenance program.

Three or four small modern kilns are generally more economical to operate than one large rotary kiln—and vastly more flexible in consumer relationships.

Fuels and Combustion

The quality and type of fuel exert a profound effect on the quality of lime produced, not just the quantity employed. Certain fuels are innately more conducive to producing porous, soft-burned, reactive limes; others tend to yield hard-burned, slow reactive limes; others are intermediate between these two extremes. Consequently, selection of the proper type is vital for optimum efficiency and poses usually a more complex decision than quantity. For example, a change in source of bituminous coal to another of presumably comparable carbon, ash, and volatiles content and calorific value may upset the delicate combustion balance just enough to yield a lime of noticeably different quality or properties. Some fuels, like stone, are simply unsuitable for lime burning; others are of marginal utility, satisfactory for some, unfeasible for others; others are suitable in virtually all plants, but still require various operating modifications for high efficiency and/or to conform to specified or desired properties. Of the usable fuels, many companies will purchase the more costly types since they have observed empirically that they yield the desired quality most consistently. As a result, fuels should be selected with considerable care for both existing kilns and new plants in the blueprint stage. In the latter case, influence of fuel on kiln design can be critical. Often considerable trial-and-error methods are necessary for final selection.

The principal fuels, together with their approximate range in calorific values for lime, are summarized as follows:

Fuel type	Calorific values	Unit
Solid fuels as—	11,000–16,000	Btu/lb.
Bituminous coal		
Anthracite coal		
Coke		
Producer gas (coal)		
Natural gas	900–1200	Btu/ft.^{3}
Fuel oil	16,000–18,000	Btu/lb.
Wood	5000–6000	Btu/lb.

There have been other fuels employed successfully, such as sawdust, propane, and even camel dung, but they do not warrant elaboration since generally they are either prohibitively expensive in unit cost per Btu or have limited availability.

Traditionally, bituminous coal was the predominant fuel for burning lime in the United States. In 1939 in a U. S. survey[43] encompassing 75% of commercial lime and 80% of dead-burned dolomite, it was reported that 55% of the fuel directly used was bituminous coal and indirectly 79%, if producer gas and coke are included. The latter two fuels were 17% and 6.5% respectively; 100% of dead-burned dolomite was produced with bituminous coal. The remainder on lime was: natural gas—9%, wood—7%, and fuel oil—6%.

However, starting with World War II, natural gas steadily encroached on coal because of its lower cost per Btu, so that the percentage distribution of fuels in the 1960's has changed, as seen in Table 8-1.

Meanwhile, wood has virtually disappeared as a fuel due to lack of availability and soaring costs. Producer gas has ebbed and oil has remained static. Actually, since 1955 coal has recouped some of its losses to natural gas due to ascending prices of the latter. In a few locations coal and oil are now on economic parity in cost per Btu. Contributing to coal's gain has been increasingly severe gas curtailments in cold-weather periods, necessitated by increasing domestic demand for gas at much higher prices. This stems from low-cost, interruptible contracts tendered by natural gas producers to lime and industrial consumers. The uncertainty of the time of curtailment and its duration is gradually forcing most companies to install stand-by fuel facilities at substantial cost with either coal or oil to provide continuity of production.

Table 8-1. 1962 percentage breakdown of fuels for
lime plants[a]

Fuel	Percent
Natural gas	36
Coke	30
Bituminous coal	22
Oil	6
Unspecified and miscellaneous[b]	6
	100

[a] Includes U. S. Bureau of Mines' survey of 226 active plants, embracing commercial, captive, and recovery plants.

[b] One plant each reported using coke oven gas, wood chips, and sawdust.

Great Britain employs solid fuels (largely bituminous coal) almost exclusively, whereas Germany's coal-derived production is primarily coke because of rigorous smoke-abatement provisions. There is very little natural gas available in western Europe. This dearth of natural gas is expected to be short-lived. A tremendous new gas field has been discovered in Holland that is being commercially developed and should make deep incursions into the industrial fuel market of northwestern Europe by the late 1960's. Meanwhile, interest is gradually increasing in Europe for fuel oil because of steadily mounting coal and coke costs. For the first time in history at or near some European port cities, the economics slightly favors fuel oil over coal, and quite a few plants are converting to oil in the 1960's. Wood is still extensively employed in a few countries, notably Brazil, for lime burning.

Assuming it is economical, *wood*[44] is considered the ideal fuel for vertical kilns for the following reasons: It produces a longer flame than solid, liquid, or gaseous fuels, enabling the heat to penetrate farther into the stone mass and creating a broader burning zone. This maximizes kiln capacity and promotes more uniform calcination and soft-burned lime. Considerable steam is generated from wood, more than any other fuel, contributing a tempering effect that lowers the flame temperature required for calcination. The resulting cooler temperature lessens the danger of overburning. In fact, it is considered almost impossible to overburn lime with wood. As a result, lime manufacturers being forced to use other fuels for economic reasons strove to modify the combustion of solid, liquid, or gaseous fuels so as to emulate the effect of wood fire. Such modifications involve primarily diverse methods of creating draft to support the most efficient combustion. With wood-burning plants usually only *natural draft* is necessary. In addition to natural draft the following methods are employed with other fuels often in combination:

1. *Forced draft* is created by blowing steam artificially under the grates of shaft kilns into the discharge or cooler end of the kiln, causing the gases to rise—or blowing air, creating pressure.

2. *Induced draft* is propelled by various capacity fans, which draws the hot gases through the preheating zone and also may be employed in recirculation of gases (suction).

3. *Eldred method* is a combination of forced and induced draft employed in vertical kilns in which the hot exhaust gases are induced from the top of the kiln by recirculation down under the grate and into the kiln again by forced draft. Usually recirculation is supplemented by partial venting of the exhaust gas. There are other modifications.

Table 8-2 covers all of the principal fuels or those that have potential for lime burning and provides a typical range of analyses on their composition.

Solid Fuels. Table 8-3 provides a range of analyses on the composition of individual bituminous coal reported for use in direct firing and producer gas. There are a number of interrelated properties for evaluating *bituminous coal* for lime burning—reactivity (free burning), caking and coking tendency, calorific value, and carbon, ash, sulfur, hydrogen, volatiles, and moisture contents. However, usually reactivity is the most important criterion, since this factor determines kiln temperature and affects both (1) the primary stage in which the coal is oxidized to CO_2 and (2) an undesirable secondary stage in which some of the CO_2 might react with additional carbon for reduction to CO.

The reason the secondary combustion stage is undesirable is that, being endothermic, it reduces kiln temperature and dissipates potential heat in the exhaust gases, thereby lowering kiln efficiency. This characteristic is generally applicable to highly reactive, free-burning fuels. Other properties of these fuels are a tendency to resist caking, high porosity, low strength, high volatiles and moisture contents, and relatively low carbon content and calorific value. The reverse is generally true of fuels of low reactivity that possess high caking power and greater density. So it is coal of *moderate to low reactivity* that is

Table 8-2. Analysis of principal fuels for lime

Combustible fuel	H$_2$O	Ash	C	H$_2^b$	H$_2^a$	S	O$_2$	N$_2$	CO$_2$	CO	CH$_4$	C$_2$H$_6$
							% Composition					
Air-dried, hardwood	20	0.2	40.4	4.2	1.0	—	33.9	0.3	—	—	—	—
Metallurgical coke	—	13.4	81.4	0.2	1.2	0.9	1.9	1.0	—	—	—	—
High-grade bituminous coal	3.2	7.0	78.1	0.6	4.3	1.0	4.7	1.2	—	—	—	—
Semibituminous coal	3.5	4.2	80.3	0.7	3.9	0.7	5.7	1.0	—	—	—	—
Natural gas (Texarkana)	—	—	—	—	—	—	—	5.4	2.1	—	92.5	—
Natural gas (Cleveland)	—	—	—	—	—	—	—	1.9	—	—	64.9	33.26
By-product coke-oven gas	—	—	—	—	9.6	—	—	32.0	3.0	15.0	40.5	—
Blast-furnace gas	—	—	—	—	0.2	—	—	56.5	18.9	24.4	—	—
Producer gas	—	—	—	—	11.0	—	—	52.0	5.0	28.0	4.0	—
Fuel oil (midcontinent)	—	—	84.7	0.1	13.4	0.4	1.1	0.3	—	—	—	—
Fuel oil (Western)	—	—	86.2	0.1	11.8	0.7	1.0	0.2	—	—	—	—

^{a} Unavailable.
^{b} Available.

Table 8-3. Range of analyses of
U. S. bituminous coals

Component	Range %
Carbon[a]	70–90
Hydrogen	4.5–5.6
Volatile	24–42
H_2O	1–15]
Sulfur	1– 5
Ash	4–12

[a] As fixed carbon values, above would be 50–65%.

most adaptable for lime burning. The quality of the *coke* is directly related to the coal from which it is derived. Thus, hard metallurgical-grade coke of low reactivity is superior for lime burning. In mixed-feed kilns the tremendous consolidated weight of alternate stone layers on coke demand high compressive strength, since if the coke is crushed, the resulting fines will filter down through the voids and cause localized heat imbalances.

Anthracite coal is an excellent fuel for lime burning since it is hard and has low reactivity, ash, and volatiles; often its use is interchangeable with coke.

For all solid fuels the ash content should be as low as possible, 8% maximum, since it possesses no heat value. In fact, it exerts a negative influence by absorbing heat; it fuses to kiln linings, accelerating deterioration; and when in direct contact with the lime it vitiates the quality by introducing silica, alumina, etc. Anthracite generally has the lowest ash content, 2 to 4%. The importance of low sulfur content is discussed on p. 155. Generally ash is independent of fuel reactivity and should be considered separately. Composition of ash fluctuates widely, but silica almost invariably predominates, followed closely by iron and alumina. In tests on fifteen different United States bituminous coals, SO_3 ranged between 2 and 7%.

Coke and anthracite produce higher average temperatures than all bituminous coals and may require closer controls for heat tempering to prevent overburning. Having little or no volatiles, these fuels waste less thermal calories in gaseous incombustibles, which are lost as they are vented with the kiln exhaust gases. As a result, a lower ratio of coke or anthracite to stone is applied than with bituminous in mixed-feed kilns by about 15 to 20%.

Low-grade solid fuels, such as lignites and peats, are totally un-suitable for lime burning because of their relatively poor heat values and high ash and volatiles contents. (A new Hungarian kiln, however, reportedly is utilizing these low-grade fuels successfully in producer gas shaft kilns, where the fuels do not contact the lime.)

The desired sizes of solid fuels are adjusted to the type of kiln and stone sizes and are discussed in other appropriate sections of this chapter.

Calorific values of coal derived *producer gas* depend upon the heat value of the coal and the furnace design and method of operation, both of which affect the concentration of the gas, but in commercial practice they range around 150 to 350 Btu/ft.^{3} of gas. The composition of *natural gas* is entirely different from producer gas, being largely methane, so-called marsh gas. In the United States natural gas has largely replaced producer gas for lime because of superior all-round performance and economy.

Most *fuel oils* utilized are Bunker C grade, and unless combustion is carefully controlled, overburning can result from excessive kiln temperatures. Oil has a potentially greater magnitude of heat genera-tion power than coke and anthracite. Invariably it is gasified before introduction in the kiln. The gasification of oil is effected in the following sequence:[46]

1. Oil is pumped and preheated to about 200°F.

2. It is injected under 200 to 250 psi pressure into a vaporization chamber where it is atomized into minute oil globules.

3. Hot recirculating gases vaporize the atomized oil.

4. The hot oil vapors are oxidized to fuel gases.

5. Rising temperature decomposes, or "cracks," the remaining heavier hydrocarbons.

6. Petroleum gas is then introduced into the calcining zone of the kiln through side ports and/or center burners, or a single large burner.

Visually it is possible to determine the desired temperature range of the gasified oil. If it is a yellowish-brown color, its combustibility is ideal; if it appears as a white fog, temperature is too low; if it is a dark brown-blackish color, temperature is too high.

Empirical Survey on Combustion. Comparative experience with dif-ferent fuels and combustion performance has been reported based on a partial survey of the U.S. lime industry.[47] Significant conclusions were:

ROTARY KILNS.

1. A majority reported slightly higher capacities are achieved (5 to 10%) with natural gas than with pulverized coal.

2. There is a general decline in thermal efficiency ranging from 2 to 20% with natural gas as compared to pulverized coal. A plausible explanation for this disparity is the difference in comparative heat values—96.5% for coal and 85% for natural gas, on an average. This is due to the former having a higher carbon and lower hydrogen content than the latter.

3. Invariably in long kilns of 300 ft. or more, greatest efficiency is obtained when natural gas is introduced into the kiln at high velocity of 100,000 to 150,000 ft./min.

4. Primary air along with natural gas is used as a combustion aid by 90% of plants.

5. A minority of plants use coal or oil as auxiliary fuels with natural gas.

6. For controlling combustion, 95% of kilns are equipped with instrumentation for analyzing stack gases for O_2, CO_2, and combustible gases. Orsat apparatus is usually employed.

7. The majority reported greater ease of combustion control with natural gas over coal and oil.

8. A majority produced superior-quality lime with natural gas.

9. Opinions were divided on whether gas, coal, or oil provided longer refractory life. Pros and cons were in balance.

VERTICAL KILNS.

1. There is preference for oil rather than coal as stand-by fuel for natural gas.

2. Marked increases in capacity were reported by 85%, ranging from 15 to 100% with natural gas rather than coal or producer gas.

3. The majority reported a 15 to 20% increase in thermal efficiency with natural gas instead of coal. Those formerly using producer gas experienced the greatest gains.

4. The majority use Orsats for stack gas analysis.

5. All plants report better combustion control with natural gas.

6. All plants report improved lime quality with natural gas.

7. Divided opinion resulted on refractory life, with a slight preponderance reporting longer life with natural gas.

Heat Balance. Next to the cost of kiln feed, fuel costs are the second most critical factor in lime production costs on an average. Due to much higher unit fuel cost, this is particularly true in Europe, much more so than the U. S. In fact, fuel costs generally equal or exceed kiln feed in Europe. As a result, it represents a fruitful area for engineering ingenuity in improving thermal efficiency, and thereby over-all costs. As evidence of this, a sizable number of the more progressive American and European companies have reduced their fuel consumption in Btu/ton of lime by 60 to 100% since 1939. As a requisite prelude to such a program, an accurate detailed heat balance must be ascertained on all existing kilns.

Throughout this chapter all references to fuel consumption are predicated on Btu/ton of lime, rather than fuel ratios. Since the former automatically allows for the varying calorific value of fuel, it is much more meaningful and precise. When coal was the universal fuel for lime burning, fuel ratios were the vogue in expressing efficiency. A 1:4 or 1:5 ratio by weight of coal to lime is not accurate enough for heat balance studies, since the Btu value of the coal might range from 11,000 to 15,000 Btu/lb. In addition to greater accuracy, Btu/ton acts as a convenient common denominator in comparing such totally divergent fuels as coal or coke with natural gas or wood or fuel oil, which are sold in tons, cubic feet, cords, and gallons respectively.

It has already been established that the theoretical minimum heat requirement to make a ton of lime from 100% pure $CaCO_3$ is 2.77 million Btu/ton and that about 1.6 million Btu is necessary to heat the stone to the dissociation temperature (p. 136). Values are less for magnesian stone and dolomite. The above preheating figure is not included as a portion of the heat requirement, since theoretically it occurs only once on warming the first charge of stone. From there the Btu recovery from exhaust gases serves predominantly to preheat the successive stone charges. Percent of heat efficiency is first obtained making allowance for the purity of the resulting lime, by the following formula:

$$\frac{\text{Theoretical Heat Requirement} \times \% \text{ Available Oxide Content}}{\text{Total Heat Requirement}}$$
$$= \% \text{ Thermal Efficiency}$$

Thus, if the lime has a 93% available CaO content and total fuel requirements are 6 million Btu/ton, then:

$$\frac{2,770,000 \times .93}{6,000,000} = 43\% \text{ (efficiency)}$$

Naturally, 100% efficiency is unattainable for three reasons. First, there is no commercial limestone available of 100% purity; second, it is impossible to calcine lime without some dissipation of heat calories; third, production of lime with zero core and recarbonation, without hard-burning, is formidable to say the least.

However, much more detailed information than provided by the above formula is necessary in order to pinpoint the most fruitful areas of heat dissipation. Table 8-4 reveals a breakdown of the specific values that must be obtained from instrumentation charts and data as well as kiln tests, preferably as realistic average values. The three examples listed are based on the following hypothetical kiln operating conditions:

Kiln A. Conventional rotary kiln without any heat recuperative equipment, but operated at optimum efficiency.

Kiln B. The same type of rotary kiln as A but possessing complete heat recuperative equipment and segmentation, and also operated at high efficiency.

Kiln C. This corresponds to the lowest fuel efficiency on record, that of the most advanced German mixed-feed kiln and which might be regarded as the ultimate in fuel efficiency that appears attainable.

These three simple heat balances are also illustrated graphically in Figs. 8-28, 8-29, and 8-30 for kilns A, B, and C respectively.

Commenting on the values in Table 8-4, obviously nothing can be saved on calcination (item 1) since this is a fixed value, regardless of kiln or efficiency of operation, assuming complete calcination. Item 2 on heat for drying stone can be reduced if the kiln feed is stored under cover and protected from rain and humidity. Usually the slight fuel savings cannot justify the cost of enclosed storage silos. The

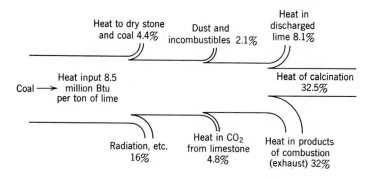

Fig. 8-28. Kiln A—standard rotary kiln—heat balance.

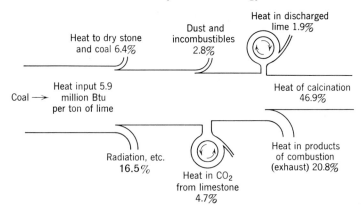

Fig. 8-29. Kiln B—rotary kiln with recuperative accessories—heat balance.

same situation also applies to coal in No. 3. In kiln C presumably the coke that the Germans used was bone dry.

Heat losses primarily from the external shell of the kiln as well as from auxiliary appurtenances, such as the cooler and preheater, are caused by radiation and convection currents (No. 4). Great savings in fuel are possible. First, the lining of the kiln (refractory and insulation) can be thickened so as to reduce conductivity of heat. The very nature of the rotary kiln does not lend itself to as thick linings as the stationary shaft kiln. In the rotary, maximizing the refractory brick from 6 to 9 in. helps, and additional insulation can be added to the calcining zone, but it can never match the low heat conductivity of the German kiln. The Germans use an average of 2 ft. 9 in. of lining in the hot zones, including five different refractories in achieving this incredibly low loss of radiated heat. From inside out in sequence are: 10 in. of a patented dolomite brick, 2 in. of chrome magnesite granules, 5 in. of fire clay brick, 5 in. of high-porosity fire brick, and 10 in. of a special brick with diatomaceous earth, called "Sterchamol"; $\frac{3}{4}$ in. steel boiler plate encases the lining.

In No. 5 the drastic minimization of heat losses with an efficient recuperative cooler is clearly evident and is substantial enough to warrant the capital cost of this equipment. In kiln A the lime is probably nearly red hot at discharge.

Usually the most prolific waste of heat is item 6, the exhaust gases. Inherently, rotary kilns cannot equal the most efficient vertical kilns because the waste exhaust gases used to preheat the stone cannot be recirculated back to the kiln as readily. Yet through segmentation (quadrants and cross-sections) much of this loss of heat can be curtailed by increasing the interior surface area of the kiln to absorb (and retain) more of this transferred heat for the continual calcination that is occurring. This results in lower temperature of exhaust gases after preheating. The same situation also applies to No. 7 on exhaust CO_2 derived from the dissociated limestone.

Losses in combustible gases—carbon monoxide, methane, oxygen, and hydrogen—from the stack can be disastrous to this heat balance factor. Their presence indicates incomplete combustion. The gravity of this should be realized, since if only 1% of CO is present in the exhaust, it is equivalent to the loss of about 200,000 to 240,000 Btu/ton of lime; in contrast, 1% of O_2 wastes only about 40,000 Btu. To guard against this eventuality, vigilant plants use continuous gas analysis recording equipment and regular spot checks with an Orsat apparatus.

The losses due to dust, degradation of stone and lime that are removed in dust-collector systems, and incombustibles in the limestone (impurities) and ash are too inconsequential to warrant much expense in curtailing Btu wastage (see Table 8-4, item 8). Avoidance of degradation by effecting a smooth flow of material through the kiln, screening off fines from the kiln feed, closer sizing of feed, and use of low ash fuel greatly minimize these losses.

Thermal efficiency can also be analyzed in a less exact but reason-

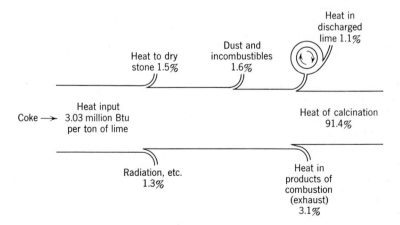

Fig. 8-30. Kiln C—most advanced German mixed-feed kiln—heat balance.

Table 8-4. Comparison of heat balances of three hypothetical kilns.

Heat consumption	Thousands of Btu/ton of lime		
	Kiln A	Kiln B	Kiln C[a]
1. Theoretical heat for calcination	2770	2770	2770
2. Heat to dry stone	295	295	45
3. Heat to dry coal	80	80	—
4. Heat of radiation and convection	1365	975	40
5. Heat in discharged lime	685	110	35
6. Heat in products of combustion (exhaust)	2720	1230	90
7. Heat in liberated CO_2	410	270	—
8. Heat in dust loss and incombustibles	175	170	50
9. Total heat requirement	8500	5900	3030
Thermal efficiency for 93% available lime	35%	50%	85%

Assume:

High calcium limestone of 98% $CaCO_3$ is calcined.

Atmospheric temperature during calcination for A, B, and C is 60°F.

Average exhaust gas temperature for A—1250°F.

Average exhaust gas temperature for B—735°F.

Average exhaust gas temperature for C—230°F.

[a] This calculation is based on low heat value.

ably accurate and more simplified manner by just an evaluation of exhaust gases. Proper interpretation of the results will reveal such impediments to calcination as incomplete combustion, use of excessive air, and a combination of factors that can indirectly provide pertinent heat balance data. To facilitate this approach Azbe[48] has developed a special chart, calibrated to include the principal variables, which are illustrated in Fig. 8-31.

The following example is depicted on this chart by broken lines between letters: Assume that in manufacturing high calcium quick-lime, burned with high-grade coal, exhaust gases contain 2.5% O_2 shown at A and 30.1% CO_2 at B. A horizontal line is drawn from A which intersects the vertical line from B at C. To allow for the 2.5% O_2 the "equivalent CO_2" is calculated from the formula inset in the chart at lower right or is drawn diagonally down to D at 34.2% in parallel with the guide lines. Next, a vertical line is extended up to E, where it intersects the gradient representing coal fuel. Adjusting for different fuels, observe the other fuel gradients for coke, oil, and natural gas, which compensate for their different thermal

values. Then a horizontal line is extended to the left ordinate at
F, indicating a fuel consumption of 5.8 million Btu/ton. Assume CaO
content is 95%. Then a diagonal line is drawn to G, where it bisects
the 95% vertical line. A horizontal line to H finally reveals net
fuel consumption of 5.6 million Btu. If gases had also contained CO,
allowance would be made at point B by adding 0.5% of CO_2 for
every percent of CO to redetermine "equivalent CO_2."

When excess air is used in combustion, CO_2 in the exit gases
may be reduced and O_2 content increased—or incomplete combustion
results with less CO_2 and formation of CO gas. Both situations dissi-
pate heat. That is why the determination of "equivalent CO_2" is
important, since it reconstructs theoretically the amount of CO_2 that
would be present if combustion was balanced and there was no surplus
of air. Generally, the higher the CO_2 is in the exhaust gases, the
higher is the thermal efficiency.

For dolomitic quicklime another chart would have to be calibrated
to compensate for the greater percent of CO_2 that $MgCO_3$ liberates.
With continuous gas sampling it is possible to make a series of plots
on Fig. 8-31, from which a more realistic average fuel consumption
figure can be computed.

A simple formula for determining "equivalent CO_2" is:

$$\text{Equivalent } CO_2 = \frac{100(\%CO_2 + \%CO)}{100 \cdot 4.78 \left(\%O_2 - \dfrac{\%CO}{2} \right) - \dfrac{\%CO}{2}}$$

(The more complete formula in Fig. 8-31 can also be used which
allows for other combustible gases, H_2, CH_4, and C_2H_6, if they are
present.)

Weisz[45] proposes some supplemental simple formulas to be used
in heat-balance and thermal efficiency calculations which consider
the quality of the limestone and lime—what its carbonate and oxide
content is, ratio of CaO to MgO in magnesian and dolomitic limes
and limestones, and core content, all of which have a bearing on
efficiency. He arbitrarily designates the stone-lime quality factor as
A. If the chemical analysis of lime is known, then:

$$A = \frac{\%CaO \times 1.785}{(\%CaO \times 1.785) + (\%MgO \times 2.092)} \times 100$$

1.785 is theoretical factor which converts CaO to $CaCO_3$.
2.092 is same factor converting MgO to $MgCO_3$.

(Values for MgO and CaO should not include any carbonates or
core found in the lime.)

If only the chemical analysis of the stone is known, then:

$$A = \frac{\%CaCO_3(100 - \%CaCO_3 \text{ in final product})}{\%CaCO_3 + \%MgCO_3(100 - \%CaCO_3 \text{ in final product})} \\ + \%MgCO_3 \\ \times 100$$

Weisz postulates that any core remaining in lime must be $CaCO_3$, not $MgCO_3$, regardless of the type of stone, since $MgCO_3$ dissociates at lower temperature than $CaCO_3$ and would be completely calcined long before the calcium component. This concept is reflected in the above formula.

Translating effect of lime quality on fuel consumption, the following formula is suggested:

$$\text{Millions of Btu/ton of product} = \text{millions of Btu/ton of oxides} \\ \times \frac{(\%MgO + \%CaO)}{100}$$

Instrumentation. Great strides in automation ("push-button" plants) have steadily progressed since World War II.[49] Presaging this advance is vastly improved instumentation, which, in addition to manpower savings, has contributed to fuel efficiency, capacity, and lime quality and uniformity as well as accident prevention. Instrumentation panels are seemingly being continually augmented with additional supplemental controls.

The same degree of progress applies to vertical kilns, but on an average they are still not quite as highly instrumented as rotary kilns and probably never will be, at least in the aggregate. Some of the new special kilns, like the Fluo-Solids and Calcimatic, are just as automated as the most advanced rotaries.

Improved instrumentation is probably just as responsible as superior kiln and plant design for the greater plant efficiency and the resulting higher quality and uniformity of lime that has been so manifest since about 1940. Large modern rotary kiln plants may invest $40,000 to $60,000 in such control, measuring, and indicator equipment, but of all lime capital investment costs, this is probably the simplest to justify, since savings and benefits are so tangible. They provide the flexibility in control necessary for "tailor-made" lime and facilitate the necessary adjustments to accommodate changes in types of kiln feed and fuel. Figure 8-32 displays a modern central control panel in a rotary kiln plant.

Fig. 8-32. Modern kiln instrumentation panel.

The following are most of the instrument controls utilized by modern efficient lime plants:

TEMPERATURE

1. In the kilns. *Recording pyrometers* connected to thermocouples are built into various (or critical) zones in the kiln, particularly the calcining zone (and preheater and cooler in rotaries). *Optical pyrometers* are also utilized which provide less accurate temperature measurement, but with experienced operators a satisfactory approximate check is obtained.

2. At exhaust gas outlet. *Thermocouples* measure temperature of waste gases; in rotary kilns such measurements are also made to to protect the *induced draft fan* from excessive temperature.

1. Primary air. *A ratio controller* automatically maintains the desired constant fuel-air ratio in rotary kilns, and a special controller actuates the induced-draft fan damper.

2. Secondary air. *A ratio controller* also maintains the desired proportion of primary and secondary air. *Oxygen analyzer* and *recorder* measures combustion air input through the cooler or cooling zone. *Recording draft and pressure gauges* measure the introduction of any source of combustion air quantitatively. *Anemometer* measures approximate amount of air entering vertical kiln under natural draft.

FUEL

1. Gaseous. Automatic *gas pressure and flow controller* meter adjust volume and velocity of gas injection into kilns. The same applies to liquid fuels that are gasified. *Flame failure controller* automatically shuts off gas if flame is extinguished in the kiln.

2. Solid fuel. Coal is *metered* into pulverizers and kilns, and the flow of solid fuel into the kiln can be controlled directly or with a *ratio controller*.

EQUIPMENT

1. Kilns. *Tachometer* measures rotation speed of rotary kiln, which is coordinated with *speed drive reducer*. Remote manual controls at central control panel are also employed to adjust kiln speeds. Kiln feed flow is measured and recorded and may be coordinated with rate of kiln rotation.

2. Electrical. Power is controlled with standard *voltmeters, watthour meters, recording demand meters, circuit breakers*. Special *alarms* are employed. *Electrically interlocking* the magnetic starters, motors, and conveying equipment at the kiln feed end is often practiced.

3. Motors. Special *controls* for all motors that can stop, start, change speed exist at central control panel, along with *indicator lights. Ammeters* indicate motors' load for such functions as charging, discharging, induced draft fan, cooler fan, etc.

EXHAUST

Various types of gas analyzers are employed—automatic, manual, and intermittent—to determine combustion efficiency—*continuous gas analyzers, oxygen analyzers, CO_2 and CO analyzers*, etc. *Orsat* type enjoys wide acceptance.

GENERAL

Supplemental coordinating controls, such as *multipointer indicator gauges* that may correlate differentials of primary and secondary air, kiln outlet draft, hood draft, and pulverized coal feed in rotary kilns are used. *Selector valves* provide for alternative manual control of key processing functions.

Readings are transmitted by fixed *electrical conductors*, usually to a central control panel.

In addition to complete instrumentation, two large German plants at Wulfrath and Dornap in the Rhineland utilize *industrial TV* as a further aid to plant control.[15] At strategic plant locations supervisory personnel can switch the TV to a series of channels where progress can be viewed at critical phases of the operation, like the primary crusher, stone classification plant, kiln charging and discharging locations, hydration, bagging, etc. In this manner, breakdowns and bottlenecks are detected and the resulting problems are communicated instantly throughout the plant area so that all departments can retard their rates of production until the problem is alleviated, thus minimizing down time.

Facilitating and supplementing these communications are a series of intricate *varicolored lights* that signal warnings of production bottlenecks to all plant and quarry departments. This enables mobile maintenance crews to be rushed to trouble spots for necessary repairs in minimum time. No other lime plants in the United States or elsewhere possess such an elaborate system, and it is doubtful if such installations can be justified except in very large, far-flung, integrated operations. These two German plants are the largest producers of lime in the world, as well as being large producers of commercial stone of many gradations, hydraulic limes, and portland cement, aggregating a prodigious total average annual production rate for both shipment and internal consumption of about 10 million tons. Their over-all operations encompass at least the equivalent of a half square mile.

Classification. Quicklime is commercially available in the following forms:

1. *Lump.* Sizes range from $2\frac{1}{2}$ to 12 in., but usually a more restricted size gradation, such as 3×6 in., 4×8 in., 6×10 in., etc., is provided. Without exception these sizes are derived from vertical kilns.

2. *Pebble.* Sizes range from $\frac{1}{4}$ to $2\frac{1}{2}$ in., but actual size distribution

Fig. 8-33. Small-sized pebble lime (0.5 in. average size).

is more precise, such as $\frac{1}{4} \times \frac{1}{2}$ in., $\frac{1}{2} \times 1$ in., $1 \times 1\frac{3}{4}$ in., etc. The pebble size was first introduced from rotary kilns, but now it is available from vertical kiln plants that crush their lump quicklime to smaller sizes. It is also supplied by the Calcimatic and some of the other special kilns as a primary product (Fig. 8-33).

3. *Ground*. Typical size gradation is about 100% passing a No. 8 sieve and 2 to 4% passing a No. 100 sieve. The only kiln that might produce this form of lime as a primary product is the Fluosolids. It is a secondary product from the other kilns as a result of screening off fines or grinding and classifying coarser sizes.

4. *Pulverized*. Typical size gradation is about 100% passing a No. 20 sieve and 85 to 95% passing a No. 100 sieve. Without exception this product is a secondary product from all kilns and is the direct result of pulverization and classification.

5. *Pelletized*. Pellet sizes are usually quite uniform (about 1 in.) and are made by compressing quicklime fines with special mechanical molding equipment. This is the newest form of quicklime, a postwar development. It is only produced by a few American companies and is only available in limited tonnage. It possesses no advantages over

the other forms except that it provides the most uniform physical size of all types.

Figures 8-34 and 8-35 show pebble and lump lime being conveyed. (The latter is being beneficiated by hand selection.)

Regardless of the kiln employed, after being discharged, quicklime is conveyed by belt conveyor to *screens* where the fines and undersized particles are removed. Either before screening or immediately afterward, some companies will upgrade their quality by *hand selection,* namely, removing overburned, underburned, or impure lumps of quicklime that contain visible amounts of silica on a picking conveyor by hand (Fig. 8-35). Usually two specially trained employees stationed on opposite sides of the conveyor discard the substandard

Fig. 8-34. Inspecting rotary kiln pebble lime.

Fig. 8-35. Hand selection of vertical kiln lump lime.

lime. This practice may be less costly than selective quarrying, or with marginal stone deposits it may be an essential supplemental beneficiation practice to selective quarrying in order to comply with exacting specifications.

The fines that are removed are either conveyed to the hydration plant or are sold, with or without further classification, as ground or pulverized quicklime or are converted to pellets. The amount of fines accumulated varies greatly, depending largely upon lime's resistivity to decrepitation during calcination and attrition during discharge and conveyance. Most United States plants are able to utilize all of their fines profitably in manufacturing hydrated lime (Chapter 10) and/or by purveying pulverized quicklime. The motivation behind the development of pelletized quicklime was caused by the inability of a few companies to market their fines as hydrate or quicklime. By pelletizing the fines to a marketable or usable form, wastage of quicklime is averted. High calcium lime is much more amenable to pelletization than dolomitic. When the MgO content exceeds 10%, compaction of the fines becomes complicated. A majority of American companies, however, cannot rely exclusively on waste quicklime fines as their only source of raw material for hydrated lime, and they will grind additional tonnage of lump or pebble quicklime as supplemental raw material.

A few West German companies have ingeniously solved the problem of an imbalance in demand between quicklime versus hydrate and pulverized quicklime by manufacturing *portland cement* from their surplus quicklime fines and waste clay or shale overburden from their quarries. In one such plant about 70% of the quicklime fines, No. 8 mesh to dust, is blended with about 30% clay overburden and milled into cement flour (or kiln feed). It is then sintered at 2600°F in a standard rotary kiln 7 ft. in diameter and 190 ft. in length, equipped with a grate cooler of 31 × 6 ft. that has a cement capacity of 350 short tons/day. The resulting clinker is then pulverized in the conventional manner into high-quality portland cement that meets basic specifications. This is undoubtedly the lowest-cost cement produced in the world, since there is only a 5% loss on ignition as opposed to conventional cement plants predicated on limestone that lose 20 to 25% of the kiln feed weight on ignition, largely as CO_2. Furthermore, fuel consumption is much less—reported to be only 1.65 million Btu/ton of cement clinker—since heat is not required for calcination of limestone. Based on the success of its initial cement kiln, one German company has added a second rotary kiln. Thus, the lime industry has entered the cement business on a by-product basis. The same attractive potentiality exists for a few other American and European lime plants.

Dust Collection. Dust control is a greater problem with rotary than modern vertical kilns, since recirculation of hot exhaust gases is not practiced after preheating the kiln feed. The higher capacities and smaller kiln feed are also contributory to this condition. Increasingly (and invariably in urban areas), dust-collection systems are installed, usually adjacent to the kiln stacks. One or a combination of the following dust-control systems[50, 51] is exercised, depending upon how rigorous atmospheric pollution is being abated:

1. *Large dust chambers,* in which the heavier dust particles settle out from exhaust gases of low velocity flow, are employed primarily for preliminary dust settling to be followed by more involved systems. This method is completely inadequate to employ alone in urban areas but may be satisfactory in some rural sections.

2. *Multiple dry cyclonic* types are much more efficient than item 1 but are still incapable of removing more than 75 to 85% of the lime dust. As a result, supplementary equipment is often necessary.

3. *The Cottrell precipitator,* or the electrostatic collector, will remove 95 to 98% of the lime plant dust. Its principal deterrent, balanced against this high collection efficiency, is high capital equipment

cost. Some such collectors are designed with preliminary collection chambers.

4. *Wet scrubbing towers* of many different designs are employed and effect high dust removal of about 95% or more. They are less costly to install than item 3. Settling basins are often employed to concentrate the waste solids that are removed; some plants feed the waste slurry into sewers or lagoons.

5. *Miscellaneous.* Other new methods are being introduced, such as filter collection in *glass bags*[52] and *sonic* methods. Both bottom and top inlet glass bags are used with dampers that collapse the bag, discharging the dust to bins below. Bag life is two to two and one-half years. Although the bag method is costly to install, it appears very promising because of its high efficiency and low maintenance.

Dust losses in rotary kiln operations generally range between 2 and 8% by weight of the limestone charge. In addition, there is ash and unoxidized carbon from the coal and spalled fines from the kiln linings. There is, of course, some variance in the analysis of dust, but the following might be regarded as typical with a high calcium lime, fired by pulverized coal in a rotary kiln:

Total oxides (CaO and MgO)	37%
Total carbonates ($CaCO_3$ and $MgCO_3$)	25
CaO chemically combined in compounds, including some $CaSO_4$	10
SiO_2	10
R_2O_3	5
Carbon	12
SO_3	1
Total	100%

The lack of uniformity of this heterogeneous dust, even from the same source, tends to render this waste by-product difficult to merchandize at any price. The wet-process sludge material from the collector has the added disadvantage of requiring drying, a cost which cannot possibly be absorbed in this usually valueless material. As a result, it is usually accumulated in segregated waste piles or lagoons on the plant premises and ultimately may constitute a disposal problem.

Dead-burned Dolomite Production. In the United States rotary kilns are almost exclusively employed for the production of dead-burned dolomite (refractory lime).[53] The sole exception is one plant that utilizes an ore-sintering machine with a down-draft traveling grate. In Europe rotary kilns also predominate, but not to the same extent.

washing, and dewatering steps, a major loss occurs in the kiln in which lime and limestone fines are swept out with the hot, turbulent exhaust gases. Because of the greater volume of fines lost in this manner, atmospheric pollution is a more vexatious problem, and as a consequence recovery kilns generally require a more elaborate dust-control system before the clarified exhaust gases are vented up the chimney. Combined losses of kiln feed in recovery plants are reported as high as 10 to 15% $CaCO_3$ equivalent as compared to 2 to 8% in the average rotary plant. This factor, together with higher fuel and equipment requirements, probably increases the average total cost of recovery lime by at least $1.50/ton over commercial lime on a comparable basis, excluding transportation cost considerations.

Rotary kilns are almost invariably employed in such plants and must be regarded as the preferred kiln. The only exception would be the Fluo-Solids kiln, which has been employed with some success. Vertical kilns would be incompatible with this type of raw material.

Hydraulic Lime. Heretofore, in this chapter and in Chapter 6 most of the discussion has been devoted to calcination of pure (or relatively pure) forms of carbonate stone, shell, and precipitated sludge. Impure carbonate materials are not utilized in modern lime manufacture—only in the manufacture of hydraulic lime and cement. No text on lime is complete without at least a cursory inclusion of this chemically impure form of lime, which has only one end use, namely, as structural lime (p. 398).

In the United States hydraulic lime has never flourished, and in 1964 there was only one such manufacturer in existence—in Virginia. However, in Europe, South America, and parts of Asia it has been produced extensively for many centuries, although in the past few decades its production has ebbed since portland and masonry cement and lime-cement mortars have captured some of its traditional market.[58]

Chemically, hydraulic lime might be broadly classified as being intermediate between lime and portland or natural cement. It is derived from argillaceous limestone and contains after calcination a rather high percentage of calcium silicates that possess varying, but generally mild, hydraulic properties, meaning that it will harden under water. At the same time, unlike cement, it possess considerable free lime (or CaO + MgO), so that the product slakes in water. Generally, CaO is valued much more than MgO, but this concept is subject to controversy. Thus, the rock from which it is made possesses much more silica (and usually alumina and iron too) than limestone for lime manufacture, but still less impurities than contained in the port-land cement raw material mix or so-called cement rock.

The degree of hydraulicity[59] of these limes varies considerably so that they have been classified into three different grades as follows: (1) *feebly hydraulic lime;* (2) *moderately hydraulic lime;* (3) *eminently hydraulic lime,* which is also called *Roman lime* and which approaches hydraulic and natural cements in hydraulicity and strength imparting qualities—however, compared to modern portland cement these properties are relatively feeble. The Germans and French recognize these three groupings in their building material specifications.

Traditionally, hydraulic limes have been evaluated by the *Cementation Index* theory, which is based on certain scientific and empirical assumptions, such as:

1. The hydraulic properties are imparted by the formation of compounds of calcium and magnesium with silica, alumina, and iron.
2. The silica combines with lime molecularly to form tricalcium silicate ($3CaO \cdot SiO_2$).
3. Alumina combines with lime to form dicalcium aluminate ($2CaO \cdot Al_2O_3$).
4. Magnesia reacts molecularly the same as lime, except at a slower rate.
5. Iron oxide has the same equivalent molecular reaction as alumina.

There is some disagreement on these assumptions, but mainly pertaining to the relative hydraulic values of these compounds. So the following Cementation Index formula[6] may be modified by some authorities slightly:

$$\text{C.I.} = \frac{2.8 \times \%SiO_2 + 1.1 \times \%Al_2O_3 + 0.7 \times \%Fe_2O_3}{\%CaO + 1.4 \times \%MgO}$$

Assume the following hydraulic lime analysis:

SiO_2	18.47
Al_2O_3	5.73
Fe_2O_3	3.29
CaO	68.19
MgO	2.66

Then:

$$\text{C.I.} = \frac{(2.8 \times 18.47) + (1.1 \times 5.73) + (0.7 \times 3.29)}{(68.19) + (1.4 \times 2.66)} = .839$$

Based on this formula the three groups of hydraulic limes are arbitrarily classified by Cementation Index as follows:

Feebly hydraulic—.30 to .50

Moderately hydraulic—.50 to .70

Eminently hydraulic—.70 to 1.10

Thus, the product in the above example is eminently hydraulic (.839). The greater hydraulic value ascribed to silica as compared to alumina and iron is clearly evident in this formula. An impure limestone of low silica but high in other impurities would be of little or no value since this is fundamentally a lime–silica reaction. The Al_2O_3 and Fe_2O_3 are of doubtful hydraulic value. They simply act as fluxes, facilitating the desired calcium silicate combination and disposing of further excess lime by also uniting with it to form aluminates and ferrites. The lime–silica ratio is the *controlling factor*.

The manufacture of these limes is quite similar to what has already been described. Lumps of impure limestone are charged into vertical kilns. (No plant is reported to employ rotaries.) The calcining temperatures are usually 1 to 200°F higher than in ordinary shaft lime kilns, since the object is to combine chemically as much of the impurities with the lime as possible, and the higher temperature stimulates these reactions. Because of this fluxing effect the lumps resemble clinkers after calcination. However, if there is still 10 to 15% free lime remaining in the clinker, the hard sintered lumps will disintegrate into a powder on addition of just the minimum amount of water to satisfy its hydration affinity. Retention of clinkers and water in silos for several days may be necessary to complete hydration, which is usually relatively sluggish. If there is little or no free lime present, then these clinkers will be unaffected by water and the hard lumps will have to be mechanically pulverized, forming a product akin to natural cement. As a generalization, the higher the Cementation Index, the less free lime is available—or conversely, the greater the free lime content, the lower the Cementation Index or hydraulic value. The free CaO content is only critical in the eminently hydraulic limes since in the feebly and moderately hydraulic materials only 20 to 60% of the total CaO is chemically combined, leaving ample amounts to hydrate and effect comminution through expansion caused by the heat of hydration.

In some highly hydraulic French plants a portion of the clinkers are almost totally chemically combined lime compounds, which will not slake. These lumps, called "Grappiers," are removed by screening from the hydrated material and are pulverized mechanically. They have a much higher Cementation Index of 1.10 to 1.70 and have earned the name of Grappiers' cement. Some of the Grappiers is then blended into the hydraulic lime, upgrading its hydraulicity, or it is sold unmixed for use as a natural cement. In the latter part of the nineteenth century, France exported this material to the United States.

Table 8-6 presents typical chemical analysis of hydraulic stones of varying hydraulicity and origin. Essentially they are derived from very impure limestones with carbonate contents ranging between 78 and 92%, none of which would be suitable for ordinary lime manufacture. For the hydraulic material the most satisfactory stone would be in the 78 to 82% carbonate range, with a maximum amount of silica. In contrast, "cement rock" is 70 to 72% carbonate.

Table 8-7 provides analyses of calcined hydraulic lime before slaking. When highly hydraulic types are slaked, the resulting pulverulent product only absorbs about 7 to 10% of water as compared to about 25% with pure limes. Note the wide range in analyses, even among hydraulic limes of the same grade.

In spite of elevated calcination temperatures, this lime frequently contains core which is removed in milling the hydrate to a consistency of 99.5% passing a No. 50 mesh sieve. Compared to conventional quicklime at similar operating efficiencies, hydraulic lime probably costs less to manufacture. The higher burning temperatures coupled often with more complex milling are more than offset by the lower loss on ignition, and, because of its lower carbonate content, a correspondingly lower theoretical heat requirement for calcination

Table 8-6. Analyses of hydraulic lime rock

Component	(1)	(2)	(3)	(4)	(5)
SiO_2	5.00	7.40	10.30	16.35	11.03
Al_2O_3	$\lbrace$ 2.23	2.70	0.65	1.00	3.75
Fe_2O_3	2.00	5.30			5.07
CaO	48.65	40.82	48.30	43.85	43.02
MgO	1.86	4.52	0.30	0.55	1.34
CO_2 and H_2O	40.26	37.06	40.45	38.25	35.27
Cementation Index[a]	0.356	0.581	0.739	1.05	0.91

(1) Feebly hydraulic lime from Holywell, England.
(2) Moderately hydraulic lime from Horb, Wurtemburg, Germany.
(3) Eminently hydraulic lime from Malain, France.
(4) Eminently hydraulic lime from Le Teil, France.
(5) Eminently hydraulic lime from Hausbergen, Germany.
[a] Speculated values based on stone analysis.

Table 8-7. Analyses of hydraulic limes (before slaking)

Component	German 1	German 2	French 3	French 4	French 5[a]
SiO_2	7.60	11.95	21.7	22.59	31.10
Al_2O_3	11.60	4.25	1.8	2.63	4.43
Fe_2O_3	0.96	8.52	0.6	0.84	2.15
CaO	79.09	65.73	74.0	65.62	58.38
MgO	—	7.25	0.7	1.54	1.09
Alkalies (Na_2O, K_2O)	—	—	—	—	0.94
SO_3	—	—	—	—	0.60
CO_2	—	—	—	—	1.28
Cementation index	0.431	0.58	0.84	0.99	1.56

[a] Grappier's Cement.

(Values above left blank do not mean that the lime did not contain any of the chemical constituent itemized. No values were reported, but undoubtedly each lime contains at least a trace, usually much more, of each component listed.)

results per ton of product. But the major factor is that when an eminently hydraulic lime is made, a ton of impure limestone only shrinks to about 0.65 ton instead of about 0.55 ton for normal lime. So the heat requirement is generally about 25 to 30% less than vertical kiln lime, assuming comparable efficiency.

A major problem confronting most hydraulic manufacturers is lack of uniformity in the finished product, namely, shipping a lime with a Cementation Index of 0.92 one day and 0.70 the next, etc. The chemical analyses in impure deposits frequently changes abruptly from one ledge and strata to another. Blending of stone to secure a composite analysis alleviates this problem to some extent but can never be pursued with the same flexibility of portland cement plants. Furthermore, the intensive analytical testing that it entails is costly. This chronic problem of quality control, more than any other factor, has forced hydraulic lime to lose position primarily to portland cement and secondarily to pure ("fat") lime, both of which have beneficiated their quality and uniformity.

Selective Calcination. Because of the differential in dissociation temperatures of $CaCO_3$ and $MgCO_3$, it is possible to selectively calcine dolomite so that the higher-level temperature component, $CaCO_3$, remains substantially uncalcined, as illustrated in the equation below:

$$2CaCO_3 \cdot MgCO_3 + \text{heat (about } 1600°F)$$
$$= 2CaCO_3 + 2MgO + 2CO_2\uparrow$$

Precise temperature control is essential for such calcination. The resulting MgO is chemically more reactive than in normal dolomitic quicklime, since with the subnormal temperature it is necessarily soft-burned. Such a pseudocalcined product has been commercially manufactured in Europe and by a few American companies on an intermittent small scale, but although the product possesses interesting properties, it has never enjoyed a widespread or continuous market. This is also the commencement of a process to produce magnesium hydroxide; MgO is slaked and is then removed from $CaCO_3$ as a hydroxide.

A patented process[60] for selectively calcining ground dolomite in the above manner in which steam is employed as an alleged catalyst has been reported. A laboratory electric induction furnace and a small simulated rotary kiln, adapted from a rotary drier, were employed in calcining $\frac{1}{8} \times \frac{1}{4}$ in. and No. 100 mesh dolomite. Considerable calcination occurred at 600°C, and complete calcination was consummated at 650°C when steam was injected; in contrast, in an atmosphere of air, there was only a trace of calcination. The reported catalytic effect of steam is illustrated in Fig. 8-36.

Pelletized Calcination. One completely revolutionary method of making lime has been reported in 1965 from Oamaru, New Zealand.

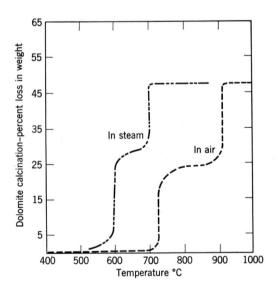

Fig. 8-36. Effect of steam on calcination temperatures with dolomitic limestone.

A very pure, soft coralline limestone, which would hopelessly decrepitate in conventional kilns, is utilized in the following unique manner. The stone is pulverized and the resulting limestone fines are dried and then ball-milled into a powder with a predetermined amount of pulverized coal. The lime-coal "flour" is conveyed to a large silo. From there it is fed into a flat revolving drum, causing pellets to form. Finally, when the pellets accrete to about $\frac{1}{2}$–$\frac{3}{4}$ in. diameter, they are fed, as kiln feed, into the kiln, forming an inverted cone. The pellets are uniformly spherical and quite hard. Heat is applied under precise control; if applied too rapidly, steam will erupt from the pellets, causing them to rupture. Lime gravitates through a series of grates at the bottom of the kiln onto an oscillating plate for discharge.

In spite of requiring meticulous control with elaborate instrumentation, it is reliably reported that lime of high quality and uniformity is produced at reasonably good thermal efficiency. It is believed that this kiln (35 ft. high and 7 ft. inside diameter) can attain a capacity of 100 tons per day. Although this kiln represents a relatively high capital investment (comparable to rotaries on an equivalent basis), it might prove an economical answer to the spalls problem and for other types of soft limestone.

Manufacturing Costs. It should be obvious from the foregoing description of lime burning that the cost of American lime manufacture varies widely. Chapter 4 (p. 70) discussed the widespread cost of obtaining stone or *kiln feed*, an important segment of over-all manufacturing costs.

Fuel cost traverses just as broad a range in both unit cost of fuel and thermal efficiency—both have a spread of about 100%. Of course, thermal cost disparity may be mitigated. With some plants purchasing the highest-cost fuel, this disadvantage is diminished by high thermal efficiency; other plants that enjoy the lowest-cost fuel may be wasteful in thermal units; still others possess high fuel costs and mediocre efficiency, a very unfavorable combination that is only tenable because of their isolated location and protection afforded by lofty freight rates.

Labor costs and fringe benefits for comparable job classifications also possess about 100% variance. Again, some plants with highest wages counterbalance this disadvantage with the ultimate in mechanization and automation; others with low wages and limited automation and appreciable hand labor may conclude that the capital investment for the requisite labor-saving equipment is unwarranted.

Of the remaining costs *depreciation* is probably the greatest factor, at least for all relatively new plants. The 350 to 400% postwar inflation of manufacturing equipment interjects a formidable capital equipment or plant amortization schedule. (See p. 432.) Other principal costs are *interest, refractory, electric power, maintenance supplies and parts, testing* (quality control), *insurance, taxes,* and *administration.*

Under the circumstances any arbitrary estimate of over-all manufacturing costs would be irresponsible. In the United States no statistical study of this kind has ever been undertaken, and what very sparse, fragmentary cost information that does exist is probably misleading and outdated. Consequently, the only reasonable solution would be an estimate of approximate ranges or magnitude in costs as of 1964 for the principal cost categories. The following figures do not necessarily represent minimal and maximal values, but they do indicate that great breadth prevails:

Costs per net short ton of quicklime

	Minimum	Maximum
1. Kiln feed (*a*)	$1.30	$5.00
2. Fuel (*b*)	1.50	4.50
3. Labor (*c*)	1.25	3.75
4. Miscellaneous (*d*)	2.00	3.00
Total	$6.05	$16.25

a. Based on $.65 to $2.50/ton of stone and assuming 2 tons of stone/ton of lime to allow for stone losses; included are plants purchasing their stone from outside sources.
b. Based on all types of fuels—natural gas, producer gas, coal, coke, and oil; widely varying thermal efficiencies of from about 4.5 to 12 million Btu/ton; and extensive breadths in unit fuel costs.
c. Based on all labor except supervisory, engineering, and chemists (salaried employees).
d. Includes interest, depreciation, insurance, taxes, power, refractory, maintenance supplies and parts, quality control and testing, and administrative. (Sales costs are excluded.)

It should not be construed from the above that *any* plant has the minimum cost indicated because they probably do not, but it is plausible that a few plants may have the maximum cost—mainly small, old plants that are captive producers. The vast divergence in total costs is far more pronounced than in other industries, like chemicals and cement. In fact, it is doubtful if many industries have as high a *percentage spread* in costs as lime. Generally unit costs decline as output increases, and larger plants enjoy substantially lower costs than smaller plants—as well as lower proportionate capital investment. But there are notable exceptions. Generally operation at 90 to 95% of capacity stimulates the lowest cost. At 100% capacity plants tend to push their plants to the limit, causing greater attrition on refractories and equipment, leading to increased kiln relining and general maintenance. Competitive relationships are discussed on p. 430. With regard to western Europe it is the writer's opinion that their 1964 lime costs are not much different from the United States, probably only 10 to 20% less on an average.

The last such lime cost estimate ventured in a reference text for the United States was by Eckel[6] for the 1903–1913 era. He found a similarly wide range in total costs then too of $1.20 to $2.90/ton. This included costs for kiln feed of $.50 to $.90; fuel of $.30 to $.75; labor of $.25 to $.80 (all figures per ton of lime). It is apparent that total modern costs are not nearly as inflated as various individual cost categories, such as (in 1903–1913) labor of $.20/hour, coal (delivered) of $1.25/ton, etc. Inflation of the latter factors is about eight-to tenfold. Plant amortization would be more than tenfold.

A more recent (1948) postulation on cost limited to vertical kilns was by Lacey.[61] He outlined how costs of vertical kiln lime could be reduced from $8.50 to $6.27/ton or to $5.55/ton by modern plant mechanization. However, considerable subsequent inflation tends to obscure the significance of such figures.

Lime Plant Safety. Unless a lime company establishes a constructive plant-wide accident prevention program and pursues it unremittingly, it can be a dangerous environment in which to work. The widely varying lime safety records of United States plants are usually in direct proportion to the amount of effort and emphasis that is placed on safety from top management through all the production echelons down to the foreman. Some companies have maintained consistently superb safety records. Probably the all-time international safety record was established by Ash Grove Lime and Portland Cement Company at their Springfield, Missouri plant from 1949 to 1962 of an amazing 4634 consecutive days without a lost-time accident. About ten other United States plants have operated 2 to 3000 consecutive days without a lost-time injury.

Since 1935 the U. S. Bureau of Mines has conducted a voluntary annual lime plant safety contest in which The National Lime Association participates by rendering awards to all injury-free plants. Figure 8-37 illustrates the results of this industry-wide safety program during

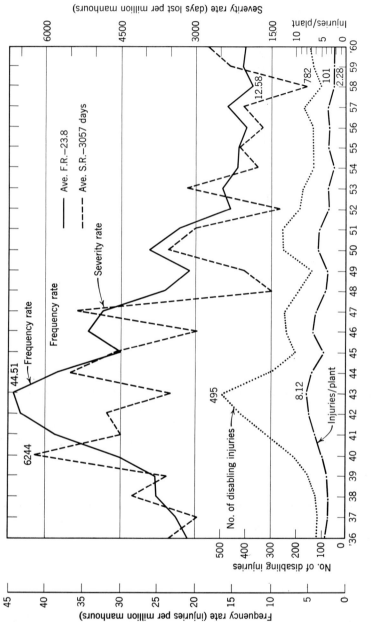

Fig. 8-37. Disabling injury experience of American lime industry (1936–1960).

most of the contest years (1936–1960).[62] It reveals that except for the World War II years, when plants were forced to use many inexperienced workers and labor supply was short, that the over-all safety record has improved, possibly stimulated by the contest. Since 1955 the industry frequency rate (in injuries/million manhours) has ranged around 12 to 15; severity rate (in days lost/million manhours) has averaged about 1700; lost-time injuries per plant has averaged between 2.2 to 2.5. Sixty-four fatalities occurred during these twenty-five years, with only one year—1958—being death free.

Causes of accidents were widespread, but the following injury categories were the most common:

Cause	*% of Total*
1. Falls of persons	15.5
2. Burns (including lime)	14.0
3. Handling materials (other than rock)	12.0
4. Haulage	10.2
5. Machinery	7.6
6. Falling objects	7.5
7. Hand tools	7.4
8. Rock slides	7.2
9. Handling rock	5.1
10. Miscellaneous	13.5

Of the above tabulation, falls of persons was by far the worst in severity rate, with a staggering total of 105 days lost per injury. Contributing to this were eight fatalities and four permanent partial disabilities. Six of item 2 involved loss of sight in one eye from lime burns. Back strain disabilities from improper lifting predominated the handling materials classification. The average injury represented forty-two days of lost time, but some of the injuries were severe, one permanent and nine permanent partial disabilities (but no deaths).

For all injuries reported, regardless of the cause, the following parts of the anatomy were most frequently affected, including percent of total:

Eyes	19.3%	Hands	11.3%
Back	17.4	Feet	10.8
Legs	16.4	Arms	7.5

The National Lime Association has developed recommended safety standards.[63] Accidents are a definite but largely avoidable cost of production. In 1960 the National Safety Council calculated that lost-time accidents (for all industry) cost $2256 as an average per injury. It is believed that the average cost in the lime industry is higher than this, so it is more true than trite to ennunciate that a lime-plant safety program pays.

REFERENCES

1. *Pit & Quarry* (on rotary kiln)—5/50, p. 96; 8/50, p. 65; 5/51, pp. 104, 118; 11/51, p. 80; 5/52, p. 108; 12/52, p. 76; 5/53, p. 110; 3/55, p. 136; 5/55, p. 116, 144; 5/56, pp. 107, 140; 5/57, p. 101, 138; 6/57, p. 100; 7/57, p. 178; 12/57, p. 144; 5/58, p. 120; 11/58, p. 122; 5/59, pp. 122, 132, 140, 146, 160, 168; 5/60, p. 104; 6/60, p. 96; 8/60, p. 96; 1/61, p. 146; 5/61, p. 104, 120, 134; 11/62, p. 108; 3/63, p. 84, 5/63, pp. 108, 148; 5/64, p. 92.
   (on vertical and other kilns)—4/44, p. 64; 5/47, p. 95; 1/48, p. 110; 5/49, p. 104; 5/55, p. 78, 92, 122; 2/56, p. 108; 5/56, p. 92; 5/57, p. 116; 8/57, p. 62; 5/58, p. 128; 7/58, p. 123; 5/60, p. 104; 11/60, p. 93; 2/62, p. 97; 5/62, pp. 126, 152; 10/62, p. 92; 5/63, p. 122, 143; 5/64, pp. 116, 136.
2. *Rock Prod.* (on rotary kilns)—2/51, p. 108; 12/51, p. 99; 1/53, p. 150; 2/56, p. 57; 6/56, p. 101; 10/56, p. 84; 11/56, p. 37; 8/58, p. 50; 11/58, p. 90; 5/59, p. 120; 2/61, p. 53; 7/64, p. 72, 81.
   (on vertical and other kilns)—11/49, p. 66; 10/51, p. 125; 2/52, p. 117; 7/54, p. 66; 10/55; p. 126; 12/56, p. 62; 1/57, p. 122; 5/57, p. 94; 10/57, p. 95; 11/57, p. 107; 2/58, p. 172; 7/58, p. 94; 9/58, p. 134; 4/59, p. 94; 8/60, p. 129; 3/64, p. 95.
3. V. J. Azbe, *The Theory & Practice of Lime Manufacture* (A compendium of *Rock Prod.* articles by the author from 1922–1945); Azbe Engineering Corp. (Clayton, Mo., 1946).
4. *Pit & Quarry Handbook,* Pit & Quarry (Chicago, 1963).
5. A. B. Searle, *Limestone and Its Products,* E. Benn, Ltd. (London, 1935), pp. 270–394; 404–447.
6. E. C. Eckel, *Cements, Limes, & Plasters,* John Wiley, 2nd Ed. (1922), pp. 99–113; 172–199.
7. N. V. S. Knibbs, *Lime & Magnesia,* E. Benn. (London, 1924), pp. 154–187; 196–201.
8. V. Azbe, *Fundamental Mechanics & Methods of Control of Calcination,* ASTM Symposium on Lime, (Mar. 8, 1939).
9. V. Azbe, *Theory & Practice of Lime Manufacture, Rock Prod.* (1953 series) —Feb., p. 100; Mar., p. 102; Apr., p. 138; May, p. 84; July, p. 80; Sept., p. 100; Dec., p. 111.
10. R. Priestley, *Vertical Kiln Advantages,* Chem. Enging. **69** (Nov. 1962), p. 330.
11. H. Erasmus and H. Leuenberger, *U. S. Patent* 2,933, 297 *Lime Kiln* (to Union Carbide Corp., Apr. 1960).
12. W. Trauffer, *Pit & Quarry* (May 1959), p. 122.
13. J. Wuhrer and G. Rademacher, Chem. Ing. Tech. **28** (1956), p. 328.
14. G. Pohl, *Tonind. Ztg.* **77** (1953), p. 165.
15. R. Boynton, *European Lime Manufacture,* Pit & Quarry (May 1960), p. 124.

16. W. Bauer, *How to Control Heat for Calciners,* Chem. Enging. (May 1954).
17. V. Azbe, *Rotary Kiln Performance Evaluation & Development, Rock Prod.* (Feb. 1953–Oct. 1955).
18. V. Azbe, *A Super Rotary Kiln, Rock Prod.* (July, Aug., Sept. 1964).
19. I. Warner, *Elements of Efficiency in Lime Calcination, Rock Prod.* (Jan. 1952).
20. L. Niems, *U. S. Patent* 2,858,123, "Apparatus for Cooling & Calcining" (to Marblehead Lime Co.) (Oct. 1958).
21. C. Ellis, *Various Methods of Improving the Efficiency of Rotary Lime Kilns,* Nat'l. Lime Ass'n. Operating Meeting (Sept. 29, 1960).
22. C. Furnas, *The Rate of Calcination of Limestone,* Ind. Enging Chem., **23** (1931), p. 534.
23. Anon., "Produces Chemical Lime with 20% Less Fuel," *Chem. Proc.* (Jan. 1961).
24. F. Kowhanowski, "Grate Kiln Systems for Lime Production," *Pit & Quarry* (May 1963), p. 143.
25. R. Kite and E. Roberts, "Fluidization in Non-Catalytic Operations," *Chem. Enging.* (Dec. 1947).
26. Anon., Install Fluo Solids System, Rock Prod. (Dec. 1956), p. 62.
27. B. Herod, "The Calcimatic, a New Concept in Lime Burning," *Pit & Quarry* (Nov. 1963).
28. Anon., "Ash Grove with Pacific Northwest," *Pit & Quarry* (June, 1964), p. 138.
29. N. Rockwood, *Industrial Minerals & Rocks,* AIME (1949), pp. 467–512.
30. R. Boynton and K. Jander, *Encyclopedia of Chemical Technology* **8**, Interscience (New York, 1952), pp. 346–382.
31. R. Boynton and K. Gutschick, *Industrial Minerals & Rocks,* AIME (1959), pp. 497–519.
32. H. Utley, "Kennecott Copper Operates Own Lime Burning Facility," *Pit & Quarry* (May 1964), p. 136.
33. B. Corson et al., *U. S. Patent* 3,022,988, "Material Treating Device" (to G. & W. H. Corson, Feb. 1962).
34. J. Maduras, "A Modern Lime Burning System," *Cement, Lime & Gravel* (May 1962), p. 131.
35. Anon., *Rock Prod.* (Feb. 1964), p. 106.
36. News releases, private correspondence, & advertising claims in *Zement, Kalk, Gips.*
37. Private correspondence (author).
38. W. Trauffer, *Pit & Quarry* (May 1961), p. 134.
39. H. Heiligenstaedt, *U. S. Patent* 2,948,521, "Apparatus for Heating Cross Stream Shaft Furnace, Particularly for Calcination" (to Rochling'sche Eisen & Stahlwerke, Aug. 1962).
40. V. Azbe, *U. S. Patent* 2,994,521, "Terminal Calcining Kiln" (to Azbe Corp., Aug. 1961).
41. Nat'l. Lime Ass'n., *Chemical Facts,* 2nd Ed. (1964), p. 28.
42. H. Robertson, "Refractory Types & Applications," *Iron & Steel Eng.* (Jan. 1954), p. 86.
43. F. Moyer, *U. S. Bu. Mines I.C. 7174* (June 1941).
44. W. Emley, "Manufacture of Lime," *U. S. Bu. Standards Circ. 337,* 104 pp. (1927).

45. W. Weisz, "Kiln Performance Charted from Studies of Gas Analysis," *Rock Prod.* (Mar. 1952), p. 88.
46. V. Azbe, "Fuel Oil & Its Application to the Combustion System," *Rock Prod.* (June 1954), p. 129.
47. I. Warner, *NLA Survey,* Nat'l Lime Ass'n Operating Meeting (Oct. 10, 1956).
48. V. Azbe, *Rotary Kiln Evaluation & Development, Rock Prod.* (Mar. 1954).
49. C. Cox, "Instrumentation at National Gypsum Co. Kimballton Plant," *Pit & Quarry* (Aug. 1948), p. 74.
50. J. Dallavalle, "Dust Collector Costs," *Chem. Eng.* (Nov. 1952), p. 177.
51. J. Orth, *Nonmetallic Minerals Proc.* **3,** 1 (1962), p. 14.
52. W. Ballard, "Glass Bags—from Batch to Bag-House," *Rock Prod.* (Oct. 1962), p. 69.
53. K. Gutschick, "Basic Refractories," *Pit & Quarry* (May 1953), p. 95.
54. H. Lee, "Refractories from Ohio Dolomite," *Ohio State U. Enging. Expt. Sta. News* **XIX,** no. 2 (Apr. 1947).
55. Anon., "U. S. Gypsum Calcines Lake Ponchartrain Clam Shells," *Rock Prod.* (July 1964), p. 72.
56. W. Shearon, Jr., "Oyster-Shell Chemistry," *Chem. & Eng. News* **29** (July 30, 1951), p. 3079.
57. W. Trauffer, "Virginia Paper Mill Operates Highly Instrumented New Kiln," *Pit & Quarry* (June 1953), p. 101.
58. N. Rockwood, "Hydraulic Cements," *Rock Prod.* (Jan. 1948).
59. M. Roberts, "Constitution of Hydraulic Lime," *J. Am. Ceram. Soc.* **39** (Sept. 1956), p. 182.
60. W. MacIntire and T. Stansell, *I. & E. Chem.* **45** (July 1953), p. 1548.
61. G. Lacy, "The Cost of Lime," *Rock Prod.* (Sept. 1948).
62. K. Gutschick, "25-Year Study of Lime Plant Injuries," Nat'l Lime Ass'n Operating Meet. (Oct. 6, 1961).
63. Nat'l Lime Ass'n Bull., "Safety Standards for the Lime Industry" (1955).
64. J. Orth, "Upgrading Limestone & Lime Products by Compaction," Non-met. Min. Proc. (Mar. 1962), p. 14.

CHAPTER NINE

Theory of Lime Hydration

The most prodigious tonnage of quicklime (as such) and that which is converted into commercial hydrated lime is actually consumed for hundreds of uses in a classical *solid–liquid phase reaction* by either dry or wet hydration (slaking with varying amounts of excess water) to produce a dry, pulverulent hydrate or a putty, slurry, milk-of-lime, or saturated hydrate solution. In the aggregate this exceeds the vast tonnage of quicklime applied in dry, thermal processes, such as sintering, smelting, fluxing, etc. Consequently, hydration or the reaction of lime and water is extremely important, both as a method of application of quicklime and in the manufacture of a commercial product, hydrated lime.

Technically the terms "hydration" and "slaking" are synonymous and interchangeable. Yet based on popular connotation hydration yields a *dry* powdered hydrate, whereas slaking involves more water, producing *wet* hydrates. All phases will be fully discussed.

Accompanying hydration is a *strong exothermic reaction,* and it is surprising that the appreciable heat of hydration, 490 and 380 (both Btu/lb. of quicklime) for high calcium and dolomitic types respectively, has not been utilized commercially to a greater extent. This means that the heat of hydration of 1 lb. of pure quicklime is sufficient to heat 2.3 lb. of water from 0°F to boiling point (212°F)—or can heat 3.4 lb of water (3.3 pints) from room temperature, 70°F, to 212°F. Including captive and regenerated limes, there are about 12 million tons of lime slaked annually in the United States which theoretically evolve 1.0 to 1.15×10^{13} Btu/lb. of heat in the aggregate or the heat equivalent of 350,000 to 400,000 tons of a high-grade bituminous coal (14,000 Btu/lb. heat value). This, of course, presupposes 100% pure lime and 100% heat generation efficiency, which is impossible. But even at 50 to 75% efficiency the magnitude of this heat total is staggering. An estimated 98 to 99% of this heat is dissipated into the atmosphere. As a result, the potential value of this wasted heat offers a challenge to the research ingenuity of chemists and engineers to harness it for useful purposes as well as to utilize this property in new applications. This characteristic appears to have been generally overlooked or ignored.

As further evidence of this pronounced heat generation, wooden box cars carrying bulk quicklime have ignited and burned up as a result of leaks into which rain penetrated. The macabre application of quicklime by gangsters in obliterating the corpus delecti in a pit by dumping quicklime, followed by water, onto the corpse of one of their victims is well documented—as well as the sanitary uses by farmers in disposing of the carcasses of dead animals and refuse from privies in a like manner.

Regardless of the function that hydrate performs, water serves strictly as a diluent, as a convenient vehicle for application, and as the necessary solvent or aqueous medium to release the calcium, magnesium, and hydroxyl ions in many chemical reactions. Other than this, it is valueless so that these slaked forms of lime are simply dilutions of the most concentrated form—quicklime. The value of the hydrates is predicated on the lime solids or oxide content ($CaO + MgO$), which varies tremendously, depending upon the amount of water added.

Chemical Reaction. The hydration of both high calcium and dolomitic quicklime is actually one stage of a reversible reaction, since dehydration, effected by subsequent heating, will recreate quicklime and volatilize the water of hydration as vapor, as displayed in the chemical equations below:

High calcium hydrate

$$\overset{56}{CaO} \text{ (h.c. quicklime)} + \overset{18}{H_2O} \rightleftharpoons \overset{74}{Ca(OH)_2} \text{ (h.c. hydrate)} + heat\uparrow$$

$$\overset{74}{Ca(OH)_2} \text{ (h.c. hydrate)} + heat \rightleftharpoons$$
$$\overset{56}{CaO} \text{ (h.c. quicklime)} + \overset{18}{H_2O} \text{ (vapor)}\uparrow$$

Normal dolomitic hydrate

$$\overset{96.4}{CaO \cdot MgO} \text{ (dol. quicklime)} + \overset{18}{H_2O} \rightleftharpoons$$
$$\overset{74.1 \quad 40.3}{Ca(OH)_2 \cdot MgO} \text{ (dol. hydrate)} + heat\uparrow$$

Highly hydrated dolomitic lime

$$\overset{96.4}{CaO \cdot MgO} \text{ (dol. quicklime)} + \overset{36}{2H_2O} + pressure \rightleftharpoons$$
$$\overset{74.1 \quad 58.3}{Ca(OH)_2 \cdot Mg(OH)_2} \text{ (dol. hydrate)} + heat\uparrow$$

These equations demonstrate that high calcium quicklime slakes much more readily than dolomitic. The dolomitic, because of its hard-burned MgO component (explained on p. 136), usually requires pressure for complete hydration. Some hard sintered types will resist complete hydration unless subjected to an inordinate amount of pressure, temperature, and time. It is possible, however, to produce a relatively softburned dolomitic quicklime from pure dolomite by calcining small-sized stone (-1-in. size) at low calcination temperature and under rigid time-temperature conditions that will substantially hydrate at atmospheric pressure. But in all instances high calcium quicklimes, even those that are hardburned, slake with greater facility, hydraulic or very impure limes excepted.

When exposed to water, regardless of form (liquid, steam, moist vapor, or ice), quicklime exhibits a strong affinity for moisture, adsorbing it into its pores. Because of this inherent hygroscopicity a chain reaction occurs, which appears to be simultaneous with rapid slaking limes. As the water penetrates into the surface pores, heat of hydration is triggered. This, in turn, exerts great internal expansive force in the lime particle and causes it to fracture, shatter, and then disintegrate completely into countless microparticles, either as crystalline "dust" or as a colloidal suspension, the difference contingent on the amount of water added. With a lump of reactive lime, it visually appears to be a miniature cataclysmic, volcano-like eruption. Accompanying this reaction is steam, evaporated water from the heat of hydration, and a hissing noise owing to the turbulence generated. Invariably the resulting hydrate is whiter than the quicklime from which it is derived. The assimilated water is *chemically combined* into one distinct hydroxide molecule, a *strong base*, as opposed to other hydrated chemicals, like Glauber's salt, in which the moisture is in the form of water of crystallization and molecularly only loosely adhering to the salt molecule.

Water Content. From the above equations the theoretical water content of hydrated lime can be calculated. Assuming complete hydration and 100% pure quicklime, the water of hydration is 24.3% for high calcium and 27.2% for true dolomitic with values of magnesian limes intermediate, depending upon MgO content. Pure $Mg(OH)_2$ contains 30.8% water. The balance of 75.7% and 72.8% for high calcium and dolomitic respectively is the total oxide or *lime solids* content. This means that there is an *increase in weight* of the original quicklime to at least the extent of the water of hydration, namely, 100 lb. of high calcium quicklime plus about 32 lb. of water equal 132 lb. of calcium hydrate; the gain in weight of a completely hydrated dolomitic lime would be proportionately greater.

Practically, an excess of moisture over the theoretical amount is essential to achieve complete hydration since some water will be lost through evaporation precipitated by the heat of hydration, and invariably there is some, at least a fractional percent of adsorbed, *free* water that is not chemically combined but that envelopes the hydrate particle, like a film or bound water. This latter moisture cannot be removed by mechanical means, only by heat, so in a sense it is not free. Consequently, if only the theoretical amount of water is added, the lime will be incompletely hydrated and unstable; oxides will still be present with the hydroxides. Practically, the minimum amount of water required for complete hydration of a high calcium quicklime is at least 52% of the lime solid's weight of the average commercial product.[17] Variance in this value is contingent upon purity and degree of reactivity of a given quicklime.

Hydrated Forms. The following is a discourse on the principal hydrated (hydroxide) forms of lime, the main distinguishing characteristic of which is the amount of excess water they contain:

1. *Dry hydrate.* This is the most concentrated form of hydrate, a dry, fine, white powder that is produced commercially as hydrated or slaked lime. Commercial products contain 72 to 75% and 69 to 72% lime solids (oxides) for high calcium and dolomitic (highly hydrated) types respectively. Most of the balance (24 to 27%) is chemically combined moisture; the remainder is impurities and a minute amount of free water. The sole exception is normal dolomitic hydrate, which may contain as little as 16% water of hydration, since only a small portion of the MgO component is hydrated; 17 to 19% chemically combined water is typical, with the balance lime solids. Generally hydrate is manufactured without too much excess water, in the range of one part by weight of quicklime to 0.5 to 0.75 parts water. Commercial hydrates, depending upon the quality of the quicklime and the method of hydration, possess divergent physical properties that influence their chemical reactivity and usefulness. This will be discussed later in the chapter.

2. *Putty.* As the name infers, this is a wet, amorphous-appearing, plastic paste form of hydrate that has "body," so that it is moldable, a thixotropic mass. It is characterized as a colloidal gel by some investigators. It contains about 30 to 45% free water in addition to the normal increment that is chemically combined. The difference in free water content determines whether it is a stiff putty or a sloppy, wetter putty with less "body" and thinner consistency. Corre-

sponding to the above free moisture content, the oxides or lime solids content range between 40 to 60%. It is produced from quicklime by adding more water during slaking than in the case of dry hydrate, usually 1 part of quicklime to 1–1.4 parts water by weight. It can also be made by adding less water to a dry hydrate. In either case intimate mixing is essential to yield a smooth, homogeneous putty.

3. *Slurry.* This colloidal suspension form of hydrate has the consistency of thick cream. Unlike the putty, it has no body or plasticity. It will flow and is pumpable, like a viscous liquid. It contains about 55 to 73% free water, corresponding to a lime-solids content of 20 to 35%. It is readily obtained by adding an extra increment of water when quicklime is slaked, usually 1 part quicklime to about 2 parts water by weight, or by adding a much lesser quantity of water to a putty. Intimate mixing is essential to achieve uniform consistency. Table 9-1 gives the varying concentrations of slurries and lime suspensions.[12]

4. *Milk-of-lime.* This form of hydrate has the consistency of whole milk, is considerably thinner than slurry, and will flow almost as readily as water. Whitewash is synonymous to milk-of-lime. This aqueous colloidal sol suspension is employed in broad concentrations, ranging from 1 to 20% lime solids, so that it is largely composed of free water. It is easily prepared from quicklime by adding extra increments of water, more than required for putty or slurry, but more often it is obtained by diluting a slurry, putty, or hydrate sufficiently and agitating thoroughly. If prepared directly from quicklime, one water plant chemist recommends a quicklime to water ratio of 1:3 to 1:4.5 at a slaking temperature of 150 to 180°F.[3]

5. *Lime water.* This is a pure, saturated or unsaturated aqueous solution of calcium or calcium-magnesium hydroxide without *any* lime solids. Depending upon the temperature, Ca $(OH)_2$ will contain about 1.4 to 0.054 CaO equivalent of hydroxide in g./l. sat. sol. (See p. 176 on the solubilities of lime hydrates.) Since solubility decreases with rising temperature, a gain in temperature will reprecipitate some of the hydroxide solution as a lime-solids sediment. Conversely, a drop in temperature will dissolve some of the sedimentation in a supersaturated solution, thereby increasing the concentration of the solute. All hydrate suspensions or supersaturated solutions contain the same percent of saturated solution at a given temperature, assuming equal intimacy of mixture. Lime water is obtained from such suspensions by filtration or sedimentation, followed by decantation.

6. *Air-slaked.* If quicklime is exposed to the atmosphere of moderate to high relative humidity, the moisture vapor in the air will cause

Table 9-1. Strength of lime suspensions of a wide range of concentrations, providing data on specific gravity and unit weight of slurries, lime increments for both CaO and Ca(OH)$_2$, and ratios of water or suspension to lime

Specific gravity @ 15°C	Baumé °	Lb. per gal.	Grams per liter	Grams CaO per liter	Grams Ca(OH)$_2$ per liter	Lb. CaO per gal.	Lb. CaO per ft.³	Lb. CaO per lb. suspension	Lb. Ca(OH)$_2$ per lb. suspension	Ratio susp. to lime	Ratio water to lime
1.010	1.44	8.41	1,010	11.7	15.46	.097	.726	.012	.016	86.70:1	85.70:1
1.020	2.84	8.50	1,020	24.4	32.24	.203	1.519	.024	.032	41.87:1	40.87:1
1.030	4.22	8.58	1,030	37.1	49.02	.309	2.312	.036	.048	27.77:1	26.77:1
1.040	5.58	8.66	1,040	49.8	65.81	.415	3.105	.048	.063	20.87:1	19.87:1
1.050	6.91	8.75	1,050	62.5	82.59	.520	3.890	.060	.079	16.83:1	15.83:1
1.060	8.21	8.83	1,060	75.2	99.37	.626	4.683	.071	.094	14.11:1	13.11:1
1.070	9.49	8.91	1,070	87.9	116.15	.732	5.476	.082	.108	12.17:1	11.17:1
1.080	10.74	8.99	1,080	100.0	132.14	.833	6.232	.093	.123	10.79:1	9.79:1
1.090	11.97	9.08	1,090	113	149.32	.941	7.040	.104	.137	9.65:1	8.65:1
1.100	13.18	9.16	1,100	126	166.50	1.05	7.855	.115	.152	8.72:1	7.72:1
1.110	14.37	9.25	1,110	138	182.35	1.15	8.603	.124	.164	8.04:1	7.04:1
1.120	15.54	9.33	1,120	152	200.85	1.27	9.501	.136	.180	7.35:1	6.35:1
1.130	16.68	9.41	1,130	164	216.71	1.37	10.249	.146	.193	6.87:1	5.87:1
1.140	17.81	9.50	1,140	177	233.89	1.47	10.997	.155	.205	6.46:1	5.46:1
1.150	18.91	9.58	1,150	190	251.07	1.58	11.820	.165	.218	6.06:1	5.06:1
1.160	20.00	9.66	1,160	203	268.24	1.69	12.643	.175	.231	5.72:1	4.72:1
1.170	21.07	9.75	1,170	216	285.42	1.80	13.466	.185	.244	5.42:1	4.42:1
1.180	22.12	9.85	1,180	229	302.60	1.91	14.289	.194	.256	5.15:1	4.15:1
1.190	23.15	9.91	1,190	242	319.78	2.02	15.112	.204	.270	4.91:1	3.91:1
1.200	24.17	10.00	1,200	255	336.96	2.12	15.860	.212	.280	4.72:1	3.72:1

Table 9-1. (Continued)

1.210	25.16	10.08	1,210	268	354.14	2.23	16.683	.221	.292	4.52:1	3.52:1
1.220	26.15	10.16	1,220	281	371.31	2.34	17.506	.230	.304	4.34:1	3.34:1
1.230	27.11	10.24	1,230	294	388.49	2.45	18.328	.239	.316	4.18:1	3.18:1
1.240	28.06	10.33	1,240	307	405.67	2.56	19.151	.248	.328	4.04:1	3.04:1
1.250	29.00	10.41	1,250	321	424.17	2.67	19.974	.256	.338	3.90:1	2.90:1
1.260	29.92	10.49	1,260	331	437.38	2.81	21.022	.268	.354	3.73:1	2.73:1
1.270	30.83	10.58	1,270	343	453.24	2.92	21.845	.276	.365	3.62:1	2.62:1
1.280	31.72	10.66	1,280	356	470.42	3.03	22.667	.284	.375	3.52:1	2.52:1
1.290	32.60	10.74	1,290	370	488.92	3.14	23.490	.292	.386	3.42:1	2.42:1
1.300	33.46	10.83	1,300	382	504.77	3.25	24.313	.300	.396	3.33:1	2.33:1
1.310	34.31	10.91	1,310	396	523.27	3.37	25.211	.309	.408	3.24:1	2.24:1
1.320	35.15	11.00	1,320	410	541.77	3.48	26.034	.316	.418	3.16:1	2.16:1
1.330	35.98	11.08	1,330	422	557.63	3.58	26.782	.323	.427	3.09:1	2.09:1
1.340	36.79	11.16	1,340	435	574.81	3.70	27.680	.332	.439	3.02:1	2.02:1
1.350	37.59	11.25	1,350	448	591.99	3.81	28.503	.339	.448	2.95:1	1.95:1
1.360	38.38	11.33	1,360	460	607.84	3.92	29.326	.346	.457	2.89:1	1.89:1
1.370	39.16	11.41	1,370	472	623.70	4.03	30.148	.353	.466	2.83:1	1.83:1
1.380	39.93	11.50	1,380	484	639.56	4.15	31.046	.361	.477	2.77:1	1.77:1
1.390	40.68	11.58	1,390	496	655.41	4.25	31.794	.367	.485	2.72:1	1.72:1
1.400	41.43	11.66	1,400	510	673.91	4.37	32.692	.375	.496	2.67:1	1.67:1
1.410	42.16	11.75	1,410	524	692.41	4.50	33.665	.383	.506	2.61:1	1.61:1
1.420	42.89	11.83	1,420	538	710.91	4.61	34.487	.390	.515	2.57:1	1.57:1
1.430	43.60	11.91	1,430	550	726.77	4.71	35.236	.395	.522	2.53:1	1.53:1
1.440	44.31	12.00	1,440	562	742.63	4.82	36.058	.402	.531	2.49:1	1.49:1
1.450	45.00	12.08	1,450	575	759.81	4.93	36.881	.408	.539	2.45:1	1.45:1

the quicklime to "air-slake," a slow but inexorable form of hydration. Because of the scant moisture present, there is very little heat of hydration generated, and days or weeks may be necessary to distintegrate the quicklime lump or particle, depending upon the relative humidity and reactivity of the lime. In air-slaking, disintegration into hydrate particles occurs very gradually.

Seemingly, accompanying hydration is absorption of CO_2 from the atmosphere so that the disintegrated particles of air-slaked lime are composed of a mixture of hydroxides, oxides, and carbonates. Presence of moisture catalyzes carbonation of both quicklime and hydrated limes, but technically, hydration always precedes carbonation.

This material is of no commercial significance. In fact, it is an undesirable reaction and simply illustrates that *all* limes are perishable unless adequately protected. The sole exception to this premise is a farmer who spreads ground burnt quicklime on the soil as the most concentrated agricultural liming material. In the absence of rain the quicklime will air-slake, and with rain it will slake and absorb CO_2 more rapidly.

Of the above hydrated forms normal dolomitic hydrate will require about 20% less water than its high calcium counterpart, since very little MgO actually hydrates.[4] The stoichiometrics of highly hydrated dolomitic lime (Type S) indicate that about 12% more water is required, but practically it is about the same as high calcium since there is less evaporation due to the slower and lower heat of hydration. The least water is required for hydraulic lime since there is only 10 to 40% free lime present to react with water (see p. 273).

Rates of Hydration. As a generalization it can be concluded that the varying rates of hydration and divergent physical properties of hydrates are directly related to its derivative quicklime, just as the quality of the quicklime to a large measure is related to its parent limestone, as explained on p. 163. However, there are also paradoxical exceptions, so that the rate of slaking and optimum hydration conditions must be empirically and individually established. That is why in American lime materials specifications there is a standard clause to the effect that "all quicklimes should be slaked according to manufacturers' directions." In dry hydration of commercial hydrate the rate of reaction may be accelerated or retarded in order to obtain certain special properties desired by different classes of customers. Slight deviations in hydration conditions may exert a marked effect on its rate of reaction.

Variances in reactivity of slaking were reported by Murray.[5] He

compared the slaking rate of six different limes (three high calcium
and three dolomitic) at the same starting temperature of 25°C with
fixed amounts of water and lime. Results are graphically illustrated
in Fig. 9-1. One lime is then slaked with varying amounts of water,
with same starting temperature, and fixed lime increment (Fig. 9-2).
Invariably he found that the greatest temperature rise occurs with
the least water.

On contact with water highly reactive high calcium quicklimes slake
with literally explosive violence. The more rapid the rate of hydration
the greater the turbulence and temperature rise. Other dense, hard-
burned high calcium or dolomitic limes hydrate very slowly. For all
limes the rate ranges from seconds to hours or even months, with
overburned, recarbonated, impure or hydraulic high calcium quicklimes or hard-burned dolomitic. In some cases with the latter types
years are required for complete hydration at atmospheric pressure
and temperature. The much more rapid rate of the soft-burned, porous

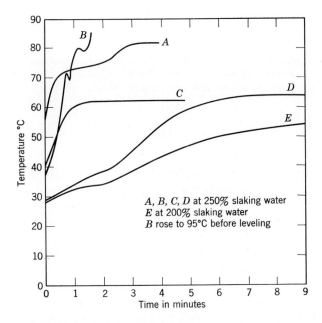

Fig. 9-1. Slaking gradients are displayed for different commercial limes, depicting the varying rates of slaking under
identical conditions, except that lime *E* was slaked in only
200% water. (A lime more reactive than any of the five
in this graph is illustrated in Fig. 9-2.)

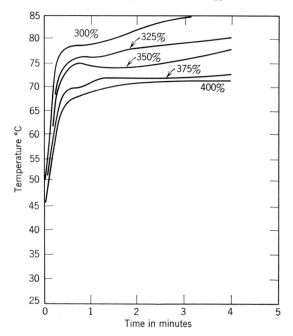

Fig. 9-2. Slaking gradient of one very reactive lime, illustrating how increasing amounts of slaking water retard hydration and temperature rise.

limes is due to greater permeability for water penetration. Other factors influencing rate of reaction are:

1. *Purity.* High-chemical purity abets rapid hydration. The greater the percent of impurities, the slower the rate. Pores are clogged, and the surface is partially coated with a slag formed by lime fluxing of such impurities as silica, alumina, and iron, rendering it more impervious to the entrance of water.

2. *MgO content.* Increasing increments of MgO have a retarding effect on the rate of reaction, as explained earlier. A very impure dolomitic lime would be the slowest, almost impossible to hydrate.

3. *Size.* Small-sized ground quicklime of $\frac{1}{4}$ to $\frac{1}{2}$ in. top size is appreciably more rapid in slaking than lump or pebble. Still finer, dustlike particles increase the rate even more.

4. *Temperature.* The rate accelerates with increasing temperature of both reactants, particularly water, and with many limes reaches its peak with steam. With some limes the rate increases so rapidly

that it may double for each 10°C rise in water temperature, thus simulating an explosion when hot water is employed. Chilled lime added to hot water reacts more slowly.

5. *Amount of water.* Increasing the amount of water retards the rate of hydration and mutes the heat evolved in the diluted mixture.

6. *Agitation.* Agitation of the lime and water increases the rate and dispersion of the lime particles markedly.

7. *Air-slaked.* A quicklime, partially air-slaked or recarbonated, hydrates sluggishly.

However, as will be revealed in the following sections, the most rapid-slaking method is not necessarily the most desirable because of other considerations; a reasonably rapid hydration rate is a more practical objective.

The slaking rate gradients, shown in Figs. 9-1 and 9-2, generally parallel each other in some respects.[5] In all cases the most rapid rise in temperature occurs instantaneously as the lime and water make contact. With reactive limes velocity is so rapid that it cannot be clocked. (It would appear in Figs. 9-1 and 9-2 that these limes start slaking at 35 to 40°C instead of the 25°C starting temperature for all limes.) This is caused by the heat of absorption of water in contact with the chemically reactive lime surface and the presence of quicklime fines. The next similarity is more obscure, but except for the most reactive lime, there is a lag in temperature rise of varying degree and duration after the initial instantaneous upsurge. Then, in the third stage the isotherm tilts sharply upward again, denoting more rapid exothermic activity. Finally the gradients plateau rather suddenly, indicating complete hydration.

If a large excess of water is added rapidly (or at one time) to lump or pebble quicklime, an adverse reaction may occur in which the lime is *"drowned."* The surface of the lime becomes hydrated but tends to be impervious to water penetration of the interior of the particle. A loss in initial temperature during the quiescent period following the initial temperature rise stifles hydration before the particle ruptures. Slaking is retarded and incomplete or is subject to delayed hydration. The resulting putties are unsound because of the danger of delayed hydration; they lack plasticity and resemble a thin, watery paste.

The other extreme from "drowning" is to add insufficient water to the lime, also resulting in incomplete hydration. This causes the hydrate to be *"burned,"* due to generation of excessively high temperature of 400 to 550°F that may even actually dehydrate some of the lime that has initially hydrated. More often a mass of hydration on the surface of the particle impedes penetration of water into the unhydrated interior. The expansive force of hydration does not fracture the crystal structure. "Burned" lime is also unsound because of the presence of unhydrated oxides and the hydrated particles are considerably coarsened, lacking the fine colloidal properties of a normal dry hydrate. "Burning" is more apt to occur with reactive quicklimes, but agitation and adequate water are preventatives. "Drowning" is much more prevalent with the hard-burned, slowly reactive limes.

In dry-hydrating a reactive quicklime it may be preferable to add excess water in several small distinct increments, not all at once or in a continual light sprinkle, so as to avoid the malpractices above. A slim majority contend that greater hydrating control is exercised by adding water to the lime than the converse of adding lime to water, but the latter practice has strong adherents. *Agitation* of the mixture permits higher temperature and faster slaking and is the best protection against localized overheating, which can lead to "burning." Grinding large particles to $\frac{1}{4}$ to $\frac{1}{2}$ in. size promotes a rapid but smooth reaction. Finer grinding is generally unfeasible. Although it increases the rate, it is more costly, and undesirable impurities will be comminuted which will be difficult to remove in the final milling.

Rate of Carbonation. Somewhat conflicting theories exist on carbonation. Staley[17] discovered that its rate is faster initially with dolomitic than high calcium hydrate, but then the rate slows and ultimately the high calcium hydrates absorb CO_2 more rapidly than dolomitic products. He and other investigators postulate that CO_2 is at first physically absorbed and then combines chemically with the lime. Knibbs[6] contends that when pure high calcium quicklimes are subjected to air-slaking, carbonation does not occur to an appreciable extent until hydration of CaO is near completion. With dolomitic, magnesian, and impure siliceous limes carbonation tends to occur more concurrently with hydration (although initial hydration precedes and catalyzes initial CO_2 absorption).

The maize of contradictory information on rate and extent of carbonation is unfortunate since lime's affinity for CO_2 is, along with its rheological properties, the basis for its historical use in mortar and plaster (see pp. 191–389).

Particle Size. Various methods of wet slaking and dry hydration along with divergent qualities of quicklime yield hydrates of varying particle size, gradation, and surface area, all of which affect the quality and utility of the resulting hydrate. For a large majority of purposes fine particle size is preferred with a reasonably restricted size distribution. All + No. 30 mesh material, largely core and impurities, should be removed. The removal of grit is discussed later in Chapter 10. Particles can be extremely minute and have been measured down to 0.1 micron and in angstrom units. The preponderant percentage, however, of commercial hydrates is in the 1 to 2–5 micron range, with about 10 to 25% below and above this range. Generally, wet hydrates yield finer particles than dry hydrate from the same quicklime.

Bishop[7] measured the particle distribution of four high calcium hydrates as they were sedimented from a suspension down to 2 micron size with butanol employed as a dispersing agent. Figure 9-3 illustrates the straight-line gradients obtained for four hydrates. The main difference was in particle size distribution since the geometric weight-mean diameters were quite similar. Like particle distribution measurements (except values are obtained down to 1 micron) were made on four wet hydrates (putties from slaked quicklime), as shown in Fig. 9-4. Maximum particle size of both classes of hydrates was similar, but the wet hydrates were much finer, possessing a much larger percentage of submicron sizes. Table 9-2 containing more data by Bishop indi-

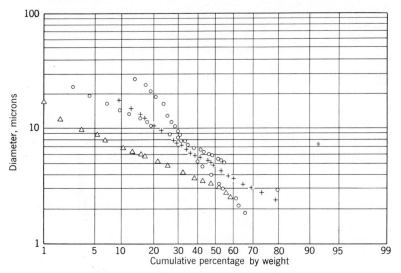

Fig. 9-3. Particle size distribution of four high calcium hydrated limes.

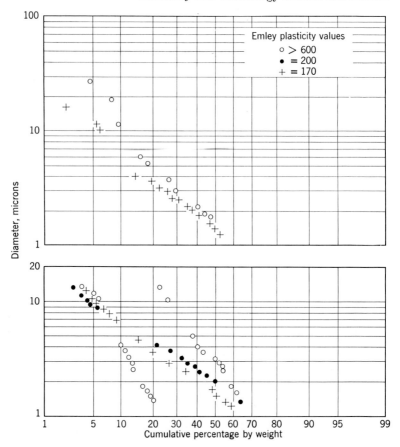

Fig. 9-4. Particle size distribution of four high calcium quicklime putties.

cates that the mean average particle diameter, particle size distribution, and specific surface areas of twenty-five hydrates do not necessarily correspond with each other. In this study it was observed that rate of settling generally adhered to size patterns; namely, 5 micron size was much slower than 10 micron; 2 micron slower than 5; and the 1 and submicron sizes were so slow that their sedimentation rate was largely immeasurable.

Adams[8] reports that average particle size of dry hydrate can be reduced if the quicklime from which it is derived is pulverized finer. He and other investigators have discovered that the putty volume is larger from soaking dry hydrates that possess the finest particle

Table 9-2. Particle size, specific surface area, and Emley plasticity data of twenty-five different commercial hydrated limes

Lime number	Emley plasticity value	Geometric weight-mean diameter	Geometric standard deviation	Specific surface	Lime number	Emley plasticity value	Geometric weight-mean diameter	Geometric-metric standard deviation	Specific surface
		Microns		cm.²/g.			*Microns*		cm.²/g.
1	60	6.0	1.9	5,000	14	300	5.0	3.4	11,000
2	80	4.2	3.1	12,000	15	310	6.6	3.6	8,000
3	80	4.6	2.7	9,000	16	320	7.2	2.8	6,000
4	80	5.6	2.2	6,000	17	400	6.3	2.8	7,000
5	80	5.8	2.4	7,000	18	400	7.2	4.9	12,000
6	80	7.0	2.2	5,000	19	410	7.5	4.5	10,000
7	110	4.4	2.5	9,000	20	420	5.2	2.9	8,000
8	120	3.3	3.0	14,000	21	420	6.8	4.4	10,000
9	120	5.4	2.2	7,000	22	480	4.1	3.0	11,000
10	140	3.7	6.1	36,000	23	490	3.0	10.5	110,000
11	220	7.4	3.9	9,000	24	550	4.9	2.6	8,000
12	240	7.8	3.4	7,000	25	600	2.9	2.1	11,000
13	280	4.0	2.5	9,000					

size. Murray[5] and other investigators observed that putty volume is always in inverse proportion to average mean particle size, increasing in volume as particle size decreases.

Many researchers[9-11] agree that slaking quicklime with a generous excess of water improves dispersion of the hydrate particles, contributing to finer particle size and slower settling qualities. Conversely, insufficient or only a slight excess of water causes an irreversible agglomeration of fine hydrate particles into coarse, rapid-settling hydrates of lower chemical reactivity. Slow rates of hydration caused by overburned quicklime, "drowning," or "burning" contribute markedly to crystal growth and flocculation of particles.

In addition to liberal amounts of water, Whitman attaches equal importance to reasonably high hydration temperature and rapid agitation in achieving fineness in particle size. There is disagreement among authorities on the desired temperature level, probably because they have experimented with different limes. A composite of the various views on recommended temperature would be for dry hydrate at boiling or slightly less and for wet slaking of quicklime about 160 to 200°F.

Murray distinguishes sharply between hydrating quicklime to form dry or wet hydrates of maximum fineness and surface area. For dry hydrates he recommends use of highly porous, soft-burned quicklime, ground to a particle size of No. 16 mesh. But for slaked quicklime putties or slurries his recommendations are diametrically opposite: use of hard-burned, relatively dense quicklime; 212°F temperature; minimum "practical" water ratio (still two to three times in excess of theoretical); and additives like sugar and ethanol are plus factors.

Whitman studied a wide variety of hydrates produced under different conditions microscopically. He concludes that the prime reason for variance in partical size is the innate tendency for hydrate particles to flocculate into agglomerates or clusters after hydration. Some particles have a greater tendency to floc than others, depending upon slaking conditions, and these provide coarser particles. He categorized hydrates produced under various conditions into the different type clusters they form as follows:

Type of cluster	*Method of slaking*
1. Large and tight	Theoretical water at 30°C
	Theoretical water at 100°C
	Vapor at 30°C[a]
	Steam at 100°C
	Vapor at 250°C
2. Small and very tight	Commercial dry hydrate
3. Small and tight	Five times theoretical water at 30°C
	Twenty times theoretical water at 100°C
	Twenty times theoretical water at 250°C
4. Small, some tight and some loose	Twenty times theoretical water at 60°C, with boiling[b]
5. Small and loose	Twenty times theoretical water at 100°C, with boiling
	Five times theoretical water at 100°C, with boiling

[a] Laboratory method: dry quicklime was placed in a desiccator over water and allowed to stand for three weeks at 30°C.

[b] Boiling was achieved by adding lime to water at 50°C in a vacuum desiccator. Pump was turned on and liquid was boiled violently.

Whitman concludes from the above that the finest particles are produced by rapid hydration, since hydrate crystals are given little opportunity to agglomerate. He feels that excess water, agitation, and increasing temperature are advantageous to employ since (1) they reduce tendency to agglomerate into large crystals and might actually promote faint deflocculation, (2) prevent local overheating ("burning") of lime, and (3) increase rate of hydration at any given temperature.

Surface Area. Probably the most significant physical property criterion of hydrates is surface area—more so than particle size. Two microscopic particles of hydrate may possess the same approximate particle size, yet one may be more rectangular and thicker as opposed to a flat plate or laminar shape containing pores. The latter type may possess several times greater surface area than the former. As a consequence, high surface area exerts a profound effect on chemical reactivity, settling rate, putty yield, plasticity, and the generally desired qualities of hydrates for most purposes—just as it does with quicklime (p. 146).

When slaked properly, surface area of all hydrates is markedly incremented over their derivative quicklimes. Based on Staley's earlier work[12] and his own research, Murray[13] illustrates this fact as follows in which five quicklimes of various surface areas, representing degrees of soft- to hard-burning, were slaked under normal conditions:

Quicklime	*HB 1*	*SB 2*	*SB 3*	*HB 4*	*HB 5*
Surface area in m.^{2}/g.	0.44	1.18	1.30	0.67	0.55
Hydrated lime					
Surface area in m.^{2}/g.	26–32	17–24	13–22	17.6	14.6

(Surface areas were obtained by the nitrogen absorption method and include both interior and exterior surfaces.)

HB = hard-burned; SB = soft-burned

This indicates that the fine quicklime crystals have been further greatly fragmented by expansive violence of the hydration reaction. Murray estimated the average hydrate particle from slaked quicklime at about 0.5 micron, smaller than Bishop found.

Staley and Murray then slaked one of the above quicklimes under varying abnormal conditions and obtained widely divergent surface areas, both lower and higher than the above values. This is reported as follows:

Conditions	*Surface area in m.^{2}/g.*
Slaking under cold conditions (5°C), ice added	6.67
Hydrating in steam at 113°C	8.05
Slaking with 2% sugar solution	37.0 to 46.3
Slaking with ethanol solution	36.0 to 45.6

This indicates how adjustments in hydration practice can markedly alter the surface area. Air-slaking, not reported above, produced the lowest surface area of all methods attempted.

Of the five different quicklimes above, the soft-burned types (as would be expected) yielded dry hydrates with one-third greater average surface area, when hydrated with a minimum of excess water, than the hard-burned types. Inexplicably, the hard-burned products, when slaked in four times excess water into suspensions, yielded greater average increases in surface area than the soft-burned quicklimes. This incongruity illustrates why the Delta T temperature rise slaking test (and other similar tests) described on p. 483 is not an equitable evaluation of chemical reactivity for *all* quicklimes, since hard-burned types invariably have lower temperature rises than their soft-burned counterparts. It should not be construed from this that all hard-burned limes yield highest surface areas, because many (a majority) do not. Only those derived from certain limestones exhibit such a property. This is simply another of lime's many anomalies that must be empirically determined until further research explains these apparent discrepancies.

By inducing maximum temperature rise, Staley[1] discovered that surface areas were correspondingly reduced in dry hydration. On slaking quicklime putties decrease in temperature below an optimum of 212°F reduced surface area. Values of all five limes gained with increasing increments of excess water. In dry hydration addition of ethanol or sugar did not increase surface area as it did in wet slaking.

Staley also calculated the pores in these microparticles. In the soft-burned derived hydrates pores were about 30 Å, with some as small as 7 Å; in hard-burned products pore sizes were no smaller than 45 Å.

Table 9-3. Changes in specific surface of a high calcium hydrate, actuated by varying temperatures and water–lime ratios

Temperature °C	4	10	20	40	60	90
°F	39	50	68	104	140	194
Ratio H_2O/CaO			All Values in cm.^{2}/g.			
2.5	50,736	54,293	52,790	56,606	57,355	58,300
4.5	—	—	48,307	—	52,260	55,255
7.5	35,246	34,534	—	47,035	49,183	53,070
10.5	29,133	29,840	—	45,203	48,920	51,126
13.5	23,166	24,419	36,520	41,080	45,967	52,658
18.0	17,833	18,968	31,556	37,620	48,307	53,925
25.0	15,314	18,597	29,405	40,910	48,244	53,295

Table 9-4. Changes in calculated particle diameter of a high calcium hydrate, actuated by varying temperatures and water–lime ratios

Temperature °C	4	10	20	40	60	90
°F	39	50	68	104	140	194
Ratio H_2O/CaO			Microns			
2.5	0.53	0.49	0.50	0.47	0.46	0.46
4.5	—	—	0.55	—	0.51	0.48
7.5	0.76	0.77	—	0.57	0.54	0.50
10.5	0.91	0.89	—	0.59	0.54	0.52
13.5	1.15	1.09	0.73	0.65	0.58	0.51
18.0	1.49	1.40	0.84	0.71	0.55	0.49
25.0	1.74	1.43	0.90	0.65	0.55	0.50

Miller[14] generally confirmed the Staley and Murray findings on specific surface area and how adjustments in slaking conditions can markedly alter this property, but he used a different method for determining surface area—the Blaine Air Permeability method (p. 486), which is standard practice in portland cement testing. He concluded that this method was the most accurate and expeditious for evaluating properties of hydrates and that changes in hydration or wet-slaking conditions were closely correlated with resulting changes in surface areas. He found that this correlation was much more dependable than with time-consuming sedimentation tests. Similarly, good correlation was also obtained with mean particle diameter determinations. Miller's specific surface area and particle diameter measurements from wet slaking high calcium quicklime under diverse water ratios and temperatures is contained in Tables 9-3 and 9-4 and reveal how these values can be adjusted to obtain hydrates possessing precise settling and reactivity characteristics for special applications.

Other conclusions from Miller's research were:

1. Specific surfaces vary less with changes in water temperatures at low water–lime ratios and with water at high temperatures for all water–lime ratios.

2. Low specific surface can be obtained with reactive soft-burned quicklimes by hydrating at low water temperatures (40 to 55°F) with high water ratios of 18–25:1. Such hydrates will react like those derived from hard-burned limes.

3. If maximum surface area is desired, there is an optimum slaking

water temperature and water ratio that can be empirically established for any lime.

Plasticity. In varying degrees the inherent rheological properties of lime putties, whether prepared by slaking quicklime or soaking dry hydrate, provide plasticity characteristics upon which the structural applications of lime depend (pp. 388–410). This putty property is difficult to measure with exactitude in the laboratory, and no one as yet has devised a test that is completely reproducible and universally acceptable. In their strictly empirical way, skilled masons and plasterers can possibly appraise the degree of plasticity in a putty more accurately than the laboratory. However, certainly the Emley plasticimeter test, described on p. 475, is the most satisfactory method yet propounded, although, except for the United States, other countries do not recognize it and rarely employ it. Indirectly, the standard water-retention test, which determines the extent at which a putty retains its absorbed free water, is a dependable indicator of plasticity. The higher the moisture retained, the higher, generally, the plasticity.

Although there is some disagreement over some aspects of what constitutes plasticity and how it is obtained, there are certain generally recognized facts. A soft-burned, porous, highly reactive quicklime is a desirable starting point, since with proper coordination it can provide rapid slaking, high initial temperature rise, and fine particle size of the resulting hydrate. Generally the finer the particle size, the greater is the putty volume, which contributes to increased plasticity. If excess water is added to such a putty, it will be a slow-settling suspension.

Yet such an explanation is perhaps oversimplified. There are many hydrated limes and quicklime putties with fine particle size that settle slowly, which paradoxically do not possess too much plasticity. This is borne out in Table 9-2 on p. 301 which shows relatively little difference in the average micron size of twenty-five different hydrated limes, which possess widely divergent Emley plasticity values of from 60 to 600. Bishop[7], who reported this data, concluded that the degree of plasticity must be determined for dry hydrates from the minus 2 micron particles and for quicklime putties from the minus 1 micron sizes. He observed that the particle distribution of the +2 and +1 micron sizes respectively exerted no apparent influence. Particle size distribution curves for four hydrates were plotted which were very similar straight-line curves; yet plasticity values (Emley) were 80, 110, 280, and 480.

It has been established by Hedin[15] and other investigators[5] that *particle shape* is at least as important as size for some uses (probably more so). Microscopically they have observed that the finest particles of plastic hydrates and putties are largely elongated and laminar in shape. The more of such miniscule, tabular-shaped particles present, the greater the plasticity. Empirically this has been proven by lime manufacturers, who commonly process their dry hydrates in tube mills to enhance plasticity. The kneading action of this mill tends to compress and flatten particles, even though in so doing many particles are loosely agglomerated. On subsequent wetting or soaking these plate-like particles disengage and are readily dispersed as colloids.

Disagreement arises on the slaking temperature and ratio of lime to water. Holmes et al.[16] contend that relatively low slaking temperatures induce particle fineness. Whitman and Davis[9] experimented in hydrating an alleged reactive high calcium quicklime under twelve conditions and concluded that the finest, slowest-settling hydrate (presumably high plasticity too) was obtained by rapid hydration with boiling water (100°C) and with only the "minimum" amount of water, which they regarded as 50% over theoretical to allow for evaporation loss. They also felt that thorough agitation was paramount. Hedin is also an advocate of rapid hydration with the smallest amount of hot water but stresses the essentiality of *cooling* the hydrate mass promptly after maximum temperature rise (completion of hydration). Cooling, he contends, counteracts the innate tendency of the hydrate molecules to migrate, owing to surface diffusion, and recrystallize with the larger crystals flocculating the most minute crystals. Such recrystallization, unless impeded, yields coarse, thicker particles instead of the desired minute, platelike particles.

Recrystallization into agglomerates and conglomerates has been generally observed by most investigators. Ray and Mather[17] characterize hydrate as a *nonreversible colloid*. They and Hedin have commented on how particles absorb a film of water around them and retain this moisture tenaciously. This lubricates the particles rendering the putty mass plastic. Hedin[18] attributes the adsorption energy of hydrate to the high surface tension that exists between $Ca(OH)_2$ and its saturated solution, which he estimates at about 600 erg/cm.^{2} Ray and Mather also hypothesize that plastic limes exhibit some degree of *cataphoresis*, namely, the power of forming charged particles on slaking or soaking through ionization that indisputably occurs (see p. 185). It is these charged particles that attract and retain moisture films that encase the hydrate particles (zeta potential principle). The extent of cataphoresis varies greatly with different limes and is nonexistent, they contend, with hydrates of low plasticity.

(In the author's opinion this is doubtful—possibly less, but not non-existent.) In comparing quicklime putties with dry hydrates they consistently discovered that charged particles were more prevalent with the former. Therefore, they conclude that on an average, putties from quicklime are inherently more plastic than dry hydrate. (However, this research preceded the advent of the quick plastic Type S hydrated limes.)

In a laboratory study[19] limes were produced from pure marble and precipitated calcium carbonate by calcining at different time-temperature conditions in air and under vacuum. Plasticities of the resulting slaked putties were measured, and the putties were subjected to X-ray spectrograph. Increasing numbers and intensities of the lines in the X-rays indicated a decrease in plasticity. Limes burned in the vacuum furnace exhibited more plasticity than limes from the same carbonate materials calcined in air in the electric furnace. Plasticities also decreased with increasing amounts of core (carbonate). Yet empirically a few lime manufacturers have observed paradoxically that small amounts of core (usually not over 5%) upgrade plasticities of their

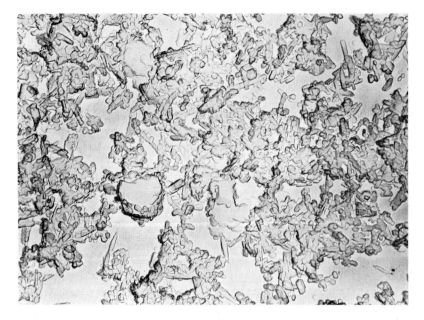

Fig. 9-5. Lime putty from dry hydrate, soaked for seven days, which raised Emley plasticity to 665 from an initial 115, with well-dispersed, laminar-shaped, fine single crystals. (31,000 ×).

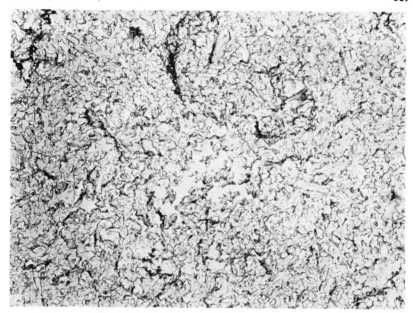

Fig. 9-6. Lime putty from slaked quicklime, soaked for seven days, that developed Emley plasticity of 650; well dispersed with very uniform, fine particle size. (15,750 ×).

dry hydrates. This core is interground and milled uniformly into the hydrate powder. Others report adverse results from core.

Highly magnified microphotographs taken by Dr. Paul Ney, German chemist from Köln, illustrate putties of different plasticities. Figures 9-5 and 9-6 reveal the microstructure of two highly plastic putties. Both are relatively well dispersed, but the laminar-shaped particles are more easily discernible in the former than the latter, probably resulting from greater magnification. In contrast, Figures 9-7 and 9-8 illustrate two hydrates of low plasticity showing less dispersion, more agglomeration into larger coarse particles. The latter (Fig. 9-8) has many of the desired tabular shapes, but they are much larger and complicated by the presence of conglomerates, large single crystals, and hydrocalcium silicate. The former are all German limes. Lastly, an American highly hydrated dolomitic putty is displayed in Fig. 9-9 after only thirty minutes of soaking. It has high plasticity in spite of evident agglomeration.

Putties from many dry hydrates and slaked quicklimes develop greater plasticity from long periods of soaking or curing putties under

Fig. 9-7. Lime putty from dry hydrate, soaked for thirty minutes; possesses Emley plasticity of 119, large conglomerates, and coarse particles. (6,100 ×).

saturated conditions.[10] Unaccountably, some are drastically improved while others exhibit only modest gains. On an average, the greatest gains in plasticity from tests on seventy-five different hydrated limes subjected to twelve to twenty-four hours soaking occurs with normal dolomitic hydrates, the least with highly hydrated dolomitic, because of its high initial plasticity. High calcium is intermediate but closer to normal dolomitic hydrate in extent of gain. Further gains in plasticity of dry hydrates by soaking for periods longer than twenty-four hours are of no consequence, but putties from many slaked quicklimes will continue to increase in plasticity up to a year. Recarbonation adversely affects plasticity.

One of the most lucid fundamental explanations of plasticity, although some of his hypotheses may be questioned by some authorities, is by Hedin,[18] which is quoted as follows:

. . . the first contact between (lime) oxide and water will result in the formation of a supersaturated hydroxide solution, which gives rise to the formation of larger or smaller quantity of crystallization nuclei. Then, the crystals begin to grow and take the form of microscopic rods or laminae. In order to insure plasticity, it is important that the quantity of laminar crystals should be as large as possible. This will be the case when the supersaturation of the solution is relatively low. The lowest supersaturation will be obtained if the crystallization is allowed to take place at a low temperature, since the saturation concentration will then be relatively high. Furthermore, it is important that the calcium oxide should be as fine grained as possible. Then, the slaking will be quick, and this will cause an instantaneous high supersaturation, which will result in a large number of crystallization nuclei. Consequently, in the course of subsequent growth, the supersaturation will rapidly become so low as to give rise to the formation of tabular crystals.

Thus, it is advisable to use a fine grained, strongly reactive lime and to insure efficient cooling of the slaking tank during the slaking process.

It follows that the properties of burned lime have extremely important effects on the quality of the hydroxide formed in the process. Therefore, it is primarily necessary that lime manufacturers produce burned lime consisting of small crystal unit cells and having great porosity so as to insure that reactivity will be as high as possible.

Water Retentivity. The ability of lime putty to hold and retain moisture, even providing appreciable resistance against suction, is

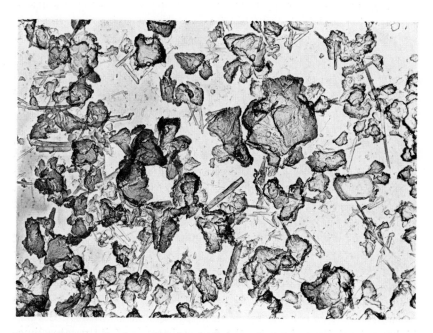

Fig. 9-8. Lime putty from slaked quicklime, soaked for seven days; has Emley plasticity of 120; possesses agglomerates, large crystals, and hydro calcium silicates. (3,400 ×)

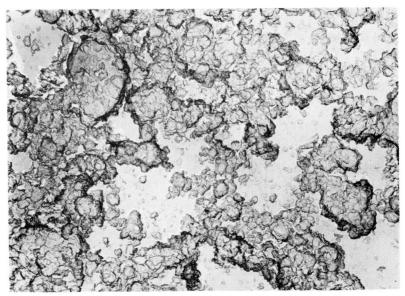

Fig. 9-9. Lime putty from highly hydrated dolomitic, soaked for thirty minutes; has Emley plasticity of 345; is agglomerated but well dispersed. (6,100 ×)

one of its principal attributes as structural lime. This characteristic prevails even when putty is mixed with sand.

Wells, Levin, and Clarke (and other researchers) have found that limes with a high plasticity, as measured by the Emley plasticimeter, invariably possess high water retentivities.[20] The converse is not always true. Usually a lime of high water retentivity of 90% or more possesses high plasticity, but there are a few inexplicable exceptions. But there are no exceptions involving highly hydrated dolomitic lime; all have uniformly high plasticity and water retentivity.

Water retentivity is invariably increased by soaking or aging a lime putty, whether derived from quicklime or hydrate, for twelve to twenty-four hours, as opposed to thirty minutes, but the extent of increase varies with different limes and is least with the highly hydrated dolomitic products. Tests comparing a 1:3 lime–sand mixture by weight as against the same ratio by volume invariably develop higher retentivity values for the former because of its greater lime–sand mixture ratio (richer mix). Thus, water retentivity is gradually lowered by the addition of sand.

Putty Volume. Another characteristic of lime hydrates is the inevitable expansion in volume that occurs when either quicklime is slaked or dry hydrate is soaked into a wet putty of standard controlled consistency. Large increases in volume or high putty yield is of interest to consumers of structural lime because of the economy involved, since lime is always proportioned in mortar and plaster mixes by *volume* instead of by weight and is usually two and a half to three times bulkier than the sand aggregate. Thus, one ton of a high-yield putty may be equivalent in volume to 1.5 tons of a low-yield putty. But of greatest importance is the fact that the large-volume putty generally possesses superior chemical and rheological properties. The increase in volume is due to lime's inherent ability to absorb and retain a considerable amount of *free* moisture in addition to the stoichiometric increment that is chemically combined. However, all limes, quick or hydrated, high calcium or dolomitic, pure or impure, differ in the amount of free water that they retain at standard putty consistency. Contributing further to this variance is the broad range of values of quicklimes and dry hydrates in apparent density or weight of a unit volume in dry form, even though the calculated true specific gravities of most limes are almost identical ($\pm 2\%$) for those derived from limestones. Chalk is much different—much lighter.

Testing by Holmes and Fink[16] on seventy quicklimes and thirty-seven hydrates (all commercial types) for the National Lime Association in the early 1920's revealed that the average quicklime putty increased in volume from a dry basis of 3.1 cu.ft., representing a 180 lb. barrel, to about 7 cu.ft., with a range of putty volumes of from 4.5 to 8.7 cu.ft. The amount of water remaining in the putty including that which is chemically combined, ranged from 123 to 275 lb. for 100 lb. of quicklime, with the average about 210 lb. On an average, quicklime putties were able to absorb more free water than putties from dry hydrates and, as a result, yielded a larger proportionate increase in volume, averaging about 50% more on an equivalent basis. High calcium quicklimes averaged a higher putty yield than dolomitic, 7.0 to 5.3 cu.ft., respectively. Apparent densities of these seventy quicklimes ranged from 1.15 to 2.1.

There were some unaccountable exceptions, but a large majority of the voluminous quicklime putties possessed the slowest sedimentation rates, highest plasticity, and reactivity, as determined by the rate at which lime causticized sodium carbonate. The reverse was true with the lowest putty volumes. There was no correlation, however, in putty volumes with the diverse bulk densities.

In contrast, the amount of free water absorbed by putties from thirty-seven dry hydrates ranged from a ratio of 68 to 102 parts water to 100 parts dry hydrate, with an average of about 90 parts water. Stated another way, the putty yield from 100 g. of dry hydrate ranged from 109 to 147.5 cm.³. The difference in putty yield between high calcium and dolomitic was not nearly as pronounced as with quicklime putties; it averaged slightly higher for high calcium. However, these tests were conducted before the advent of Type S hydrated limes. Apparent densities of these dry hydrates ranged from 0.749 to 0.985, but there was very little correlation between this value, putty yield, and settling rate.

Sedimentation. The settling rates of both wet and dry hydrates are of critical importance to most chemical users of lime. A large majority prefer slow settling characteristics; a minority desire fast settling; others are intermediate in their requirements.

Probably the most exhaustive study on sedimentation of lime suspensions is by Holmes, Fink, and Mather,[21] who tested settling rates of 125 quicklimes and 40 hydrates under many conditions. Tables 9-5 and 9-6 offer a partial tabulation of their results on quick and hydrated limes respectively, showing the tremendous variance in sedimentation rates. In the tests, 7.5 g. and 10.0 g. of each lime was either slaked or soaked and allowed to remain in putty form for twenty-four hours. Each putty was then added to a 100 cm.³ graduated cylinder of water, shaken thoroughly so as to disintegrate the putty and disperse it uniformly in the water. After each five-minute interval the top of the lime suspension was recorded as it settled. After twenty-four hours the final volume was recorded. Tables 9-5 and 9-6 demonstrate the time required to settle the lime half-way in the graduate to the 50 cc. level, the volume of the sediment after completion of settling in twenty-four hours, and the available lime content of the quicklime. There were some disconcerting exceptions in results, but the following is a summarization of principal conclusions:

1. All magnesian and dolomitic quicklimes and their normal hydrated limes settle more rapidly than their high calcium counterparts owing to the necessarily harder burned MgO component.

2. Generally the purer the lime, based on available CaO content, the slower the sedimentation rate.

3. Most of the high calcium quicklime putties settled more slowly than their derivative hydrates—some appreciably so.

4. Excess water yields more fineness and slower settling than theoretical amount of water for slaking.

5. Settling rates of all commercial hydrates are more uniform, with less deviation, than slaked quicklime.

6. There was less difference in settling rates between dolomitic quicklimes and hydrates than with the high calcium types.

7. Generally the supernatant lime solution above the suspension

Table 9-5. Settling rates of twenty-three commercial quicklimes

NLA lime no.	Type of lime	Time in min. required to settle to 50 cm.^{3}	Final volume 24 hrs.	Available CaO % (nonvolatile)
2	High calcium	67	30	94.02
3	High calcium	18	27	90.12
5	High calcium	1195	45	96.44
7	High calcium	472	35	91.36
18	High calcium	2080	50	86.12
27	High calcium	175	40	93.19
38	High calcium	89	35	93.29
44	High calcium	315	38	91.95
45	High calcium	6	21	83.88
63	High calcium	16	23	90.15
76	High calcium	1288	43	93.18
81	High calcium	649	40	87.03
86	High calcium	2	17	79.05
93	High calcium	2269	45	95.67
23	Dolomitic and magnesian	8	16	—
25	Dolomitic and magnesian	160	35	81.63
65	Dolomitic and magnesian	58	24	—
71	Dolomitic and magnesian	2	16	—
103	Dolomitic and magnesian	90	29	—
111	Dolomitic and magnesian	22	20	—
115	Dolomitic and magnesian	156	30	—
117	Dolomitic and magnesian	111	24	—
119	Dolomitic and magnesian	87	28	—

Table 9-6. Settling rates of twenty commercial hydrated limes

NLA lime no.	Type of hydrate	Time in min. required to settle to 50 cm.[3]	Final volume 24 hrs.
2a	High calcium	26	29
3a	High calcium	34	30
5a	High calcium	63	28
35a	High calcium	44	25
45a	High calcium	33	39
46a	High calcium	50	24
59a	High calcium	77	28
79a	High calcium	78	29
88a	High calcium	8	18
94a	High calcium	29	24
98a	High calcium	45	23
105a	High calcium	7	21
23a	Magnesian and dolomitic	58	25
25a	Magnesian and dolomitic	27	25
52a	Magnesian and dolomitic	39	24
65a	Magnesian and dolomitic	15	21
71a	Magnesian and dolomitic	40	30
74a	Magnesian and dolomitic	44	28
75a	Magnesian and dolomitic	30	27
103a	Magnesian and dolomitic	39	28

line was clear. With some of the finest hydrates the supernatant was faintly milky white, indicating presence of minute colloids.

8. Addition of sugar retards sedimentation.

9. Addition of calcium chloride acts as a depressant on lime suspensions.

10. Hydrated limes tend to settle more slowly as hydration temperature is reduced.

11. Soaking hydrates for a few days before test retards settling rate.

12. Maintenance of temperature after hydration is completed tends to increase settling rate of commercial hydrates.

13. The slowest settling limes produced the greatest putty volumes.

(The values cited in Table 9-5 and 9-6 are based on U. S. limes in the 1920's. The average quality of lime has been definitely improved during the past forty years, so that the poorest and average values would be higher in the 1960's.)

In slaking quicklimes with increasing increments of water of 225%, 250%, 275%, and 300% over theoretical, Murray[13] observed that settling rates became progressively more rapid with increased dilution. This is at odds with conclusion item 4 above.

All investigators recognize that the settling rates of limes can be readily altered by changes in the hydration process. In striving for extremely slow settling qualities, many researchers have experimented with chemical additives as possible dispersion agents in order to impede the tendency of hydrates to flocculate. Murray, after experiments with about fifty different additives, was unable to find a conclusively satisfactory dispersant.[13] Yet Staley[1] was moderately successful in deflocculating hydrates with ethanol or sugar. Haslam and others have confirmed the dispersing effect of sugar. Several lime companies in the 1960's are using fractional percents of special proprietary organic air-entraining agents as dispersants in their hydrated limes for mortar and have encountered enthusiastic trade acceptance in providing apparently improved workability for mortars.

If a putty is subsequently dried without recarbonation occurring and then reintroduced to water to form a milk-of-lime, the sedimentation rate is markedly increased. This demonstrates that agglomeration of fine hydrate particles occur during drying and that subsequent wetting does not disperse or break up the resulting coarse agglomerates.

Reactivity. The concensus among authorities is that surface area is the most reliable criterion on reactivity of hydrates; the higher this value, the greater the reactivity.

This has been confirmed by various chemical tests. In one instance[22] a given quicklime was slaked into a wet hydrate slurry, with both the theoretical amount of water (30% excess allowed for evaporation) and with ten times the theoretical water content. Both hydrates were then diluted to the same concentration of milk-of-lime, and identical quantities were added to separate graduates where they were agitated equally and allowed to settle to 50% of the original volume. Rate of reactivity of both hydrates was ascertained by mixing the theoretical quantity of sodium carbonate for causticization with both specimens, which hydrolized the $Ca(OH)_2$-Na_2CO_3 solution into sodium hydroxide solution and $CaCO_3$ precipitate. Time of complete causticization was recorded. Results of this experiment were as follows:

	Settlement in minutes	Time of causticizing (minutes)
Slaked with theoretical water	10	330
Slaked with ten times the theoretical water	440	10

It was concluded from the above that the superior reactivity and slower settling rate were due to greater particle fineness and surface area obtained from slaking this particular quicklime in considerable excess water.

In all probability the above is an extreme case, and usually the difference in reactivity and settling is not so pronounced. Fineness, as determined by particle size or mean diameter, is indicative of the degree of reactivity, but is not as reliable as surface area.

Whitman[9] found that highest reactivity and fastest rate of disolution of lime hydrate occurred with slowest settling, finest hydrates of largest surface area. He measured the rate of reactivity by titrating aqueous hydrates with 10 cc. of $1N$ hydrochloric acid with phenolphthalein as an indicator and timing the reaction for the reappearance of the pink color with a decimal stopwatch. The aqueous lime suspension contained 3.7 g. of hydrate (0.1 equivalent), diluted with water to 10 cc. and the 10 cc. dose of acid was repeated ten times to achieve complete neutralization.

Dehydration of Hydroxides. Research has been conducted quietly by a number of organizations on preparing calcium oxide from hydrated lime that reportedly possesses some unique properties, including extremely high chemical reactivity and surface area. Such a material might be classified as double calcined lime or dehydrated lime, since it is prepared by conventional calcination of limestone, hydrating the resulting CaO to a dry powder and then dehydrating the hydroxide so that it reverts back to quicklime. The time and temperature control of each of these three steps must be precise in order to produce this bizarre type of lime; otherwise, the final product will be similar to an ordinary pulverulent quicklime.

Since careful milling of hydrates invariably and substantially improves the purity of lime, the final product will possess purity unequaled by any commercial quicklime. A CP grade of 99% + CaO is believed attainable. This coupled with explosivelike reactivity, submicron particle fineness, and ultrahigh surface area offers the potentiality of a relatively high-priced specialty lime product. Conjecture inevitably follows that the double hydrate resulting from slaking this unique quicklime might also possess similarly intriguing qualities. With imaginative research and development its properties could very possibly be utilized as a catalyst, dessicant, or absorbant in new chemical applications of lime. This is strictly supposition since such a material has never been commercially sold and only produced experimentally in a small pilot plant. Research to date has been largely shrouded in mystery, and only briefest mention of such material has appeared in the literature.

One researcher, General Motors Company,[23] obtained a patent on such a material, which they describe as "Especially Reactive Lime." They contend that maximum reactivity is obtained at about 85% recalcination of the hydrate so that the final product contains 15% hydroxide and 85% oxide. At 100% calcination they claim that reactivity is reduced to one-third of that attained by the 85 to 15% proportion. Chemical and physical properties are claimed to be significantly different from commercial limes.

Final calcination (dehydration) occurs at much lower temperature than in conventional lime burning; 600°C is believed to be the optimum calcining range. Rotary and Fluo-Solids kilns, muffle furnaces, or rotary driers could probably be adapted with modifications for this unconventional process.

Experimentally hydrates in putty form have been recalcined to an oxide, and the resulting quicklime is nodular in shape and No. 8 to 10 mesh in size. Reactivity is reported to be very high if dehydration is consummated at minimum temperature.

A related process[24] is a lime recovery program developed by Air Reduction Company of recalcining a waste hydrate sludge, following pelletization of the hydrate, into quicklime for reuse in manufacturing calcium carbide. The properties of this form of quicklime, however, are totally different from the above.

Preparation of CP-grade $Ca(OH)_2$. The oxalate reprecipitation method will yield $Ca(OH)_2$ of virtual 100% purity. $Ca(OH)_2$ is reacted with oxalic acid producing a calcium oxalate precipitate. After this highly insoluble precipitate is thoroughly washed and dewatered, it is calcined in a muffle furnace to yield CaO. The oxide is then hydrated in distilled water. Recalcination of the filtered hydrate follows and then final hydration with distilled water. By this method Hedin[15] obtained a hydrate of zero impurities, but titration revealed that the Ca^{++} ion was 0.7% greater than the OH^- ion than theoretical molecular calculation indicates.

A corollary of this process would be preparation of CP grade calcium oxide.

A less pure but probably a CP grade of hydrate can be made by hydrating a double-calcined quicklime (or dehydrated hydroxide), since the quicklime is relatively so pure (preceding section).

Conclusion. Some contradictions exist in the foregoing sections, but not nearly as many as occur in the literature. An earnest attempt has been made by the author to largely present the consensus of the best authorities on all factors affecting slaking. In so doing some repetition was unavoidable.

Some of the discordance that exists in the literature probably stems from unreliable experimentation, but most is believed due to the following variables, the full significance of which was minimized by some investigators and, as a result, was not clearly delineated: (1) quality of quicklime and (2) quality of water. It would be repetitious to elaborate on (1), but obviously the widely varying porosities, particle or crystallite sizes and spacings, hardness, core content or overburned effect, impurities (qualitatively and quantitatively) have all a marked effect on the conditions of slaking and the hydrate properties.

But the quality of water possibly exerts nearly as great an influence. Hydration tests with distilled CP water may yield results entirely different from industrial process waters or solutions that contain even minute amounts of other electrolytes, like chloride, sulfate, and nitrate ions, etc. Slightly brackish waters inject uncertainty; even potable waters that are relatively "hard" can influence the results. Note the changes in solubilities of limes exerted by fractional percents or parts per million of extraneous inorganic and organic substances (pp. 177–9).

Therefore, use of this information should be tempered by a certain amount of empirical testing under the *exact conditions* of its contemplated application and should not be accepted as conclusive.

REFERENCES

1. H. Staley, "Micrometrics of Lime," (Research Rep't to Nat'l Lime Ass'n, 1946).
2. National Lime Association, *Chemical Lime Facts* (1949), p. 34.
3. H. Lordley, "Lime and Lime Slaking," Water & Sewage Wks, *Ref. & Data* (1955), p. R-214.

4. L. Wells and K. Taylor, "Hydration of MgO in Dolomitic Hydrated Limes & Putties," *U. S. Bu. Stand's RP 1022* (Aug. 1937).

5. J. Murray, Research Repts to Nat'l Lime Ass'n (1953, 1954).

6. N. Knibbs, *Lime & Magnesia,* E. Benn, Ltd. (London, 1924), pp. 105–111.

7. D. Bishop, *Particle Size & Plasticity of Lime, U. S. Bu. Stand's RP 1232* (Aug. 1939).

8. F. Adams, *I. & E. Chem.* **19** (May 1927), p. 589.

9. W. Whitman and G. Davis, "The Hydration of Lime," *I. E. Chem.* **18** (Feb. 1926), p. 118.

10. D. Bonnell, *J. Soc. Chem. Ind.* **53** (Sept. 1934), p. 279.

11. M. Holmes, G. Fink, and F. Mathers, *Chem. & Met.* **27,** no. 25 (Dec. 20, 1922).

12. H. Staley and S. Greenfield, *Proc. ASTM* (1947), p. 47.

13. J. Murray, "Summary of Fundamental Research on Lime," Research Rep't to Nat'l Lime Ass'n (1956).

14. T. Miller, "A Study of Reaction Between CaO & Water," Azbe Award 1, Nat'l Lime Ass'n (1961).

15. R. Hedin, *Swed. Cement & Conc. Res. Inst. Bull. 33* (1962), pp. 57–67.

16. M. Holmes, G. Fink, *Chem & Met.* **27,** (1922), pp. 347–349.

17. K. Ray and F. Mathers, *I. E. Chem.* **20,** (May 1928), p. 475.

18. R. Hedin, "Plasticity of Lime Mortars," Azbe Award 3, Nat'l Lime Ass'n (1963).

19. M. Farnsworth, *I. E. Chem.* **19** (May 1927), p. 583.

20. E. Levin, W. Clarke, and L. Wells, *U. S. Bu. Stand's Rep't. 146* (1956).

21. M. Holmes, G. Fink, and F. Mathers, *Chem. & Met.* **27,** (1922), p. 1212.

22. *U. S. Bu. Mines I. C. 6423* (1931), pp. 32–33.

23. W. Lovell, V. Miller, M. Mulligan, and H. Lechenwalner, U. S. Patent 2,474,207 (to General Motors Co.), "Especially Reactive Lime" (1949).

24. B. Herod, *Pit & Quarry* (May 1964), p. 98.

CHAPTER TEN

Methods of Hydration

Slaking Methods

In slaking quicklime into a putty, slurry, or milk-of-lime, various batch and continuous processes are employed, but regardless of the method the correct proportion and distribution of the slaking water for a given lime should be established in order to produce the desired quality of hydrate.[1-3]

The *batch method*[4] is the oldest and simplest method of slaking lime, involving a minimum of equipment. However, since about **1925** it has been largely replaced, at least in large industrial plants, with continuous, automated slaking machines. The reason for its decline is its high labor cost, since the lime and water are fed, mixed, and discharged by hand. There is also more of a safety problem from lime burns, and the very nature of the process is conducive to a waste of lime.

Generally, the mixing of quicklime and water is achieved in large metal vats or open tanks in which the laborer agitates the mixture with a hoe or paddlelike device. Because various quicklimes slake differently, the recommended slaking procedure of the lime manufacturer should be closely observed. There is some difference of opinion on whether water should be added to the lime or vice versa. In a majority of cases the former is preferable, particularly with relatively slow slaking limes, so as to prevent "drowning" (p. 297). If a quick slaking lime is added to the water, then it is paramount that sufficient water is present to avoid localized overheating of the mass that may cause "burning" (p. 297). Control of slaking temperature is more difficult in a manual operation, but maintenance of a temperature below boiling and above 180°F. is generally desirable.

In such slaking most precise control can be maintained by slaking to a thick slurry suspension or thin putty consistency with a lime-solids content of 35 to 40%. This will require an approximate water to lime ratio for high calcium of $3:1 \pm 25\%$ and for dolomitic of $2:1 \pm 25\%$, depending upon the reactivity and purity of the quicklime. At moderate to warm temperatures the resulting slurry is barely pumpable; at cold temperatures it usually is too viscous to pump. It may be used in this form or may be diluted further into a "milk" in another tank before use. To assure complete hydration the lime should be slaked at least thirty minutes before application. Aging the putty for twelve hours or a few days generally improves the efficiency of the product, if it is covered to prevent carbonation.

Modern *continuous slakers* are typified by the drawings in Figs. 10-1 and 10-2. They consist of a dry feeding hopper; a large enclosed rectangular tank equipped with paddles to agitate the mixture; often, a second connecting tank to dilute the slurry to a specific concentration of "milk"; water jets; a classifier for degritting the slurry; a hood for dust and vapor removal; pumps and pipelines for discharging the slurry; and auxiliary control equipment. Since maintenance of temperature below but near the boiling point (about 180 to 190°F) is impor-

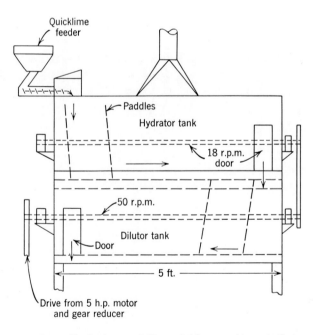

Fig. 10-1. Basic type of lime slaking equipment that possesses two main compartments for hydration and dilution of the resulting lime slurry. (Courtesy, Dorr-Oliver Co.)

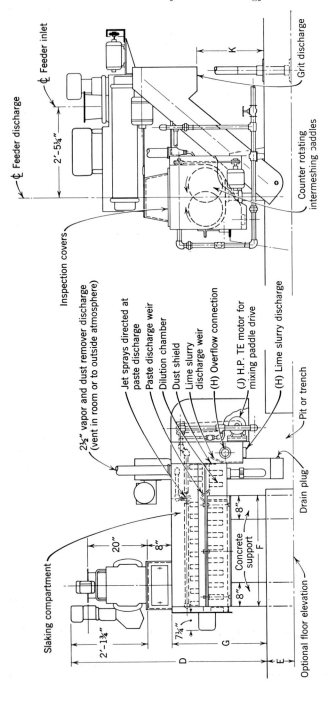

Fig. 10-2. Dual-drawings of a paste lime slaker that first hydrates the lime into a thick slurry or paste and then dilutes it into a milk-of-lime. (Courtesy, Wallace and Tiernan, Inc.)

tant, a thermostatically controlled water valve is installed which responds to a 5°F temperature change. The flow of lime and water are adjustable and uniform, and the concentration of the suspension is checked with a hydrometer for the desired specific gravity, which is calibrated with the percent of lime solids. (See Table 9-1 on strength of lime suspensions.) The automatic agitator accelerates the rate of hydration. Frequently, modern slakers are encased with insulation to retain the heat of hydration. Operationally this is sound as well as practical in reducing heat radiation in hot summer months. Capacities of commercial slakers range from 0.5 to about 5 tons per hour.

Most slakers are adapted to quicklime no larger than 2 in. top size, and probably ½ in. size material feeds more smoothly in most machines than any size. However, ground quicklime has also been successfully employed, but pulverized material may cause arching in the feed. The objective is to introduce the quicklime and water continuously and uniformly with sufficient contact time to assure complete hydration.

When a dilute lime suspension is desired, generally it is achieved in two stages by first slaking to a slurry and then diluting the slurry or putty in a second tank with agitation at the slaker's discharge end or in a separate settling tank. It can be produced in one step, but the concensus is that particle size and reactivity may be degradated. In secondary dilution hot water should be employed and the mass should be diluted, if possible, while it still retains some of its slaking heat.

In addition to improving the quality of the lime suspension, *degritting equipment* conserves the life of pumps and tanks against abrasion. *Grit* is defined as any material retained on a No. 100 mesh screen and is composed of carbonate core, silica, alumina, and insoluble calcium compounds. With relatively pure quicklimes 1.5 to 3% of grit is removed with various classifiers based on initial dilution to 5 to 15% CaO content and then rapid sedimentation of the heavier particles. A small secondary lime recovery, which is usually inconsequential, is possible by tumbling or scrubbing the grits to remove hydrate that adheres to the core of some products.

After degritting, if it is desired to reconcentrate the suspension, the milk-of-lime is pumped to *settling tanks* or *thickeners* where the suspension is dewatered. The principle of sedimentation, expounded on p. 314 applies here. Rudolfs[1, 5] also reports an enormous difference in the settling rates of lime slurries in chemical process plants. He claims that on an average dolomitic hydrate has twice the free settling rate of high calcium and two to five times as great a thickening capacity. Depending upon the temperature and severity of calcination of a given derivative quicklime, its hydrate settling rate varied from 0.6 to 8.5 ft./hr. Increasing the temperature of slaked lime, decreases the viscosity of the water phase and accelerates the settling rate. The effect of wind on the open settling basins retards settling rates in two ways: by its cooling effect and by causing eddy currents that agitate and disturb normal settling. His experience indicates that when a continuous thickener is fed a lime suspension of 4 lb. $CaO/ft.^3$ at 175°F, a free settling rate of 7.5 to 8.5 ft./hr. can normally be anticipated.

In *pumping* lime slurries centrifugal pumps are usually the most efficient. With degritted slurry cast iron pumps are satisfactory, but for slurries containing grit the NiHard pump is advisable. To minimize *scale* deposition in pipelines in which lime slurry is pumped or fed by gravity, the velocity of the slurry flow should exceed 150 ft./min. Ironically, degritted slurry tends to scale more readily than slurry with grit, since the coarse particles exert a scouring action on the pipe linings. Frequent high-velocity rinsing by pumping the "clears" from the settling basin back to the slaker, rather than fresh water for slaking, tends to remove freshly precipitated $CaCO_3$ and calcium bicarbonate from the pipelines. Yet acidizing the pipelines with corrosion-inhibited hydrochloric acid periodically may be the only solution to scale formation.

Commercial (Dry) Hydration

A vast majority of lime consumers do not purchase quicklime and slake it for their own consumption. Indeed, they cannot possibly justify the cost of capital (slaking) equipment and the inevitable problems attendant to another processing step that slaking entails. Their lime requirements are simply too small. Consequently, they prefer to purchase commercial hydrated lime; only large consumers slake their own. Invariably commercial hydrate is a dry powder, representing the *most concentrated* form of hydrate that exists. The purchase of hydrate in such dilute forms as slurry or putty would be uneconomical except in rare instances because of the cost of transporting so much free water as well as the cost of suitable containers. Dry hydrates can be readily mixed with water and reduced to any desired consistency. Furthermore, many applications are predicated on the use of a dry material. (Determination on whether to purchase quicklime for slaking or hydrated lime is discussed on p. 341.) Thus, dry hydrated lime is the *second major product* of the lime industry.

It is obvious that dry hydration is more complicated than wet slaking, since the final product must be literally dry for packing in paper bags, whereas the water concentration of putties and slurries resulting from wet slaking is not critical since these materials are usually subsequently diluted further before application. All commercial hydrate of predictable quality must be produced entirely in *closed-circuit systems* to prevent recarbonation. Originally *batch* processes[6] were exclusively employed for hydration, but *continuous* systems[7, 8] have steadily replaced most batch systems because of greater capacity, automation, and superior dust control. Yet batch systems may still be economical where hydrate production is small or sporadic and where only one quality of hydrate is required. Some continuous plants may maintain batch hydrators as stand-by capacity or for sudden surges in demand.

Hydration *plant layout,* operation, and size and type of equipment are probably as highly individualized as its derivative quarry, stone, and lime-burning plant layouts. Figures 10-3 and 10-4 present flow diagrams of two typical plant processes. Influencing modifications in plant design and equipment are the following variables confronting a hydration plant:

1. Porosity and slaking rate of the quicklime.
2. Chemical purity of the quicklime (% of impurities and core).
3. Physical size and gradation of quicklime to be fed into the hydration system.
4. Temperature of hydration water.
5. Particle size requirement of resulting hydrate, predicated on markets for which hydrate will be sold.

The significance of these and related factors has been described in Chapter 9. Yet, while externally the various plants and equipment appear to be, in some instances, radically different, essentially most operate on approximately the same principle. Case history type articles on commercial hydration plants are contained in past issues of *Pit & Quarry*[9] and *Rock Products.*[10] Generally equipment in commercial hydration plants, utilizing atmospheric pressure, is comprised of:

1. *Grinding equipment* is usually employed in most plants to reduce lump or pebble quicklime to at least $\frac{1}{4}$ to $\frac{1}{2}$ in. or smaller size even though waste quicklime fines that are screened off at the kiln discharge are also normally fed to the hydrator. In many plants the demand for hydrate exceeds the supply of by-product fines, and any excess

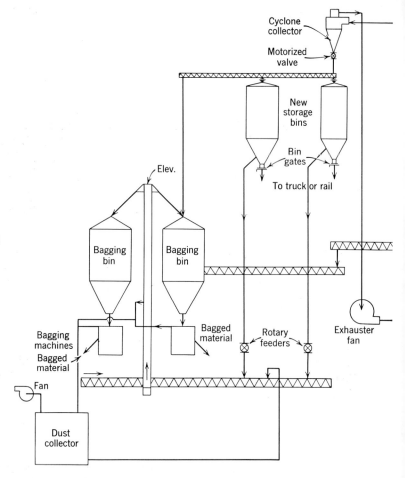

Fig. 10-3. Flow diagram of a modern hydrated lime plant from ground quick (Courtesy, *Pit & Quarry*.)

is necessarily obtained from "virgin" lime by grinding, often from the most undesirable odd sizes or from poorest quality (under- and over-burned). In other plants where hydrate is the major commercial product, the best quality of quicklime is selected. Finishing hydrates that command a premium price are invariably derived from the best quality of quicklime available. A uniform and reasonably restricted gradation is desirable for consistent and high-quality hydrate rather than variable fractions.

In some plants the fines are hydrated separately from the large

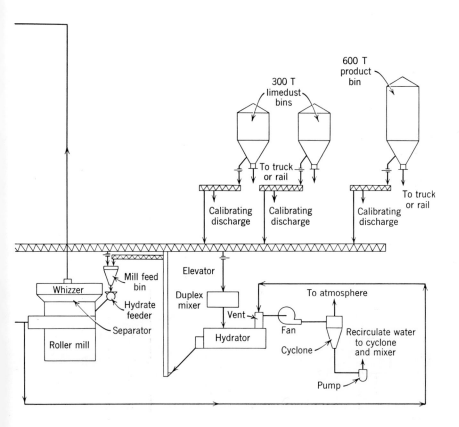

lime feed silos through to bulk hydrate storage silos and bagging department.

crushed and screened material. With reactive quicklime the heat of hydration with coarser fractions can be controlled more easily than with fines; with quicklimes of low reactivity finely divided material of Nos. 8 to 100 mesh are more efficient to hydrate.

The most adaptable type of grinders are the hammer mill, impact breaker, small gyratory, and cone mill. The comminuted material is conveyed to a raw-lime storage silo or bin.

2. *Storage bins and conveyors* are required for the classified quicklime raw feed and to convey the quicklime to the hydrator. Storage

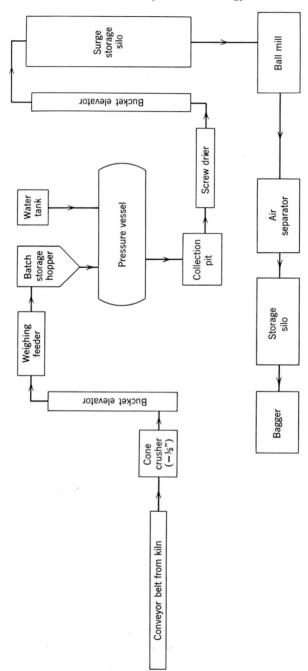

Fig. 10-4. Simple flow diagram of a pressure hydrated lime process. (Courtesy, *Pit & Quarry*.)

facilities are of a silo type with hopper bottom and with capacities of 100 to 600 tons, depending upon the capacity of the plant. The conveyor is usually an enclosed circuit bucket elevator since most storage silos are elevated.

3. *Feeder* of constant weight or volumetric type introduces quicklime from the silo-hopper to the hydrator or a premixing chamber.

4. *A hydrator machine,* equipped with agitation devices, that will intimately mix a predetermined proportion of lime and water is the "keystone" of a hydration plant. Temperature-control equipment and provision for venting off some of the heat of hydration and lime dust through a stack with dust-control apparatus are necessary auxiliary equipment in modern plants.

Hydrator chambers or troughs are both vertical and horizontal types with paddles, blades, or rabble arms for agitation and for impelling the mixture through, like a screw conveyor. Water is sprayed through an inlet into the stack and, as it descends to the hydrator, ascending exhaust heat preheats the water for more rapid hydration. The lime dust that is emitted up the stack is recovered in a washer-scrubber collector, and the resulting milk-of-lime is recycled back to the hydrator or to a *premixer chamber,* where it is initially intermixed with the quicklime before entering the hydrator. Hot, clean air is vented from the stack into the atmosphere, thus averting atmospheric pollution and lime wastage.

Meanwhile in the hydrator the fine, fluffy hydrate particles form and rise above the heavier unhydrated particles, core, and impurities and overflow a dam or weir and are discharged. Thus, feeding and discharging occur continuously. Much of the core and impurities are collected in a sump and removed as "tailings." Some hydrate models contain *sealed door openings* that permit observation of the process in operation and convenient accessibility for interior adjustment or repair.

5. *Enclosed circuit conveyor* for transporting the semiprocessed hydrate from the hydrator to the air separator, cyclone, or storage or curing compartment is of many diverse types, depending upon plant design. Usually a system of horizontal screw conveyors in tandem and bucket elevators are employed.

6. Enclosed, mechanical *air separators* of the centrifugal type that are conical shaped with hopper bottom are virtually universally employed in the final milling of the hydrate in order to classify the material to a fine state of subdivision. With highly refined hydrate this may be all — No. 200 mesh material, or even a more stringent mesh requirement, if the final product must contain 98% passing

a No. 325 mesh. Some grit escaping from the hydrator is easily ejected at this point.

Briefly stated, the principle behind the centrifugal separator is that the fine particles are lifted by strong ascending air currents, generated by a fan, operating countercurrent to the combined effect of gravity and centrifugal force. The coarser, heavier particles settle out so that this operation resembles a dry version of sedimentation.

These mills are adjustable, so that in some plants a series of several separators may be employed to produce hydrates of several different particle sizes and gradations. Hydrates that are not air separated would be considered crude, unrefined products, like agricultural hydrate, and of less value. This operation also further dries the moist, crude hydrate powder that enters the separator.

Rejects are usually conveyed to the waste pile along with "tailings" from the hydrator, providing the percent of waste is small, as it should be if a well-burned quicklime is efficiently hydrated. But much of this rejected material may contain lime, often pure lime agglomerated in granular form and also clinging to core and impurities. So if there is an appreciable amount of this waste material, there is often economy in reprocessing these "tailings" to recover usable hydrate in special machines that violently agitate the rejects in water, separating it from the grit, which is removed by sedimentation, like wet sand. The recovered lime in "milk" form is then reintroduced into the hydrator with the mixing water or into the premixer.

7. After milling, the finished hydrate is conveyed or fed by gravity to *storage silos,* located above the automatic bag-filling machines or special silos designed for bulk shipment.

8. *Bag-packing machines* encompass quite a number of highly automated patented types that can accommodate various types of multiwall bags, like the sewn-valve, sewn open-mouth, pasted open-mouth, and pasted valve, most of which are standardized at 50 lb. net weight. The cylindrical machines are made with four, six, eight, or twelve filling spouts that can be operated by one man. The nozzle of the spout is inserted through the valve opening and the bag is filled and weighed automatically, with the valve of the bag closing automatically when the bag is full. Bag spouts are located in tandem or circularly around which the filling cylinders rotate.

Productivity of hydrated lime bag packers is about 12–15 50-lb. bags per minute with the most modern machine and with one bagging operator.

Loading of bags into boxcar or truck is accomplished with push-button, "power-curve" conveyors that transport bags around curves and into small corners. This flexible conveyor consists of a series of endless spring belts, operating over grooved rollers.

9. Many plants utilize initially a *premixer* that partially hydrates the quicklime in a smaller chamber before discharging the mixture as a slurry to the hydrator. Agitation is necessary, and such mixing is achieved in equipment resembling a small pug mill. The high initial heat of hydration stimulates violent agitation and intimacy of contact between quicklime and water.

10. Other supplemental equipment are *aging bins*, which are similar to *retention silos*. These are usually only employed in plants that use hard-burned, slow-slaking quicklime or dolomitic quicklime. The raw hydrate is fed into these tanks directly from the hydrator in a wet condition (15 to 20% excess water), permitting the hydration to be completed with varying retention periods. After hydration is complete, the crude hydrate is fed to the air separator for milling. Such mills may be intermittently used to contribute flexibility to the operation in the event that slaking time of the lime lengthens and to abet coordination of the flow of material through the comminuter, hydrator, and milling equipment.

One patented process[11] claims to secure improved plasticity by initially slaking quicklime in considerable excess water and then feeding dry pulverized quicklime into the "milk," which is hydrated and reconcentrates all of the lime slurry into a dry powder. Such a process would require meticulous proportioning of lime and water.

Photographs of two atmospheric hydrators that are radically different in appearance are shown in Figs. 10-5 and 10-6. These are the most widely employed in the United States—the Schaffer and Kritzer hydrators, respectively. The Knibbs hydrator is used on an international basis.[12, 13] Capacities of hydrators naturally vary considerably (1 to 15 tons per hour), but a few will produce up to 25 tons per hour. Some plants prefer to operate two or three separate small-to-medium-sized hydrators for different quality products. The combined hourly capacity of the small units may exceed a single large hydrator and provide greater flexibility in operation.

Pressure Hydration. At atmospheric pressure high-calcium hydrates, unless they are severely overburned, completely hydrate into a hydroxide within minutes to an hour without much difficulty, but this does not occur with dolomitic hydrate (p. 289). Very few can be even substantially hydrated at atmospheric pressure in any practical length of time, and those few that can must be retained in silos (steeped in excess water) for ten to twenty-four hours. The *silo reten-*

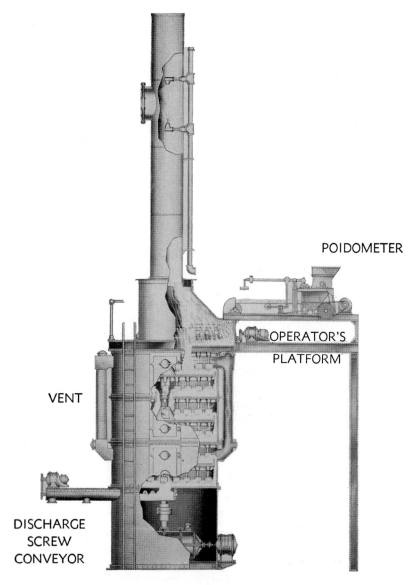

POIDOMETER

OPERATOR'S

PLATFORM

VENT

DISCHARGE
SCREW
CONVEYOR

Fig. 10-5. Schematic drawing of a modern hydrator equipped with stack and sprays for wet dust collection. Capacities of this type range from 2 to 15 tons per hour. (Courtesy, L. H. Eberhart.)

tion method[14] appears to be limited in application to only a few types of coarse-grained dolomitic quicklimes. The MgO component of dolomitic hydrate that was necessarily overburned resists hydration while the CaO portion hydrates with the same facility as high calcium lime.

Consequently, most normal dolomitic hydrates contain a paucity of $Mg(OH)_2$, generally only about 10 to 25% of the MgO is actually hydrated. A few soft-burned products from rotary kilns are more responsive, and those utilizing the silo retention principle are about 95% hydrated. As a result of the problem with U.S. federal authorities of suspected unsoundness in dolomitic hydrates because of expansion from delayed hydration (described on p. 407), commercial production

Fig. 10-6. 6-cylinder hydrator with pre-hydration chamber at top; capacity: 10–15 ton/hr. (Courtesy, Kritzer Co.)

of pressure-hydrated lime was commenced by a few U. S. producers in the late 1930's. By subjecting the dolomitic hydrate to pressure and high temperature, it was discovered that all or most of the MgO could be hydrated. This led to the development of highly hydrated dolomitic lime, which was soon recognized by ASTM in their materials specification as Type S (Special) hydrated lime (p. 397).

A pioneer of this type of lime was G. & W. H. Corson, Inc., which developed patents on their batch and continuous processes of pressure hydration and licensed some U. S. and foreign companies.[15-17]

Other patents were issued to other dolomitic companies involving continuous pressure hydration.[18] Trade acceptance in the United States of this superior-quality hydrate for structural purposes was rapid, although its growth in other countries has been slow. Since the patents appear to have some overlapping claims and since some details surrounding these processes are proprietary, this process will be described rather generally as follows:

Its principal distinguishing feature from atmospheric hydration is that an *autoclave* cylinder is used as the hydration chamber into which lime and water are introduced at a constant weight ratio or volumetric proportion. In other plants the lime and water is fed to a premixer so that the resulting partially hydrated lime slurry is conveyed into the autoclave. Pressures vary among different plants from 25 psi to 100 psi; 80 psi is frequently employed. Higher temperatures are also stimulated by the pressure (250 to 400°F range), much higher than in atmospheric hydration plants, so that hydration occurs in a vaporous atmosphere. Generally the time of hydration is shortened by the highest pressure and temperature. Size of the quicklime feed varies but is generally smaller than in conventional hydration. Most plants use ground or pulverized sizes. Proportion of water is also greater, so that hydrate at time of discharge from the autoclave is quite wet. Sizes of autoclaves vary, but cylinders 7 ft. in diameter by 20 ft. in length are rather typical.

A few companies use an explosion method[15] for discharging the hydrate from the autoclave through a small pipe into a cyclone collector. The explosive ejection of the lime dries it and markedly decreases its particle size, a large percentage of which is submicron.

From the collector the lime is conveyed to a *tube mill*. Due to the kneading action of this mill, the submicron particles are loosely agglomerated and compressed into plate-like shapes, improving the product's plasticity. From the tube mill it is conveyed to an air separator for further milling or to screens prior to bagging.

With the additional processing steps and higher capital equipment cost, manufacturing costs are higher than conventional hydration, but the product also commands a premium price by virtue of its higher and more uniform plasticity, development of instantaneous plasticity that eliminates soaking, and higher mortar strengths. Capacity of most of these pressure plants are 6 to 15 tons per hour.

High calcium producers do not employ this process, but several are reported to be employing some similar steps, such as tube milling,[19] to improve plasticity (and attainment of it quickly) in their products. So it appears that some aspects of this process are beneficial to all building lime hydrates. One such producer uses a hydroclave.[20]

The only other equipment that produces a highly hydrated, quick plastic dolomitic lime is the *Kennedy Van Saun hydrator* that operates without an autoclave.[22-23] This process is more similar to conventional hydration. Its main innovation being that a heat exchanger is uniquely employed to recover the initial heat from hydrating the CaO component and then recycling the same heat back to catalyze the hydration of sluggish MgO at much higher temperature. A high-speed mixer is used to premix the material; a rod mill, called a "plasticitor," is employed in lieu of a tube mill. It appears to be a simpler process than pressure hydration, but it is not nearly as widely employed as the latter. Possibly this is because of its later introduction, but also it is still unknown if it is applicable to all dolomitic products.

A patented process for making "adhesive lime"[24] evoked considerable excitement because of claims that the resulting hydrate develops much higher tensile strength and adhesiveness than normal limes. The process consisted of hydrating a high calcium quicklime with 1.5 to 4% HCl, forming some $CaCl_2$ in the lime. But no company has commercialized on the process to any extent.

Another patented process involved slaking quicklime in water containing a high molecular weight organic compound (wetting agent), such as sodium lauryl sulfate, in the proportion of 4 oz. of additive to a ton of quicklime. Unique properties were claimed, such as a greatly reduced surface tension of an aqueous lime solution, but it has never been commercially developed.

Purity. Qualitatively there is inconsequential difference in the chemical analyses of hydrates and their antecedent quicklimes, except for their water content. But quantitatively there may be an enormous difference. *Invariably, hydrates are purer than their derivative quicklimes.*

Let us assume that prior to hydration a high calcium quicklime contains the following impurities and that the resulting hydrate is efficiently produced and finely milled:

CO_2	2.5% (as $CaCO_3$ core)
SiO_2	1.5%
Al_2O_3	.5%
R_2O_3	.5%
	5.0% Total

Speculating on the anticipated analyses of the hydrate, the following would be typical:

On a nonvolatile basis (oxide)		On a volatile basis[a] (hydroxide)	
CO_2	0.5%	CO_2	0.375%
SiO_2	0.3%	SiO_2	0.225%
Al_2O_3	0.15%	Al_2O_3	0.113%
R_2O_3	0.25%	R_2O_3	0.193%
Total	1.20	Total	.906

[a] Calculations assume 25% total H_2O.

Most of the core and fused calcium silicate and aluminate are readily removed because of their relatively coarse gradation. Most of the residues of these impurities remaining are probably quite fine, approximating the largest hydrate particles in size. As a result, they cannot be economically removed with conventional mechanical milling equipment. Probably most of these fines are derived from attrition, i.e., grinding the quicklime and from agitation and friction of lime particles in the hydration process. Invariably, all dry hydrates will contain a fractional percent of free (or "bound") water, not chemically combined as the hydrate, of 0.2 to 1.0%. The presence of such water is not regarded as an impurity, just as an inconsequential diluent.

Thus, as a generalization, commercial hydrated lime is a purer source of CaO or CaO·MgO than quicklime, although appreciably less concentrated, due to its water content.

On an average, hydrated lime is a more profitable product to manufacture than quicklime. About 125 tons of hydrate are made from 100 tons of quicklime, and the hydrate is generally sold at about 15% higher price per ton on a comparable bulk basis. Even allowing for the loss of 5% of its weight as "tailings," the total sales return is 40 to 50% greater than its equivalent as quicklime, exceeding the additional processing cost entailed. This applies to normal hydrate; with pressure and special hydrates, profit will be still greater. Bagging adds about $3.50/ton more to costs and/or prices.

REFERENCES

1. W. Rudolfs, *Lime Handling, Application, & Storage in Treatment Processes,* Nat'l Lime Ass'n (1949), pp. 59–69.
2. E. Harper, "Lime Slaking," *J. Am. Water Wks. Ass'n* **26,** (June 1934), p. 752.
3. H. Lordley, "Lime & Lime Slaking," *Water & Sewage Wks. Ref. Data* (1955), p. R-214.
4. R. Locke, "Slaking Lime—An Art," *Plastering Inds.,* (May 1948), p. 14.
5. H. Hartung, *Water Wks. Enging* **89,** no. 8 (Apr. 1936).
6. A. Searle, *Limestone & Its Products,* E. Benn, Ltd. (London, 1935), pp. 487–516.
7. G. Lacy, "Hydrating Systems," *Rock Prod.* (Sept. 1946), p. 68.
8. W. Cliffe, "Methods Employed in Hydration & Slaking Lime," *Pit & Quarry* (June 1946), p. 77.
9. *Pit & Quarry* (articles on hydration)—1/53, p. 155; 10/57, p. 106; 11/57, p. 136.
10. *Rock Prod.* (articles on hydration)—12/53, p. 118; 6/55, p. 84; 7/55, p. 70.
11. W. Carson, U. S. Patent 2,147,191, *Plastic Powdered Hydrated Lime* (Feb. 1939).
12. N. Knibbs, Brit. Patent 480,215, *Apparatus for Slaking Lime* (Feb. 1938).
13. R. Adams, *Pit & Quarry* (May 1953), p. 103.
14. C. Loomis and W. Barrett, U. S. Patent 2,408,324, *Hydrated Dolomitic Limes* (Sept. 1941).
15. Anon., "Corson Explosion Method of Continuous Pressure Hydration," *Pit & Quarry* (Dec. 1949), p. 85.
16. B. Corson, U. S. Patent 2,309,168, *Powdered Dry Hydrated Lime* (to G. & W. H. Corson, Inc., May 1943).
17. B. Corson, U. S. Patent 2,409,546, *Methods of Conditioning & Treating Lime* (Oct. 1946).
18. W. Garvin, U. S. Patent 2,356,760 (to Standard Lime & Stone Co., Aug. '44). H. Huntzicker, U. S. Patent 2,408,647 (to U. S. Gypsum Co., Oct. 1946). H. Huntzicker, U. S. Patent 2,489,033 (to U. S. Gypsum Co., Nov. 1949). J. Volk, U. S. Patent 2,902,346 (to National Gypsum Co., Sept. 1959). J. Volk, U. S. Patent 2,957,776 (to National Gypsum Co., Oct. 1960).
19. M. Rikard, U. S. Patent 2,894,820 (to American Marietta Corp., Aug. 1959).
20. B. Herod, *Pit & Quarry* (Dec. 1956), p. 92.
21. A. Allen, U. S. Patent 2,888,324 (to Kennedy Van Saun Mfging, May 1959).
22. Anon., "Rockwell Lime Co. Operating Non-pressure Hydrator," *Pit & Quarry* (May 1956), p. 136.
23. Anon., *Pit & Quarry* (Dec. 1953), p. 114.
24. A. Pozzi, U. S. Patent 2,230,761, *Hydrating Lime* (to Adhesive Lime, Ltd., Feb. 1941).
25. H. Huntzicker, U. S. Patent 2,201,667, *Hydrated Lime* (to U. S. Gypsum Co., May 1940).

CHAPTER ELEVEN

Uses of Lime

Undoubtedly, there is no other material in commerce that has a greater myriad of diverse uses and varied functions (if as many) as lime; it uses are almost countless. In total tonnage it cannot be compared with such giants as oil or coal, but these materials are consumed primarily for one purpose—combustion—and for a narrow group of lesser applications. In contrast, there are *many* functions of lime.

In *industry* it is the second largest "heavy" (or basic) chemical, a close second to sulfuric acid and is used:

For	As
Neutralization	Flux in metallurgy
Coagulation	Specialized lubricant
Causticization	Bonding agent
Dehydration	Filler
Hydrolyzation	Raw material
Absorption	Refractory

Then it is employed *in building construction* as a cementitious material and plasticizer; *in agriculture* to supply calcium and magnesium as plant nutrients and for neutralization; and lastly (and most recently) *in highway construction* for soil stabilization and as a filler and antistripping agent in asphalt. For further elaboration on lime's uses and functions see Fig. 11-1 for diagrammatic chart.

Both high calcium and dolomitic limes are employed interchangeably for many of the above functions. However, for certain purposes and uses there are general preferences in varying degrees for one type over the other, and for some purposes only one type can be used. Similarly, in most of the above functions both quick and hydrated limes can be employed, but again, for a few purposes only one or the other is specified. Where the use is interchangeable, the

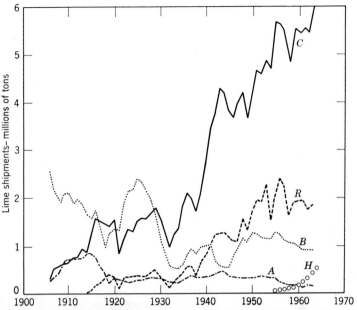

(U. S. Bureau of Mines statistics modified from 1953 to 1962 by eliminating captive lime tonnage to make end-use pattern consistent with method of compilation used from 1906 to 1953.)

LEGEND:
  C   chemical (industrial)   ———————
  B   building (structural)   . . . . . .
  R   refractory   – – – – – – –
  A   agricultural   – · – · – · – ·
  H   highway construction   ° ° ° ° ° °

Fig. 11-2. Changing end use of lime.

preference for quick or hydrated lime depends largely on the volume of consumption and to a lesser extent on the type of storage facilities. Quicklime, being an anhydrous form of lime, is more concentrated than hydrate and also costs less per ton, offering appreciable savings in raw material cost (averaging about 50 to 60% less on an equivalent basis). However, in most cases, the quicklime must be slaked into a milk-of-lime or paste by the consumer before it can be fed into the process. This necessitates an investment for slaking equipment and an added manufacturing step for the consumer to absorb. With hydrated lime the slaking is performed by the lime manufacturer. Generally if the volume of usage exceeds three to four carloads a

Table 11-1. U. S. Bureau of Mine's statistics—end-use breakdown of lime sold or used

Use	1962 Open-market	1962 Captive	1962 Total	1963 Open-market	1963 Captive	1963 Total
Agriculture	192,000	—	192,000	176,000	[a]	176,000
Construction:						
Finishing lime	446,204	—	446,204	463,838	—	463,838
Mason's lime	469,746	74,165	543,911	457,376	—	457,376
Soil stabilization	275,914	73	275,987	395,577	83	395,660
Other	24,127	—	24,127	39,392	78,903	118,295
Total[e]	1,216,000	74,000	1,290,000	1,356,000	79,000	1,435,000
Chemical and other industrial:						
Alkalies (ammonium, potassium, and sodium compounds)	26,693	3,068,236	3,094,929	11,064	3,129,125	3,140,189
Brick, sand-lime, slag and silica	25,457	—	25,457	32,267	—	32,267
Calcium carbide	565,375	361,339	926,714	583,869	368,147	952,016
Glass	255,861	—	255,861	243,717	—	243,717
Other chemical uses[b]	728,756	530,165	1,258,921	632,131	838,327	1,470,458
Metallurgical uses:						
Aluminum	50,164	280,253	330,417	91,559	90,284	181,843
Copper smelting	106,503	199,689	306,192	107,574	178,457	286,031
Magnesium	26,802	336,770	363,572	17,814	93,539	111,353
Other nonferrous	3,814	35,957	39,771	4,720	—	4,720
Ore concentration[c]	52,856	2,427	55,283	62,226	2,840	65,066
Steel flux	1,574,867	74,536	1,649,403	1,842,789	71,643	1,914,432
Miscellaneous steel processing (wire drawing, etc.)[d]	32,283	35	32,318	19,190	41,664	60,854

Table 11-1 (Continued)

Paper and pulp	610,429	33,944	644,373	747,896	40,212	788,108
Sewage and trade-wastes treatment	117,709	15,698	133,407	202,478	43,798	246,276
Sugar	30,152	515,008	545,160	25,497	585,735	611,232
Water softening and treatment	741,593	7,130	748,723	852,672	342	853,014
Total[e]	4,949,000	5,461,000	10,411,000	5,477,000	5,484,000	10,961,000
Refractory lime (dead-burned dolomite)	1,787,000	71,000	1,858,000	1,880,000	69,000	1,949,000
Grand total[e]	8,145,000	5,608,000	13,753,000	8,889,000	5,632,000	14,521,000

[a] Included with open-market agriculture lime to avoid disclosing confidential data.

[b] Includes alcohol, calcium carbonate (precipitated), coke and gas, food and food byproducts, insecticides, medicine and drugs, explosives, oil-well drilling, paint, petrochemicals, petroleum refining, rubber, tanning, salt, miscellaneous and unspecified uses.

[c] Includes flotation, cyanidation, bauxite purification, and magnesia manufacture.

[d] Includes wire drawing, and various metallurgical uses.

[e] Data may not add to totals shown because of rounding.

month, economy dictates the use of quicklime; below this consumption level hydrated lime is indicated. Such preferences and limitations will be mentioned specifically in the succeeding use descriptions.

Another peculiarity in lime's widespread use is its drastically changing end-use pattern over the past fifty years.[1] (See Fig. 11-2). In the 1900–1910 period about 75% was employed as a building material, with the balance divided about equally between agricultural and chemical or industrial uses. In 1963 only about 7% was used as structural lime, with most of the balance being used as chemical and refractory. Agricultural lime has ebbed to about 1.5% of the total, and a new use in highways now accounts for about 5%. Chemical or industrial lime is a strong first with 75.5% of the total, whereas refractory lime is 11%. Few materials have undergone such a radical change in use, and there are indications that further changes in lime's fluid end-use pattern will occur in the next two decades as certain uses expand and others ebb. Strangely and ironically, in spite of its unquestioned insurgence as primarily a chemical,[7, 10] the connotation of the word "lime" still means mortar, plaster, or agricultural liming to the layman, and even to many technical people who should know better. See Table 11-1 for the U.S. Bureau of Mines' compilation of 1962–1963 statistics on lime consumption and shipments by end use.

CHEMICAL AND INDUSTRIAL USES

Metallurgy

There are numerous applications of lime in this broad end-use category, but the most important is for fluxing steel. Actually, tonnage of refractory lime (dead-burned dolomite) has exceeded fluxing lime for many years, but the latter is growing steadily, whereas the former is declining. Certainly in 1964 fluxing lime exceeded refractory in volume for the first time in the United States. Other smaller but essential uses occur in nonferrous metallurgy.

FERROUS USES.

Steel. In steel manufacture lime acts as a *flux* in purifying steel during the "heat" by promoting fusion of the slag and assisting in the removal of phosphorus, silica, and sulfur as calcium phosphates, silicates, and sulfides in the slag that is tapped off from the molten metal. Its use, of course, is confined to the basic steel-making processes, which are utilized in most steel plants. Lime-consuming steel furnaces would be the *basic open hearth, basic Bessemer* (Thomas converter and related types), *electric furnace,* and, last, the most recently developed *basic oxygen converter* (LD, LD-AC, LD-Pompey, Kaldo, OLP, and related types). These converters are also characterized generally as "basic oxygen furnaces" or BOF.

Before introduction of the basic oxygen converter in the mid-1950's, average consumption of quicklime per ton of ingot steel in the United States was about 28 lb. In open hearths in particular, most of the flux employed is limestone (see p. 83), to which lime is added later in the "heat" in much smaller increments to achieve greater phosphorus removal or to accelerate or finish off the heat."[11, 12] In fact, some open hearths in the United States use only limestone as the flux and others only small amounts of lime when they are striving for maximum output during times of heavy demand. As a generalization, the rate of lime usage in open hearths increases in direct proportion to the amount of steel scrap in the furnace charge, so-called cold-metal shops. With high scrap charges lime is commonly applied in the initial charge with limestone as well as later in the "heat". Furnaces employing largely hot-metal, molten pig iron from contiguous blast furnaces, use much less lime on an average. These are "hot-metal" shops. Electric furnaces have been relatively high lime consumers—30 to 80 lb./ton of ingot steel—but their production capacity is too comparatively small to increase the lime factor in steel appreciably because of the dominance of the open hearth.

The greatest stimulation toward increasing the lime factor in steel is the BOF furnace, which requires *only* quicklime (and no limestone) for fluxing, and the rate of usage is 100 to 160 lb. (130 lb. average) per ton of ingot steel—over six times the average consumption rate of the open hearth. Consequently, as more oxygen-converter types are placed in operation, usually replacing open hearths, the lime factor in steel is gradually increasing, so that in 1962 it was 32 lb., a 17% increase since 1957; in 1964 it was about 38 lb.; and by 1970 the factor should approach 45 to 48 lb. This is predicated on the economy of the BOF being conclusively proven and that all major steel companies (as of 1965) are either operating these furnaces or are constructing new BOF plants in spite of possessing considerable idle capacity. It has proven to be one of the greatest technological revolutions in steel making. Steel "heats" that formerly required four to eight hours in the basic open hearth are obtained in thirty to sixty minutes in the BOF at less capital investment per ingot ton. Most steel authorities predict that no more open hearths will ever be built. Attempts to modify open hearths with oxygen injection through lances has generally failed to equal the BOF in all-round performance.

The *chemical reactions*[15] that occur in a steel furnace, regardless of type, are quite complex, but the following equations indicate the principal reactions that occur in those processes using massive oxygen injection:

1. The gaseous oxygen penetrates into the molten metal bath (various pig iron–scrap combinations) and oxidizes such impurities as carbon, silicon, and manganese into a slag phase, and inevitably some of the phosphorus and iron is also oxidized.

$$2C + O_2 = 2CO\uparrow \qquad\qquad 2Fe + O_2 = 2FeO$$

$$Si + O_2 = SiO_2 \qquad\qquad 4P + 5O_2 = 2P_2O_5$$

$$2Mn + O_2 = 2MnO \qquad 2Fe_3P + 5FeO = P_2O_5 + 11Fe$$

2. To enhance the fusibility of the formative slag, quicklime is added in an amount approximating a CaO to SiO_2 ratio of 3:1, forming a strong basic slag. The lime then combines with the silica from the slag and the phosphorus and sulfur from the metal phase.

$$SiO_2 + 2CaO = 2CaO \cdot SiO_2$$

$$Fe_3(PO_4)_2 + 3CaO = Ca_3(PO_4)_2 + 3FeO$$

$$FeS + CaO = CaS + FeO$$

$$MnS + CaO = CaS + MnO$$

Phosphorus and sulfur reactions invariably occur after the silica is removed. Since silica possesses a stronger affinity for CaO than phosphorus pentoxide, it can easily displace the combined phosphorus, as indicated below, causing a retrogressive reaction, unless there is ample free lime available, as characterized by a slag of high basicity.

$$Ca_3(PO_4)_2 + 3SiO_2 = 3CaSiO_3 + P_2O_5$$

If this occurs, more lime must be added to recombine with the P_2O_5 for removal in the slag.

In the basic open-hearth process considerable value is manifest by some metallurgists on the "lime boil," achieved through the exclusive use of limestone as flux in the initial charge. As metal is heated to a molten state, limestone gradually calcines and evolves CO_2, contributing a turbulence to the molten steel "heat." Lumps of lime or decomposing limestone float on top of the molten metal in the slag that has started to form. The CO_2 evolvement causes the slag to froth and foam, and it does oxidize some of the iron $(Fe + CO_2 = FeO + CO\uparrow)$. Meanwhile, the CaO that forms is combining with the free SiO_2 and partially displaces some of the FeO and MnO that has combined with other silica in the slag as follows:

$$2CaO + SiO_2 = 2CaO \cdot SiO_2$$

$$CaO + FeO \cdot SiO_2 = FeO + CaO \cdot SiO_2$$

$$CaO + MnO \cdot SiO_2 = MnO + CaO \cdot SiO_2$$

The small charges of quicklime are introduced toward the end of the "heat" solely for more complete dephosphorization.

Ranges and approximate average flux charges for the principal U.S. steel furnaces in lb./ton of ingot steel would be:

Basic open-hearth	
"Hot-metal" furnace	*"Cold-metal" furnace*
Limestone 45–190	0–125
(150 average)	(50 average)
Lime 0–80	25–120
(15 average)	(50 average)
Electric furnace	*Basic oxygen converter*
Limestone 0–100	0[a]
Lime 30–80	100–160
(50 average)	(130 average)

[a] A few plants use a small amount of limestone as a coolant.

The amount of flux on a CaO equivalent basis varies widely, even in the same plant in different "heats," depending upon the amount of impurities in the charge (ferrous burden and fuel) as well as the quality tolerances in the finished steel (low carbon, alloy, etc).

Independent of the fluxes employed, fluorspar and/or iron ore are added to most steel "heats" to increase the fluidity of the slag. Generally the fluorspar addition does not exceed 10 lb./ton of steel. The cooling effect[16] of materials used in steelmaking to heat to 2875°F has been calculated as follows:

Material	*Btu/lb.*
Steel scrap	592
Oxygen	722
Fluorspar	1000
Quicklime	1189
Iron ore, mill scale	1280
Steam (400 F)	1415
Limestone	1914
Water	2537

There is utter lack of uniformity among American steel companies on their *lime requirements*. Originally only a high calcium quicklime was used, preferably low in MgO, SiO_2 and S. This situation still prevails with older traditional processes, although there are numerous differences in their maximum tolerances on impurities. However, in the BOF processes there is considerable experimentation by a number of steel companies on the use of dolomitic quicklime, usually as a 10 to 20% substitution for high calcium, as a flux. The attrition on furnace refractory linings is most severe in the rapid BOF process, and it appears that this partial use of dolomitic quicklime will extend the life of expensive refractory linings and still be efficacious as a flux. But further experience is necessary before conclusive evidence of this is established.

The advent of the BOF has posed another question. Since speed is the keynote of this process, should the lime be of as high reactivity as possible? Such limes would presumably go into solution faster and perform their fluxing role with greater efficiency. This would indicate need of a very soft-burned lime. Previously in the relatively slow, cumbersome open-hearth process scant attention was given to the efficiency of a soft- or hard-burned lime. While there are exponents for a soft-burned lime among steel consumers, others contend that lime must be physically strong so that it does not decrepitate and crush easily during the severe attrition of conveyance and charging into the furnace. To satisfy this requisite a hard-burned lime is necessary. In order to reconcile these diametrically opposed viewpoints, an intermediate "burn" may be the logical solution.

Some plants specify lime of very low core content of 0.5% maximum, since they contend that CO_2 evolvement is deleterious or retarding to the process; others are not concerned with core content within reasonable limits of 2.5% maximum; a few plants are known to prefer a CO_2 content of 4 to 5%, for inexplicable reasons. Only time may resolve these various conflicting theories on optimum lime quality for BOF steel.

However, most steel companies are emphatic that they desire lime of low silica and sulfur of about 1.5% and 0.03% maximum, respectively (and less if economically possible). Tolerances on sulfur have become increasingly stringent since the late 1950's, when coal companies for economic reasons were forced to use new, highly mechanized coal mining equipment that yields coal of higher sulfur content.

The preferred physical size of lime has traditionally been lump or pebble, usually the latter, of 1×2 in. But again, the BOF may influence another change, although most BOF plants utilize pebble lime in the 1960's. In Europe some BOF plants are employing process modifications, known as the OLP and LD-AC, in which pulverized quicklime is introduced along with the stream of oxygen through injection lances into the converter in varying concentrations. One company stipulates an extremely fine grind of substantially no. 200 mesh material, a veritable "dust." Some knowledgeable metallurgists predict expanded use of pulverized quicklime in the BOF. Possibly the minute physical size will accelerate dissolution of CaO in this process more than a soft burned pebble lime.

Before the BOF, European steel plants always consumed proportionately much more lime for steel than the United States. An example is West Germany, which used 165 lb./ingot ton of steel in 1959. Similar or higher factors also prevail in other western and "iron curtain" countries. This stems largely from the poorer quality of foreign iron ores, which contain in particular a higher phosphorus content than most American ores. Europeans also utilize the Thomas converter (Bessemer) extensively in making steel—more than the open hearth. In many of these Bessemer plants only lime is employed as flux at consumption rates equal or higher than American BOF practice. It is doubtful if the BOF will increase lime consumption for steel in Europe because of their high historical rate of use. In fact, in Germany and a few other nations there is a perceptible trend in a reduced consumption rate; for example, in Germany the lime factor in 1963 had declined to 150 lb. The major reason for this decline is that these nations are constantly increasing the use of ore agglomerates (self-fluxing sinters, etc.) in blast-furnace charges which are more concentrated in iron and which contain less impurities than their traditional unbeneficiated lump iron ores. The pig irons which result are purer and in turn reduce fluxing requirements in their steel processes.

Pig Iron. Lime is not employed as a direct flux in blast furnaces in pig iron manufacture. This practice is limestone's exclusive province (p. 79). There has been experimental use of pulverized quicklime introduced into tuyeres of the blast furnace as a partial replacement of flux stone, but the results have been inconclusive.

DESULFURIZATION. Desulfurization of pig iron with pulverized quicklime[14] of 0.5 mm. size is practiced in Europe in rotary cupolas; 90 to 95% reduction in sulfur is effected within 15 min. of rotation and with a lime increment that is about 10 times the sulfur content of the unrefined pig iron, usually a 1 to $2\frac{1}{2}$% addition. Extremely low sulphur contents, below 0.005%, can be achieved with this process which is reported to be more effective and economical than desulfurization with soda ash. The process is useful when pig iron or steel of very low sulfur content is required.

Steel Refractory. Virtually all refractory lime (dead-burned dolomite) is consumed for one purpose—lining *open-hearth* steel furnaces.[20, 21] As standard practice it is blown by special machines or shoveled or rammed by hand into furnaces, covering the costly refractory brick. Thus, it serves to maintain hearths by filling and facing tap holes and facing door banks. In this manner it conserves the life of refractory brick so that furnaces do not require relining nearly as often.

This dense, granular material is stabilized against hydration by addition of iron. On being charged into open hearths the high temperature melts its dicalcium ferrite component, causing the granules to coalesce or set into a monolithic mass, which forms part of the hearth's interior structure. The basic lime component of this product then resists the erosive effect of the high-silica (acid) slag in the early stages of the "heat." The MgO periclase constituent then minimizes subsequent erosion to the lining caused by strong basic slag that forms during the latter part of the heat. After each heat, the hearth lining is inspected and additional dead-burned dolomite is added to resurface eroded sections of the hearth, usually to a depth of a few inches.

Its rate of consumption increased substantially after World War II until in 1956 it reached 50 lb./ton of open-hearth steel. The European rate of consumption parallels that of the United States, e.g., Germany's factor was 44 lb./ton in 1963. But as the new BOF converter began replacing the open hearth, consumption has gradually ebbed, since the BOF does not require this type of refractory. So the future of this market is bleak. (For more information on this material, see p. 268 on its manufacture and p. 357 on related refractories.)

Minor uses of lime in steel plants are: *neutralization* of waste pickle liquors (for further information see p. 364 on acid neutralization); *wire drawing*[19] in which the steel rod is coated with a milk-of-lime solution, which acts as a lubricant, as the rod is drawn through successively smaller dies in forming wire; *"pig casting,"* in which the pig molds are coated with milk-of-lime to prevent the pig iron from sticking; water softening of *boiler feed water* with lime-soda softening process; *coke-oven by-product recovery* of ammonia and benzol; and in whitewashing steel parts to prevent *corrosion*.

Recent research has revealed some potential use for lime in the beneficiation of certain types of *low-grade iron ores,* like nonmagnetic taconite. It is mentioned as a flotation agent along with other additives, like starch, in concentrating the pulverized taconite and also as a binder in pelletizing this type of ore.

In the production of *self-fluxing sinters* (p. 83) there is slight usage, and apparently more prospective use, of hydrated lime and pulverized quicklime, usually with pulverized limestone in amounts ranging from 1 to 6% added to the sinter mix (pulverized iron ore and coke) on sintering strands.[17, 18] It appears to possess advantages over less costly limestone with certain types of iron ore. The criteria on the selection of lime or limestone or combinations has never been established. Plant trial runs appear to be the only reliable criterion. Experimental tests with a high calcium quicklime as a partial and complete replacement for dolomitic limestone with a Mesabi and a Canadian ore and various ore blends are summarized as follows:

1. Quicklime by readily hydrating was useful in absorbing excessive moisture in the sinter mix, thereby minimizing bed collapse.

2. The resulting slaked quicklime acts as a plasticizer or binder and promotes increased formation of desirable micropellets in the sinter mix, yet preserving satisfactory permeability.

3. The heat of hydration is beneficial by elevating the preignition temperature of the sinter mix.

4. The lime reacts more rapidly with the acid oxide constituents (gangue) in the sinter mix than limestone due to its finer particle size that permits it to envelope the other materials in the mix more intimately. There is also no delay for calcination.

5. But most important, the capacity of the sintering strands was increased, in some cases markedly (70%), with lime additions up to 5%.

6. In most tests the physical strength of the sinter was increased by lime additions.

Experience in Europe with self-fluxing sinters parallels that in the United States with respect to the use of lime and limestone. Most of the tonnage has been limestone, with interest in lime being of most recent origin.

Another related use with iron ore preparation is the patented process of the Blocked Iron Corporation, Philadelphia, in which hydrated lime is used as a binder in making *iron ore briquettes* for use in blast furnaces. An undisclosed percentage of hydrated lime (believed to be 4 to 8% by weight) is thoroughly mixed with the iron ore fines, molded, and then fed into special tunnel-like ovens where carbon dioxide is introduced which carbonates the lime and hardens the briquettes.

Small amounts of a special type of dolomitic quicklime, called "Vienna lime," is used for buffing and polishing metal parts.

Nonferrous Uses. Nearly every commercial process that has been developed for the production of *magnesium metal* or *magnesia* requires either a high calcium or dolomitic lime as the basic raw material.[22] Several of these magnesia-magnesium processes are described chemically as follows:

Dow Seawater Process[24]

$$Ca(OH)_2 \text{ (hydrated lime)} + MgCl_2 \rightarrow Mg(OH)_2 + CaCl_2$$

$$Mg(OH)_2 + 2\ HCl \rightarrow MgCl_2 + 2H_2O$$

$$MgCl_2 + \text{electrolysis} \rightarrow Mg + Cl_2$$

Dow Natural Brine Process

$$CaCl_2 \cdot MgCl_2 + MgO \cdot CaO \text{ (dolomitic quicklime)} + 2CO_2 \rightarrow$$
$$2CaCO_3 + 2MgCl_2$$

$$MgCl_2 + \text{electrolysis} \rightarrow Mg + Cl_2$$

Ferro-silicon (Pidgeon) Process[23]

$$2CaO \cdot MgO \text{ (dolomitic quicklime)} + Si \text{ (or FeSi)} \rightarrow Ca_2SiO_4 \text{ (or}$$
$$Ca_2SiO_4 + Fe) + 2Mg$$

Seawater Chemical Magnesian Process

$$MgCl_2 \text{ (from seawater)} + Ca(OH)_2 \rightarrow Mg(OH)_2 + CaCl_2$$

$$Mg(OH)_2 + \text{heat} \rightarrow MgO + H_2O$$

In the *flotation* (recovery or beneficiation) of nonferrous ores lime is widely used.[25] In *copper ore flotation* it acts as a depressant (settling aid) in addition to maintaining the proper degree of alkalinity. Similarly, it is employed in recovery of *gold* and *silver* in the cyanide flotation process[26] in order to curtail the loss of sodium or calcium cyanide, costly chemical reagents, and for pH control.

In *uranium ore beneficiation* lime's greatest use is in South Africa, where as much as 300,000 tons of quicklime have been consumed annually in a flotation process to recover uranium from huge piles of gangue (called "gold slimes") accumulated from gold mining. In Canada and the United States it is employed to neutralize sulfuric acid waste liquors in acid-type uranium ore extraction plants.

There has been limited use of lime to remove sulfur in the recovery of *mercury* from "mercury blends"; in the flotation of some types of *zinc, nickel,* and *lead ores;* and as a conserving agent to assist in the recovery of xanthates, another flotation chemical reagent.

In nonferrous metal *smelting,* it is used as a flux in the manufacture of *low-carbon chrome.* After *nickel*[27] is smelted, it is precipitated in a boiling solution of milk-of-lime. Often in *copper smelting* plants, noxious gas fumes of H_2S and SO_2 are removed or neutralized by venting the fumes through a scrubbing tower containing milk-of-lime, thus preventing atmospheric pollution and corrosion to plant equipment. To a lesser extent, this is also practiced in zinc and lead refineries.

In the manufacture of *alumina*[29] by the *Bayer process,* lime is employed in huge quantities in a few plants to causticize sodium carbonate solutions in generating sodium hydroxide, the key chemical reagent used in this process.

High-grade imported bauxite, the aluminum-bearing ore, is ground and digested in a NaOH solution, resulting from the causticization of soda ash with ground quicklime. The alumina content of the ore dissolves as sodium aluminate liquor, and the gangue, mainly silica, is precipitated as sodium alumino silicate. The gangue, known as "red mud," is removed by sedimentation from the sodium aluminate liquor after filtering and washing the filter cake; the clarified sodium aluminate is reintroduced into the process. When this cooled liquor is mixed with "seed" aluminum trihydrate and agitated, about half of the sodium aluminate decomposes into aluminum trihydrate, which is precipitated and removed. The remaining unreacted solution is recycled back into the process into digesters. After several washings, the hydrate precipitate is dewatered and calcined in rotary kilns at 2000°F, yielding concentrated anhydrous alumina of about $98.5\% + Al_2O_3$.

A few plants do not employ lime in this process. As an alternative, they use commerical caustic soda directly instead of causticizing soda ash with lime. Individual plant economics determine which alternative to employ. However, all plants will use at least a small amount of lime for *secondary desilification* in the process.

Raw-material requirements are predicated on the quality of the bauxite ore, but the following was reported as an average requirement for the alumina industry in 1951–1952:

| Bauxite | 2 ton | Quicklime | 175 | lb. |
| Soda ash | 280 lb. | Starch | 2.4 lb. |

The combination Bayer-lime-sinter process is described on p. 85.

LIME REQUIREMENT. The lime requirements for steel fluxing have been presented.

For refractory use there is no proven substitute for dead-burned dolomite among other types of limes. No high calcium lime is used, although there are a few patents covering the use of a very hard-burned high calcium quicklime for this purpose, notably U. S. Patent No. 2,916,389; Russian patent No. 119,117.

For some of the other metallurgical uses either high calcium or dolomitic quick and hydrated limes can be used. The exceptions would be for water softening, coke oven by-product recovery, iron ore beneficiation and self-fluxing sinters, alumina manufacture where high calcium of similar quality to open-hearth steel is preferred. Vienna lime is only dolomitic because of unique physical qualities that only certain fossiliferous dolomitic limestone seems to possess. Since some of these miscellaneous uses are relatively small in tonnage, hydrated lime may be preferred because of convenience.

Pulp and Paper

Most of the lime consumed in the paper industry at present is by the *Sulfate Process* (Kraft) pulp plants, where quicklime is employed to causticize the "black liquor" (a waste sodium carbonate solution).[30, 31] Lime reacts with the sodium carbonate to regenerate sodium hydroxide (caustic soda) for reuse in the Sulfate Process ($Na_2CO_3 + CaO + H_2O \rightarrow CaCO_3 + 2NaOH$). Most sulfate mills now recover the calcium carbonate sludge after the causticizing operation. The sludge is dewatered in centrifuges, dried, pelletized, and fed into captive rotary kilns, where it is calcined into quicklime for reuse in the above process. In other words, the economy-minded pulp mills recover the chemical—lime—that is used to regenerate their key reagent—caustic soda. About 90 to 96% recovery of lime is effected in this manner; the remainder that is lost in the process is purchased as "make-up" lime from commercial producers. If this recovery is considered as consumption, then the pulp-and-paper industry would be the largest of all lime-consuming industries, with a total annual consumption of about 5 million tons in the United States alone. However, only about 700,000 tons are purchased.

Historically, *sulfite pulp* plants used considerable lime, but today relatively little is consumed, since most bisulfite liquors are prepared by reacting SO_2 with limestone or other alkaline bases, like ammonia, magnesia, and soda ash.

In fact, the trend is away from limestone or lime because calcium bisulfite waste liquors create a much more costly waste disposal problem and this alkaline base cannot be recovered. In the Barker Process dolomitic quicklime was at one time widely used as the alkaline medium for reaction with SO_2, forming calcium-magnesium bisulfite as the pulp-cooking liquor. In the *Soda Process* lime's use would be similar to its use in the Sulfate Process.

Most all pulp plants still require some lime for calcium hypochlorite *bleaching* of paper pulp in addition to such other bleaches as sodium hypochlorite, liquid chlorine, and peroxide chemicals. They bubble chlorine gas into a milk-of-lime solution in which the lime absorbs the chlorine, making a reasonably stable bleach

$$(Cl_2 + Ca(OH)_2 \rightarrow Ca(OCl)_2 + H_2)$$

Some paper mills will also use lime in the chemical coagulation or softening of their plant *process water*. There has been limited use of lime as a coagulant aid or for pH control in treating pulp mill waste effluents to abate stream pollution. For specialized papers, which utilize rags or straw as a source of cellulosic pulp alone or in combination with wood pulp, quicklime is often employed instead of caustic soda as the "cooking agent" in converting such cellulose to pulp.

A high calcium lime is required for sulfate pulp and bleaching; both types are provided for in ASTM specification C46-27 (Quicklime for Sulphite Pulp Manufacture). Other ASTM specifications covering paper are: C433-59T (Quicklime and Hydrated Lime for Hypochlorite Bleach) and C45-25 (Quicklime and Hydrated Lime for Cooking Rags in Paper Manufacture).

Chemicals Manufacture

Alkalis. The largest consumers of lime in the chemical industry are alkali companies, which employ it as one of their main raw materials in the manufacture of soda ash, bicarbonate of soda, and caustic soda under the Solvay (or ammonia-soda) Process. However, without exception, this is strictly a captive use. For this process to be economical large quantities of low-cost CO_2 are required. This is obtained as a co-product in the burning of lime by capturing the waste gases from the captive "on site" lime kiln stacks. Its function is to causticize the sodium carbonate to sodium hydroxide, similar to the Sulfate pulp process in preceding section and to recover ammonia in the manufacture of soda ash. The latter function is stated chemically as follows:

$$2NH_4Cl + Ca(OH)_2 \rightarrow CaCl_2 + 2NH_3 + 2H_2O$$

$$2NH_3 + H_2O + CO_2 \rightarrow (NH_4)_2CO_3$$

$$(NH_4)_2CO_3 + H_2O + CO_2 \rightarrow 2NH_4HCO_3$$

$$NH_4HCO_3 + NaCl \rightarrow NaHCO_3 + NH_4Cl$$

$$2NaHCO_3 + heat \rightarrow Na_2CO_3 + CO_2 + H_2O$$

Lime factors are approximately 1350 lb. and 1400 lb. per short tons of soda ash and caustic soda (ammonia-soda process), respectively.

Carbide. In making calcium carbide, an important source of acetylene, quicklime is mixed with coke and heated in electric furnaces to 2000°C ($CaO + 3C \rightarrow CaC_2 + CO$). Molten carbide is removed from the furnace and upon solidifying the carbide is crushed and ground for shipment or generated into gaseous acetylene. About one ton of quicklime is required to make a ton of calcium carbide.

One carbide producer, Air Reduction Company, is now recovering lime for reuse in the above process. When acetylene is generated from calcium carbide, a waste calcium hydroxide (hydrated lime) is obtained [$CaC_2 + 2H_2O \rightarrow C_2H_2 + Ca(OH)_2$]. The waste hydrated lime is then dried, pelletized, and charged into kilns where the chemically combined water is expelled, forming quicklime for reuse in the carbide process.

A derivative of calcium carbide is calcium cyanimide, an important nitrogen fertilizer. This is manufactured by heating calcium carbide in the presence of nitrogen ($CaC_2 + N_2 \rightarrow CaCN_2 + C$).

The high calcium lime required for carbide manufacture must have a low phosphorus content of not over 0.02%, and is covered in ASTM specification C-258-52 (Quicklime for Calcium Carbide Manufacture).

Insecticides. Although the new postwar organic pesticides have cut sharp inroads into inorganic pesticides, there still is considerable hydrated lime used in making arsenicals, Bordeaux mixtures, and lime-sulfur spray and powders.

In the manufacture of the insecticide, *calcium arsenate* (renowned for killing the cotton boll weevil), arsenic acid is reacted in a milk-of-lime, forming calcium arsenate. After the calcium arsenate is dried into a powder, additional *free* dry hydrated lime is added as a diluent or filler. Another arsenical, *lead arsenate,* also employs "spray" hydrated lime as a carrier. When water is added to these dry insecticide mixtures for spraying on vegetation, hydrated lime provides an adhesive or "tacky" effect, enabling the insecticide to adhere to plant foliage longer.

In the manufacture of such common fungicides as *lime-sulfur* sprays and dusts and *Bordeaux mixtures,* it is again employed as a reactant and diluent. Lime-sulfur solutions are prepared by heating sulfur with milk-of-lime (2 parts sulfur to 1 part dry quicklime by weight), whereas lime-sulfur dusts are formed by mixing hydrated lime and sulfur together in a dry state. Finely divided hydrated lime is the preferred carrier for Bordeaux mixtures, which is a blend of copper sulfate and lime.

Bleaches. Lime occupies the role of absorbent and vehicle for chlorine in such commercial dry bleaches as high-test *calcium hypochlorites* (70% available chlorine) and *chloride-of-lime* (25 to 35% available chlorine). The latter product has virtually disappeared in the United States in favor of the former more stable, more concentrated product. Stated chemically, its reaction is:

High-test Calcium Hypochlorite

$$Cl_2 + Ca(OH)_2 \rightarrow Ca(OCl)_2 + H_2$$

Chloride of Lime

$$2Cl_2 + 2Ca(OH)_2 \rightarrow Ca(OCl)_2 \cdot CaCl_2 + H_2$$

Inorganic Salts and Bases. Dolomitic lime is the raw material for the production of *magnesium oxide* and *hydroxide* under several processes (see p. 358). Lime is the base used in making most calcium or magnesium *inorganic salts* through interaction with the accompanying inorganic acids, like *calcium phosphates*[32] (mono, di, and tri), fluoride, bromide, ferrocyanide, and nitrate. In the recovery of ammonia in the ammonia-soda process (alkali), the presence of lime reacting with the ammonium chloride forms *calcium chloride,* a valuable waste chemical used widely in commerce. It is also used as standard practice as a neutralizing agent in the manufacture of *chrome chemicals* (sodium bichromate).

Magnesia Products. One of the most important and indispensable uses of dolomitic stone and dolomitic and high calcium quicklime is production of synthetic *dead-burned magnesite* and *caustic calcined magnesite* by various processes, closely akin to some of the intermediate steps in the magnesium metal processes described on p. 352. Chemically both of these materials are magnesium oxide, but their properties are entirely different.

In most of these processes *magnesium chloride* is obtained from either seawater, natural salt brines, or bitterns and provides all of the magnesium for the reaction when high calcium lime is used and about half of it when dolomitic quicklime is employed. Depending on which type of lime is used, theoretically 2.5 tons of high calcium limestone or 2.3 tons of dolomitic limestone is required to make one ton of MgO. The CaO is removed as either calcium chloride or precipitated calcium carbonate in these processes. However, the prime objective is to precipitate *magnesium hydroxide* (milk-of-magnesia) from which, after thickening, the MgO is derived by calcination in kilns.

The dead-burned product is sintered at such high temperatures that chemically this form of magnesia, which is also referred to as *periclase*, is inactive, and because of its high stability in presence of moisture, CO_2, silica, and high temperature, it is widely used as a refractory raw material in lieu of natural magnesia derived from magnesite and brucite minerals. Small amounts of iron oxide and silica may be added to the charge in the sintering process to enhance its stability. It is used as dense granules, usually with a binder, in lining furnaces and cupolas in metallurgical operations after each heat. Probably its widest application is in monolithic bottoms of basic open-hearth steel furnaces. It competes with lower-cost dead-burned dolomite for such purposes. It is also the principal raw material from which a form of refractory brick, called "basic brick," is made. The periclase granules are bonded by firing the brick or by compaction with tar or chemical bonding agents. Some of this brick is also sold encased in metal. Periclase is also used in making chrome and forsterite (magnesium silicate) refractory brick.

The other form of magnesia, caustic calcined magnesite, is not used as a refractory, being active chemically. It is the main raw material from which some special cements, *Sorel cement* (magnesium oxychloride) and *oxysulfate cement* are derived by mixing MgO with either a $MgCl_2$ or $MgSO_4$ solution, respectively. Other uses are in sulfite pulp mills, fertilizers, rayon, insulation, and in the manufacture of many magnesium-based chemicals and salts.

These two synthetic forms of magnesia have replaced much of the domestic, natural MgO from magnesite and brucite as well as imported material and dominate about 70% of the total United States MgO production.

For refractory purposes CaO theoretically possesses greater thermodynamic stability than most refractory oxides, but its use has been limited due to its penchant to absorb moisture slowly from the air and hydrate, even when sintered or dead-burned. Apparently this limitation has been circumvented in a new *slip-casting* process developed, whereby precision crucibles of $1\frac{1}{2}$ in. wall thickness composed entirely of CaO are made for holding plutonium and uranium. The latter nuclear metals are then subjected to a thermal bomb reduction process. Density of 3.15 g./cm.^{3}, or about 90% of theoretical, has been achieved for the CaO crucible.

Several commercial *salt* producers employ lime in the purification of sodium chloride brines in the refining process necessary to produce USP salt for foods. Lime precipitates out iron and traces of soluble calcium and magnesium sulfate from the brine. In the destructive distillation of hardwoods for charcoal and wood chemicals recovery, it is used to neutralize complex organic acids (pyroligneous acids) in which gray *acetate of lime* is recovered for use as such or a source of acetic acid.

Organic Chemicals. It is required in the chlorohydrin process for making *ethylene glycol* (permanent antifreeze). Chlorine and water are combined to form hypochlorous acid. Ethylene gas is then bubbled through the HClO to form ethylene chlorohydrin, with some ethylene dichloride unavoidably formed. Lime is then added to remove free chlorine and HCl from the hypochlorous acid solution in the chlorohydrin tower.

It is widely used in the manufacturing of most calcium-based *organic salts* and *pharmaceuticals*, such as calcium acetate, stearate, oleate, tartrate, lactate, citrate, benzoate, and gluconate with the lime reacting with the corresponding organic acid. It is employed in the concentration of *citric acid, glucose,* and *dextrin*.

Certain types of *dyes* and *dyestuff intermediates* require lime in some processes. In the production of *azo dyes*, it provides the necessary alkaline medium and accelerates the reaction rate by removing hydrogen chloride as it is formed during the process; in the production of *alpha* and *beta napthol*, it neutralizes excess sulfuric acid that is used in the process; it acts as a hydrolyzing agent in the manufacture of *benzaldehyde* from dichlorotoluene and *benzoic acid* from trichlorotoluene.

Miscellaneous. As an adjunct of the Ferro-Silicon Magnesium metal process, *metallic calcium*, an extremely interesting alkaline earth metal of great chemical reactivity, is produced indirectly from the dolomitic quicklime used in the process. Metallic calcium is used as an exotic flux or scavenger in the production of expensive alloys and metals made to exacting specifications and to produce calcium hydride, a dry source of hydrogen for barrage balloons, etc. Another adsorbent use is the production of *soda-lime*, used to adsorb poisonous gases (in gas masks) during wars. Lumps of quicklime have many small, scattered applications as a *dessicant* in laboratory work and in isolated chemical processes.

Precipitated calcium carbonate[35] is, of course, derived directly from lime. It is recovered in crude form as a waste by alkali companies in the ammonia-soda process of making soda ash. After refining, it is widely used as a pigment in paint, to coat paper, and fillers in rubber products, putties, dentifrices, and pharmaceuticals. However, the most direct method of manufacture, which is employed by several lime companies, is by carbonation of milk-of-lime $(Ca(OH)_2 + H_2O + CO_2 \rightarrow CaCO_3$ (precipitated) $+ 2H_2O)$. This material is also known as synthetic chalk since it resembles and has replaced English natural chalk and whiting from the White Cliffs of Dover, which were formerly exported in large tonnage to the United States. The most versatile characteristics of this product is its extreme whiteness and its uniformly minute particle size (including submicrons). See p. 92 for related information.

Water and Sanitation

Water Treatment. In total tonnage lime ranks as the prime water-treatment chemical; however, chlorine is more widely employed. It is used by over 2000 U. S. municipalities, military cantonments, and industrial plants to improve the quality of their water for potable or industrial process purposes.[36]

SOFTENING. In water softening the function of lime is to remove the temporary (bicarbonate) hardness. Where only temporary hardness exists, it is used by itself. However, usually permanent (sulfate) hardness is also present, and in such cases generally the lime-soda softening process is employed. The other reagent, soda ash, removes the permanent hardness. In a few plants, zeolite softening (base exchange minerals) is employed in place of soda ash since the raw material, salt, which is used to regenerate the zeolite mineral, is less costly than soda ash. This is called "split lime-zeolite treatment." Zeolite softening is also widely employed alone, particularly in small softening systems in private homes, institutions, and small municipalities. It removes both temporary and permanent hardness. Redundantly and paradoxically, lime is used to remove lime (Ca and

Mg.) The chemical reactions[37, 38] that occur when lime is added to hard waters that contain both calcium and magnesium are:

$$Ca(HCO_3)_2 + Ca(OH)_2 \rightarrow 2CaCO_3 + 2H_2O$$

$$Mg(HCO_3)_2 + Ca(OH)_2 \rightarrow MgCO_3 + CaCO_3 + 2H_2O$$

$$MgCO_3 + Ca(OH)_2 \rightarrow Mg(OH)_2 + CaCO_3$$

$$MgCl_2 + Ca(OH)_2 \rightarrow Mg(OH)_2 + CaCl_2$$

$$Mg(NO_3)_2 + Ca(OH)_2 \rightarrow Mg(OH)_2 + Ca(NO_3)_2$$

$$MgSO_4 + Ca(OH)_2 \rightarrow Mg(OH)_2 + CaSO_4$$

Reactions involving soda ash are:

$$CaSO_4 + Na_2CO_3 \rightarrow CaCO_3 + Na_2SO_4$$

$$CaCl_2 + Na_2CO_3 \rightarrow CaCO_3 + NaCl$$

$$Ca(NO_3)_2 + Na_2CO_3 \rightarrow CaCO_3 + 2NaNO_3$$

Note that in all of the above equations, a filterable insoluble precipitant is produced—largely calcium carbonate. Such insoluble residue that is removed by filtration is called "softening sludge." In a large softening plant such sludges are voluminous and often present a disposal problem. Where land is available and not too costly, the sludge is lagooned. Since the sludge is rich in calcium, it can be applied to farmers' fields that require soil sweetening. Some industrial and sewage plants will use this waste material in neutralization treatment processes. A few large municipal plants have successfully dewatered and dried the sludge and then calcined the waste calcium carbonate into quicklime for reuse in their softening process— a recycling operation.

PURIFICATION. By adding an excess of lime to water in retention tanks for twenty-four to forty-eight hours, it is possible to purify the water against bacteria in addition to the reduction of temporary hardness (softening). This application of lime is often employed where "phenolic water" exists, usually in areas where steel and certain types of chemical plants abound, since chlorine treatment tends to produce an unpalatable water due to the presence of phenol. This process is called "excess alkalinity treatment." It is rather well established that the high pH, about 11+, produced by the lime will kill most types of bacteria. This method was employed in Bombay when a serious epidemic of infectious hepatitis was traced to the potable water supply, and this treatment method was largely credited with stopping the epidemic. After retention the water must be recarbonated with CO_2 to precipitate out some of the excess lime and lower the pH to between 8 and 9.

COAGULATION. Lime is employed in conjunction with such coagulants as alum and iron salts (ferrous sulfate and ferric chloride) for coagulating suspended solids incident to the removal of turbidity from the "raw" water, usually from rivers and lakes (not wells). It serves to maintain the proper pH for optimum coagulation efficiency. Cities, like New Orleans, St. Louis, and Oklahoma City, that are located on muddy commercial rivers, necessarily have large coagulation facilities.

Many industries require huge volumes of water and in many areas the water supply is becoming increasingly scarce and costly. For reasons of economy or necessity, many industrial plants are reusing their waste waters (normally plant effluents), and are coagulating the suspended solids with lime and a coagulant before recycling the water back into their process. (See coagulation reactions in Appendix B.)

NEUTRALIZATION OF ACID WATER. Lime is used to combat "red water" conditions by neutralizing the acid water—thereby impeding further corrosion of pipes and mains. Further, it is employed to correct corrosive waters containing excessive amounts of free carbon dioxide by absorption of the aggressive CO_2 with lime to form calcium carbonate, which provides a protective coating on the inside of water conduits.

SILICA REMOVAL. One of the most common methods of removing silica from water used in high-pressure steam turbines is through the use of dolomitic lime. The magnesium hydroxide, derived from the dolomitic lime, has a strong affinity for silica absorption, which should be reduced to zero to avoid deposition on the precise turbine blades.

OTHER IMPURITIES. Lime also removes manganese, fluorides, organic tannins, and to a lesser extent, iron from water supplies. Oftentimes, supporting chemicals are required. Such impurity removal is often incidental to the main purpose for using lime. For example, Miami, Florida requires huge amounts of lime to soften its water from the Everglades. In addition to containing much temporary hardness, the raw water is the color of tea due to the presence of tannin which is derived from decayed vegetation. The lime-softening system automatically removes 80 to 90% of this color; the remaining color is further reduced with activated carbon.

Specifications on lime for water treatment are covered by ASTM Specification C-53 and American Water Works Association Specifica-

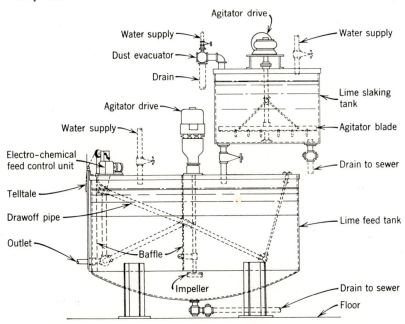

Fig. 11-3. Diagrammatic sketch of typical lime-feeding equipment used in water or neutralization treatment plants with auxiliary lime-slaking apparatus.

tion B202. Figure 11-3 contains a sketch of lime feeding and slaking apparatus used in some water plants.

Sewage Treatment. Lime has a rich heritage as a sanitation aid before the universal advent of modern plumbing. Probably anyone forty-five years old or older is aware of its legendary use in "Chick Sale" type privies. However, many of the modern sewage treatment plants also employ it, but much more scientifically.[39] It is added at the sludge digester to neutralize acidity, thereby maintaining the proper pH for biological oxidation sewage processes. Some large sewage plants will permit local industries to discharge acid waste effluents into the municipal sewage system, providing these plants pay an additional charge for the added treatment expense that is entailed, such as adding lime to neutralize the acid.

While most sewage plants employ biological processes, a few smaller plants employ chemical treatment and coagulate the nonsettleable solids that resist removal by mechanical methods or by sedimentation.[40] In such cases lime is always employed as a coagulant aid, usually with iron salts. Some sewage plants will add small amounts of lime to condition the sludge to facilitate the dewatering and compaction of the resulting filter "cake" for more efficient disposal.

Trade Wastes. Due to increasing agitation by local, state, and federal authorities in recent years for abating stream pollution by industry, hundreds of waste-treatment plants have been built and placed in operation. Many of these plants employ only biological and/or mechanical treatment processes; however, some require supplementary chemical treatment. Many of those employing chemical treatment use lime. In fact, in the waste-treatment literature, lime has been mentioned in the treatment of about fifty totally different types of industrial trade wastes. The most important of such applications are categorized and described briefly as follows:

STEEL AND METALS. Lime is being used to neutralize the free sulfuric acid and precipitate the iron salts from "pickle liquor" wastes from steel and metal-fabricating plants.[41] A voluminous calcium sulfate sludge, containing ferric or ferrous hydroxide, is precipitated. In metal-plating plants lime is invariably employed as a neutralizing agent in precipitating toxic cyanide, chrome, and copper from waste-process waters.[42, 43] The above two chemical applications stated chemically are:

For pickle liquor

(1) $H_2SO_4 + Ca(OH)_2$ (high calcium lime) $\rightarrow$
$$CaSO_4 + 2H_2O$$

(2) $2H_2SO_4 + 2Ca(OH)_2 \cdot MgO$ (dolomitic lime) $+ H_2O \rightarrow$
$$Ca(SO)_4 + Mg(SO_4) + 4H_2O$$

(3) $FeSO_4$ (iron salt) $+ Ca(OH)_2 \rightarrow$
$$Fe(OH)_2 + CaSO_4$$

(4) $Fe_2(SO_4)_3 + 3Ca(OH)_2 \rightarrow$
$$3CaSO_4 + 2Fe(OH)_3$$

For plating wastes (cyanide and chrome removal)

(1) $2NaCN + Ca(OH)_2 + Cl_2 \rightarrow Ca(CNO)_2 + 2NaCl$

(2) $3SO_2 + 2H_2CrO_4 \rightarrow Cr_2(SO_4)_3 + 2H_2O$

(3) $3Ca(OH)_2 + Cr_2(SO_4)_3 \rightarrow 2Cr(OH)_3 + 3CaSO_4$

CHEMICAL AND EXPLOSIVE PLANTS. Various and mixed inorganic and organic acid wastes from *chemical* and *pharmaceutical* plants are

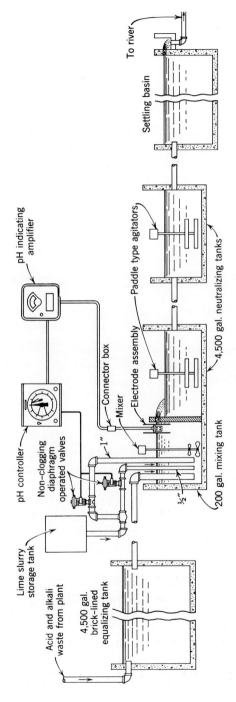

Fig. 11-4. Flow diagram of lime neutralization plant. (Courtesy, Eng. News Record.)

Fig. 11-5. Funnel feeding apparatus for lime into mixing chamber below in acid waste neutralization plant. (Courtesy, American Cyanimid Company.)

neutralized with it preparatory to their disposal into streams or municipal sewage systems. During times of war smokeless powder *ordnance plants* consume enormous tonnages of lime for neutralizing concentrated sulfuric acid wastes.[44] Fluoride waste waters and gases from the manufacture of agricultural *superphosphates* are treated with lime to precipitate the toxic fluorides as calcium fluoride.[45]

A typical chemical waste-treatment neutralization process is diagrammatically depicted in Fig. 11-4 and photographically displayed in Fig. 11-5.

CELLULOSIC PLANTS. In the *textile* industry lime is used to neutralize sulfuric acid wastes from some rayon plants and to neutralize and precipitate dissolved solids from the wastes of cotton textile finishing plants (particularly dyeworks).

In the *pulp-and-paper* industry it has limited and potential use in some pulp plants as a coagulant aid, usually with supporting chemicals, in flocculating solids like "white waters," and in printing inks that resist mechanical removal methods. In sulfate plants it assists in the recovery of valuable by-products from waste effluents, like calcium lignosulfonate, alcohol, yeast, sugars, and lactic acid from the lignin and wood sugars. In the recovery of the latter, close pH control is essential.

FOOD PLANTS. Lime has been used to coagulate solids from vegetable and fruit *canning wastes*, both alone and with other coagulants, but most canneries now lagoon their wastes and add sodium nitrate to hasten decomposition and control odors. In the *citrus industry* it is used to recover and treat citrus pulp, which is a valuable by-product used as cattle feed. *Beet sugar* wastes have been treated with lime to precipitate solids, and the waste effluent is recovered for reuse as process water.

MISCELLANEOUS. Process water in anthracite *coal-washing plants* at the mines is neutralized with lime to reduce corrosion of steel equipment and reclaim the water for reuse.[46] There has been slight usage and considerable attention given to its use in neutralizing acid *mine drainage* from abandoned coal mines.[47] Some *utility* plants clarify "water gas" effluents with lime. Then there are many highly specialized wastes where it plays a part, like the treatment of waste wash waters from cleaning out railroad tank cars; the wash waters from industrial laundries; leather tanning; breaking petroleum oil emulsions; etc.

Neutralization Considerations

In most of the foregoing uses in industrial trade waste treatment, either a high calcium or dolomitic lime can be employed. Both types have strong proponents, even in treating the same type of waste. Usually the selection of either type is determined *only* after careful analysis and before the treatment plant is constructed, since the selected lime will influence the plant design.[48-50]

In determining the most efficient and economical chemical reagent to employ in acid neutralization, there are four major considerations to evaluate.

1. *Material cost per ton* (Table 11-2).
2. *Available basicity*, which determines neutralization power or the gravimetric amount required to neutralize a given acid (Table 11-2).
3. *Reactivity* or the rate of neutralization (Table 11-3).
4. *Sludge volume and disposal.* What is the degree of solubility of reactants (Appendix A), and is the resulting waste sludge readily disposable?

Table 11-2. Cost of alkaline neutralization agents

Reagent	Form	Package	Equivalent basicity[a] factor	Approximate CL price/ton[b]	Cost/ton basicity[b]
High calcium quicklime	Pebble	Bulk	1.0	$14.00	$14.00
High calcium hydrate	Powder	Bagged	1.32	18.00	23.76
Dolomitic quicklime	Pebble	Bulk	.86	14.00	12.04
Dolomitic hydrate	Type N	Bagged	1.02	18.00	18.36
High calcium limestone	Coarse	Bulk	1.79	2.00	3.58
	Flour	Bulk	1.79	5.00	8.95
Dolomitic limestone	Flour	Bulk	1.65	5.00	8.25
Caustic soda	73 % Liquid	Tank car	1.86	60.00	111.60
	Flake	Drum	1.43	104.00	148.60
Soda ash	Dense	Bulk	1.89	31.00	58.59
	Dense	Bagged	1.89	37.00	69.93
Magnesium oxide 100 %	Powder	Bagged	.72	65.00	46.80

[a] Compares all other reagents with high calcium quicklime, given an arbitrary factor of 1. Thus, 1.32 lb. of high calcium hydrate is equal in theoretical neutralizing power to 1 lb. of high calcium quicklime, etc. Calculations assume all reagents are 100 % pure.

[b] Figures shown are F.O.B. plant prices and do not include transportation costs of chemicals, which vary greatly with destination.

Of course, the type of acid, its concentration, and the various forms of extraneous matter, such as salts, that it contains is another critical variable. But in the neutralization of sulfuric acid wastes that predominate in industrial trade wastes, some general hypotheses can be rendered on waste steel pickle liquor, as an example in which the pros and cons of various neutralizing agents in Table 11-2 can be appraised. (Many of the following criteria are applicable to other acid wastes.)

Comparison. *Caustic soda* and *magnesia* are easily the most costly neutralizers per ton, followed by soda ash. But caustic soda rates quite high in basicity and is the most reactive of all of these reagents, and since its reactant, sodium sulfate, is very soluble, it yields minimum sludge. *Soda ash,* which also possesses an equally low sludge formative capacity, lacks the basicity and reactivity of NaOH and is less effective in overneutralization than all of the stronger alkalis. Furthermore, its release of CO_2 may create a frothing problem. Magnesia possesses the highest basicity of all reagents cited and yields a soluble reactant, epsom salt, but it is relatively sluggish in reaction and is not as widely procurable. All of these alkalis will yield some sludge in pickle liquor neutralization because of the inevitable presence of iron salts.

In comparison to these lower sludge-forming neutralizers, *lime* is far less costly in F.O.B. or delivered price, but it does yield a much heavier and more voluminous sludge because of the highly insoluble reactant, calcium sulfate, that is formed. Quicklimes possess more basicity than any of the costly reagents except magnesia. Even the hydrates are slightly superior to caustic soda in neutralizing power and far stronger than soda ash. But no lime equals caustic in reactivity. The high calcium limes are slightly more reactive than soda ash, but dolomitic types are considerably slower, including MgO.

The offsetting advantages of *dolomitic* and *high calcium* limes also manifest themselves. Dolomitic possesses greater basicity, about 1.2 times as much as its high calcium counterpart. Also less sludge by weight is precipitated, since one of its reactants, $MgSO_4$, is nearly 150 times more soluble than the other insoluble reactant, $CaSO_4$, which theoretically simplifies sludge disposal. However, in some processes this advantage has been obviated by dolomitic derived sludge being fluffy and more voluminous. The settleability of bulky sludges is generally also slower and retards the treatment process.

In contrast, high calcium is far more reactive, neutralizing acids in minutes instead of hours. This factor can be translated into more economical plant design with smaller tanks, equipment, and less space—a lower capital cost—since long retention of the acid waste and alkali is unnecessary. This advantage can be minimized if agita-

Table 11-3. Reactivity of alkaline agents with pickle liquor
(Iron unprecipitated after six hours, g./l.)

	No aeration		With aeration	
Agent	Room temperature	60°C	Room temperature	60°C
NaOH	*a*	*a*	*a*	*a*
Na₂CO₃	0 in 0.75 hr.	0 in 0.75 hr.	0 in 0.75 hr.	0 in 0.5 hr.
MgO (reactive)	0 in 3 hr.	0.72	0 in 0.5 hr.	0 in 0.25 hr.
CaO	0 in 0.25 hr.	0 in 0.25 hr.	0 in 5 min.	0 in 5 min.
CaO.MgO	1.88	3.14	1.04	0.30
Ca(OH)₂	0 in 0.5 hr.	0 in 0.5 hr.	0 in 0.5 hr.	0 in 0.5 hr.
Ca(OH)₂.MgO	1.23	1.53	0.55	0 in 3.5 hr.
Cement dust (unreactive)	13.81	14.00	8.02	3.49
CaCO₃, precipitated	8.95	5.17	0 in 2 hr.	0 in 1.5 hr.
CaCO₃, limestone	20.40	18.80	2.95	0.03

a Reaction practically instantaneous.

tion and heat is applied to dolomitic up to 170°F at slightly greater cost. Adding excess dolomitic lime also helps accelerate its inherent slower reactivity. Another advantage of high calcium is that it is reactive at elevated pH levels, whereas dolomitic is only efficient in neutralizing up to a pH of 6.5. Beyond this point dolomitic's reactivity declines rapidly, and if additional lime is continued to be added, much of it, all of the MgO component, is unreacted and wasted, accentuating the sludge problem and negating its basicity advantage. Thus, in pickle liquor treatment dolomitic is at a disadvantage, since for complete treatment it is necessary to achieve a pH of 8.5 to 9.5 before all of the iron salts are precipitated. Generally dolomitic is most efficacious with strong acids and where complete neutralization or overneutralization are unnecessary. The reverse is true with high calcium; it is most efficient with dilute acids and in cases where overneutralization is required. But this may be oversimplified, since exceptions exist.

Either *quicklime* type is more economical than its *hydrate* counterpart due to greater basicity, lower price, and slightly greater reactivity. The exceptions would be medium-to-small operations where the quantity of lime consumed is too small to justify an investment for slaking equipment and expense of another processing step that is a prerequisite for quicklime. Slaked quicklime is introduced into the reaction as a slurry, usually of 10 to 15% lime-solids concentration; hydrate may be fed as a slurry or dry powder. Waste limes are also widely employed.

Of course, of all alkaline reagents *limestone* is invariably the least costly in both price and available basicity. But it possesses many disadvantages that usually negate this imposing attribute. Its reactivity is by far the poorest and its sludge is generally more voluminous and difficult to handle than with lime. When utilized in the upflow process in which an acid is percolated through a bed of limestone lumps or granules, there is a great waste of limestone and available basicity, since a thin impervious coating of $CaSO_4$ forms around the limestone particle impeding neutralization. The $CaCO_3$ and $MgCO_3$ in the center of the granule will not react and is thereby wasted. Furthermore, iron salts, which precipitate as hydrates, tend to occlude the passage of the acid wastes seeping through the bed. Depending upon the size of the granules, concentration of acid, and other factors, the unreacted portion of stone may be as much as 60 to 80%. It is even appreciable under most circumstances with small granules in the No. 4 to 10 mesh range, unless heat and vigorous agitation are applied at added cost. Consequently, this form of limestone for neutralization is generally unsatisfactory and may easily prove to be the most costly of all reagents. Reactivity and completeness of reaction increases in proportion to limestone's degree of fineness. At No. 200 mesh it is reasonably efficient and there is no waste of material due to unreacted particles. However, there are even problems with this material. The release of CO_2 from the carbonate may create a frothing problem, and unless CO_2 is freely vented during neutralization, the reaction will cease at a pH of only 4.3. So aeration or agitation (or both) is required. Similar to Na_2CO_3, limestone, being a relatively weak alkali, will not neutralize to a high pH or to where overneutralization is necessary, as in the case of pickle liquor. As the pH rises, its rate of reaction becomes progressively slower, and it is never as reactive as a reasonably well-burned lime.

Limestone flour is effective up to a pH of between 6 and 7 and will precipitate most or all of the ferric iron but not the ferrous iron, which requires a pH of at least 8.5 and often 9.5 for complete removal. Massive oxidation applied with the reaction will partially alleviate this problem by oxidizing much of the ferrous iron to the ferric state, but then the reaction is often too time-consuming for high efficiency.

A few waste-treatment plants have employed successfully a split limestone-lime method with a finely pulverized limestone utilized initially in neutralizing up to a pH of 5.5 to 6.5, and then slaked quicklime is added until the reaction is complete at pH of 8.5 to 9.5. Consequently, limestone is generally only feasible to employ with weak acids or where partial neutralization is the objective. Waste $CaCO_3$ precipitates, because of their low material cost and physical fineness, are employed economically at a number of plants.

A comparison of high calcium and dolomitic limestone again evidences pros and cons. Dolomitic possesses greater basicity, but unless it is finely pulverized, its reactivity is so hopelessly slow that it would be impractical to employ. When finely pulverized, it will also yield a lower sludge volume on an average than the faster-reacting high calcium. In spite of these disadvantages high calcium is much more widely used because its more widespread availability frequently provides a lower delivered material cost, and its relatively more rapid reaction is also a factor.

Then there are differences in the basicity and reactivity of limes and limestones of even the same type and physical form due to the geology and crystallinity of the stone and degree of calcination (for limes), as discussed on p. 143 and p. 158. The above comparisons are based on average qualities and values. So limes and limestones from potential sources should be evaluated in the laboratory before use.

Therefore, *the lowest cost per ton of available basicity is often not synonymous with the most economical reagent to use;* each case is different and can only be evaluated on an individual basis in laboratory tests or, preferably, pilot plant runs if a neutralization plant is to be designed.

Sludge Formation. The sludge problem encountered in lime or limestone neutralization can be alleviated somewhat by various modifications that improve settleability, dewatering, and compaction of sludge into a denser, less voluminous form that facilitates disposal.[51, 52] These methods are summarized briefly below. Generally with pickle liquor neutralization the hydrous iron oxide precipitate, rather than $CaSO_4$, exercises the greatest control over sludge characteristics, regardless of the alkali employed. It is the "bound" water of these hydrous oxides that determines the settleability and final volume of the hydrous oxide sludge.

1. Increasing acid concentration tends to reduce sludge volume; dilution increases equivalent volume.

2. Diluting acid concentration tends to increase settling rate.

3. Settling rate of sludge generally increases steadily as neutralization proceeds up to a pH of 10.

4. Sludge volume for an equivalent weight decreases as pH is raised up to 9.5 optimum.

5. Oxidation of either untreated pickle liquor or neutralized sludge is the most effective means of achieving improved settling and reduction in sludge volume. Direct oxidation with air is most effective with lime and limestone; with NaOH auto-oxidation with SO_2 and air is most effective.

6. At each percent of oxidation there is an optimum pH at which maximum sludge volume reduction is achieved, e.g., experimentally it has been demonstrated that a maximum 79% volume reduction was achieved at 65% oxidation at pH 9.0, but only 37% reduction at 100% oxidation and at pH of 10 for one lime and an acid waste.

Descaling. In the sedimentation and removal of calcium sulfate sludge following lime neutralization of sulfuric acid process solutions or wastes, a problem may arise with the remaining waste effluent in which after-precipitation or scaling occur in conduits and tanks of the treatment plant. This is caused by the presence of a supersaturated solution of $CaSO_4$ in the effluent which, unless precautionary measures are pursued, may slowly but inexorably accumulate into a deposit or scale of rocklike hardness, rendering the process equipment inoperative by narrowing the apertures and clogging pipes. This problem[52,53] can, however, be eliminated or greatly alleviated by the following practices:

1. *Dilution* of the effluent with clear water, followed by mild agitation, dispels the supersaturated condition of the $CaSO_4$ effluent. The extent of dilution will vary, but oftentimes very little additional water is necessary. The addition of only 35% clear water eliminated 93% of the after-precipitate in one large treatment plant.

2. A crystal-seeding technique can be employed in which some of the *return (neutralized) $CaSO_4$ sludge* is reintroduced into the effluent. The supersaturation of the $CaSO_4$ is relieved through a controlled deposition of sulfate crystals on the return sludge instead of accumulating in the process equipment on metal. This is attracted by a strong affinity that exists between sulfate sludge and its kindred supersaturated solution. This affinity is only pronounced when $CaSO_4$ sludge is derived from neutralization of concentrated sulfuric acid solutions of 20,000 ppm H_2SO_4 or more. Experimentally, 97% of after-precipitation has been eliminated in this manner. Descaling is much less when the return sludge is derived from dilute sulfuric solutions. Greater reduction in after-precipitation results when return sludge is added to the neutralized mixture rather than the unneutralized acid. This method is synonymous to double sedimentation or filtration in order to yield a trouble-free effluent.

3. When a sufficiently concentrated acid sludge is unavailable for preparing a return sludge, the same crystal-seeding technique can be performed by the addition of *commercial gypsum powder* ($CaSO_4 \cdot 2H_2O$).

4. Generally, *use of dolomitic lime* causes less after-precipitation than high calcium because of the presence of highly soluble $MgSO_4$ as one of the neutralization reaction products. The scale is almost entirely derived from $CaSO_4$, and that which is formed by high calcium lime grows more rapidly and becomes much harder and more difficult to remove than dolomitic.

5. Scaling is most extensive following neutralization of the more concentrated acids. *Dilution of the original acid* solution before lime neutralization may reduce resultant after-precipitation to an inconsequential amount.

Ceramics and Building Materials

Glass. Lime's competitive and interchangeable use with limestone in glass manufacture has been described on p. 87. Since many glass products require as much as 5 to 10% of magnesium oxide in the complex sodium calcium magnesium silicate formed in glass making, dolomitic lime is almost invariably employed instead of the high calcium type when burned lime is used.[54] The latter type could be utilized, but then glass plants would have to purchase a magnesium chemical like magnesium oxide, at greater cost to achieve the desired balance of calcium and magnesium in the glass batch. However, not every type of dolomitic lime can be used for this purpose. There are strict tolerances on iron content that average about 0.5% for standard glass bottles and range down to 0.1 to 0.2% or less for optical glass. Some glass may tolerate 0.8% iron, but generally such levels adversely affect the opacity of the glass. Most of the lime used is in the form of precisely screened ground quicklime of the approximate particle size of the sand and dense (granular) soda ash. (No. 20 to 200 mesh sizes). Sizes under No. 200 mesh are taboo. However, a few plants have used dolomitic hydrate lime, in spite of its extremely fine particle size, for special types of glass, along with light soda ash and fine sand. At the high temperatures that glass furnaces are fired, the hydrated lime quickly reverts to quicklime with the chemically combined water expelled as steam. The United States is the only country that employs burned lime products to any extent in glass making; most of the other countries use just dolomitic or high calcium limestone (or blends of both) as their source of calcium and magnesium.

Other Ceramics. A widely used refractory brick that is used for lining industrial furnaces, called *silica brick*, is made by thoroughly mixing finely ground silica (usually quartzite) with 2 to 3% lime solids as milk-of-lime. This mixture is molded, dried, and then burned in kilns. The resulting calcium silicate acts as the bonding agent. A high calcium quicklime or hydrated lime is required, which is covered by ASTM specification C49-57.

Other high-temperature, specialized refractories utilize lime as a bonding and stabilizing agent, namely, *silicon carbide* and *zirconia* refractories.

Basic brick is generally derived from dolomitic quicklime as an intermediate in making magnesium oxide or periclase with the well-established magnesia sea water and brine processes. Another basic refractory is an Italian patented product in which dead-burned dolomite is mixed with tar and then molded by hand compactors into large semicircular sectional shapes to fit a given furnace.

In the production of *whiteware pottery*, lime is sometimes employed to bind the kaolin and ball clays present and contribute to the desired whiteness of the final product. Also several formulas for *vitrified enamel* require the addition of a small amount of lime in the frit.

Sand-lime Brick. In Europe a widely used masonry unit for building construction is brick or block made out of sand and lime; 7 to 10% hydrated lime or pulverized quicklime is intimately mixed with well-graded sand, water is added, the mixture is molded at high pressure, and the "green" brick is fed onto pallets or cars into horizontal cylindrical autoclaves where these masonry units are subjected to heat and pressure. The heat and pressure varies, depending on the plant, from about 300 to 400°F+ and 115 psi to 270 psi (8–18° atm. pressure), respectively. The time of autoclaving varies also, depending upon the amount of heat and pressure, from four to eight hours. Some European plants will employ 10 to 15% silica flour in place of the corresponding amount of sand which tends to accelerate the reaction between the lime and the silica pozzolanically and produces higher ultimate strengths. Also some European plants will blend into the mix 5 to 10% clay which aids in molding the units. Chemistry of lime-silica reaction is explained on p. 194.

Regardless of these modifications in the mix, a stable hydrocalcium silicate is formed during autoclaving that can develop surprisingly high strengths (3,500 to 7,500 psi is typical). Compressive strengths for such products are much higher than so-called lightweight concrete block (cinder or pumicite block) and approximate most standard portland cement or clay brick. However, strength claims for a new patent applied-for German sand-lime product, called "Poreen," are extremely high—8 to 12,000 psi. Improved strengths are being attained through new advanced, automated equipment and know-how. Thus, the lime-silica reaction that takes place is a real cementing action, somewhat comparable to portland cement, with steam and pressure acting as catalysts. A high calcium lime is specified covered by ASTM specification C415-58T (Quicklime and Hydrated Lime for Sand-Lime Products).

The resulting products, after autoclaving, have a distinctive white color (or very light gray) and are produced in sizes of $2 \times 4 \times 8$ in., the same as clay brick. Large-sized block, comparable to ordinary concrete block, is also produced as both solid and hollow units; other products are large "Norman" brick, split stone, and slabs. Pastel units used as face brick are produced by adding pigments to the mix. At one time there was considerable production of sand-lime brick in the United States; however, this industry declined sharply after the start of World War II[55]. In contrast, sand-lime brick has grown steadily in most European countries since World War II. In 1964 Western Germany alone had 180 plants, producing over 4 billion brick equivalents, and requiring about 1.5 million tons of lime. In contrast, United States production in 1964 was less than one-hundredth of this size. Canada is a relatively large producer of sand-lime brick.

Cellular-concrete Products. A related product to sand-lime brick is a lightweight building unit called "cellular concrete block"[56] or "Ytong"[57] and "Durox," patented Swedish products. These products are mixtures of quicklime; usually a small amount of portland cement; pozzolans, like silica flour, certain oil shales, volcanic ash, or fly ash; and an aggregate, sand. A foaming agent is added to the mix—usually aluminum powder—in the presence of water. The quicklime attacks the aluminum, creating an aerating action which causes the mix to rise. When the resulting slurry starts to solidify, abetted by the exothermic (hydrating) action of the quicklime, it is cut into large-sized blocks or slabs and fed into autoclaves for curing and hardening, similar to sand-lime brick. The aluminum powder in effect acts as a leavening agent in the process. By varying the mix it is possible to produce units of various degrees of density, ranging between about 20 to 60 lb./cu. ft. The lightest material is used as insulation material, and is somewhat similar to the patented American product, Kalo. The heavier material is used as large-sized block for masonry, roof decks, and even in slabs for partition walls. While the product appears porous and spongelike, it is actually relatively impervious to moisture penetration and possesses great dimensional stability and adequate strength.

The products are more costly to produce than other types of masonry and related materials. Very little of it has been made in North America; however, it is widely used in some European countries, particularly Scandinavia.

Miscellaneous Concrete Products. Some producers of autoclaved portland cement concrete products in the United States will use lime and pozzolan (silica flour or fly ash) as a partial substitute for portland cement in the mix in varying amounts, usually ranging between 10 and 50% by volume of the cement content.[58] Economy is the main reason for such a substitution since the combination of lime and pozzolan, a cementing mixture, is less costly than portland cement. Another reason for using lime is to obtain a lighter-colored block and/or denser mix (with less voids) due to the extremely fine particle size of hydrated lime.

For the same reason a few United States producers of concrete block that do not employ autoclaving will use small amounts of hydrated lime (5 to 10% by volume) with or without pozzolans. Such products are cured at atmospheric pressure. By improving the workability of such concrete mixes these users contend that products with more precise edges and contours can be produced with less breakage.

In a few backward countries *crude brick* (or adobelike brick) is made by mixtures of clay and 5% lime. A complex lime-silica reaction occurs on curing at atmospheric temperature and pressure after twenty to thirty days, developing about 400 to 500 psi compressive strength.

Most producers of *gypsum wallboard* use a small percentage of lime in preparing the gypsum plaster used in this paperboard, laminated interior wall and ceiling material.

Food and Food By-products

Sugar. In the production of sugar from both cane and beets the crude sugar juice (sucrose) is treated with lime.[59, 60] This forms insoluble calcium organic compounds which are filtered to remove phosphatic materials and organic acids. The limed juice is then carbonated, transforming the mixture into a waste calcium carbonate and semirefined soluble sucrose. The process may be repeated several times to achieve greater purity of the sugar solution, which ultimately is crystallized and packaged. This process is known as defecation and clarification. The rate of lime usage for sugar beets is much greater than for cane sugar, averaging about 0.25 of a ton/ton of beet sugar. In contrast, cane sugar only requires about 4 to 8 lb. of lime/ton. Beet sugar's lime requirements are bolstered by the use of the Steffen's Process, in which secondary sugar is extracted from waste molasses. Only pulverized quicklime is used, unlike the other sugar processes that require milk-of-lime.

Other forms of sugar—sorghum, levulose, dextrose—also require lime in their refinement.

Since beet sugar plants must have large quantities of lime and CO_2 for economical operation, like alkali plants, all refineries burn their own lime in "on-site" kilns and obtain CO_2 as a co-product.

High calcium lime is generally used, although a few cane sugar plants specify dolomitic lime.

Packing House By-products. Waste bones and hides from slaughter houses are cooked in a lime slurry. This process swells the collagen, thereby facilitating subsequent hydrolysis. After liming, the treated stock is washed to remove lime, albumin, and mucin. The washed stock is then dried and the final product is sold by meat packers as *glue* or *gelatin*, the difference being that in the latter case government regulations specify the raw materials to be used.

Dairy Industry. When cream is separated from whole milk, lime is one of the alkalis that is employed to neutralize or reduce the acidity in the cream prior to pasteurization in *butter* making. Other alkalis employed are modified soda products.

In producing *casein* the skimmed milk is acidified, which separates the casein. Lime and a small amount of sodium fluoride are then mixed with the casein to produce calcium caseinate, used in glues and adhesives.

Fermentation of the remaining skimmed milk (whey) and the addition of lime forms calcium lactate, which is marketed as a medicinal or acidified to produce *lactic acid.*

Baking Industry. In the preparation of one type of *baking powder,* monocalcium phosphate is required as an ingredient. This is manufactured by reacting USP phosphoric acid with a very pure high calcium lime.

Fruit Industry. In *citrus* fruit processing lime is employed as a waste-treatment chemical. Canning residues, which include peel, pulp, seeds, and core, are ground and mixed with lime.[61] The resulting pulp is dried and widely marketed as *cattle feed.* In the process lime also reduces corrosion to plant equipment by neutralizing waste citric acids. Also in the canning of some fruit juices it is used to partially neutralize fruit acids in order to help stabilize the taste and color of the juice.

The refuse from *wineries*—grape "lees"—is treated with lime to precipitate calcium tartrate, which can be sold as such or converted to tartaric acid.

In the controlled atmospheric (*CA*) *storage* of freshly picked *apples,* a novel new use of hydrated lime is developing as an *absorbant of CO_2* that exudes from apples.[62] This permits the apples to retain their natural firmness and juice much longer, thereby contributing to their value and salability over a longer period. The emission of CO_2 increases the longer the fruit is stored and accelerates greatly when decomposition commences. A bushel of apples evolves about 5 cu. ft. of CO_2 per storage season. The CO_2 is not deleterious, but it so saturates the atmosphere that the normal oxygen content of the air is diluted, thus shortening the duration that apples can keep.

With its strong affinity for CO_2, hydrated lime is proving to be the most economical absorbant and is replacing wet scrubbing methods with caustic soda or refrigerated water. The most efficient application with lime is to place unopened paper bags of hydrate horizontally on racks in an air-tight box into which CO_2 is drawn from the CA room through ducts by induced draft or diffusion. Depending upon temperature and atmospheric conditions, lime consumption ranges from 0.4 to 4.0 lb./bu. apples for a six-month storage season (1 to $1\frac{1}{2}$ lb. average). Normally no recharging with fresh lime is required during a season. Lime consumption ranges between 15 to 75 ton/season for typical CA installations.

This process was developed in Canada and England but is spreading rapidly in the United States. It is utilized mainly for MacIntosh and Northern Spy apples. Experimental use of this process with cherries, Bartlett pears, plums, and tomatoes has been generally satisfactory. The same principle has been successfully employed in transatlantic shipments of pears by boat.

Miscellaneous Industrial Uses

Petroleum. Lime has a number of diversified uses in the petroleum industry. First, it is used in petroleum *refining* (Jenkin's Process). Organic sulfur impurities, known generally as "mercaptans," are neutralized. Some refineries will also vent noxious fumes of H_2S and SO_2 *waste gases* through absorption towers containing milk-of-lime in order to minimize atmospheric pollution and corrosion to plant equipment.[63, 64] However, caustic soda is much more widely used for these purposes in several different processes.

One of the most common types of *lubricating grease* is a lime-based grease made by saponifying petroleum oils with lime in a process somewhat related to soap making. A high calcium hydrated lime of rather strict tolerances is required for grease making with 1% maximum silica and 95% and 98% passing No. 200 and 325 mesh sieves, respectively. This is embraced in ASTM specification C259-52.

In drilling for oil, hydrated lime is one of the materials frequently used to condition the *drilling mud* in conjunction with such other materials as caustic soda and quebracho in producing "red lime muds," a type of drilling mud.[65] This is a highly specialized field in which the composition of the muds is frequently altered.

The Halliburton Oil Well Cementing Company has reported on using lime and a pozzolan (fly ash or volcanic ash) in a proprietary manner, as a replacement for portland cement in sealing producing *oil wells.*[66] A chemical activator is also added to accelerate the set of the lime-pozzolan mixture. It is used in Gulf coastal and tideland wells of 6,000 to 18,000 ft. in depth, where minimum temperatures of 140°F and higher are encountered. Such temperatures tend to accelerate the set of standard oil-well cements too quickly. There is a reported saving in raw-material cost in using the lime-pozzolan-activator mixture in place of portland cement mixtures. This mixture, however, would set too slowly for use in many other types of wells.

In the secondary recovery of petroleum from oil fields by a process known as "water flooding," lime is often required to condition the water, which is injected into the oil-bearing sand. In treating *plant process water* lime is frequently employed for softening, coagulation, or corrosion control. There has been limited use of lime and ferrous sulfate for *breaking oil emulsions* in petroleum waste treatment.[67]

Leather. Since time immemorial lime suspensions have been used for dehairing and plumping hides preparatory to leather tanning. To accelerate the action of the lime, "sharpening agents," usually sodium sulfide or other alkalis, are often employed as well.

Paints and Pigments. As previously mentioned on page 93 all precipitated calcium carbonate, widely employed as a paint pigment, is derived from lime. Other pigments requiring it in their manufacture include:

> *Satin white (a white pigment);* lime is reacted with alum to form calcium aluminate, which in turn is mixed with hydrated calcium sulfate to form the pigment.
> *Red iron oxide* is often made by heating ferrous sulfate in the presence of lime. A *yellow* iron oxide pigment is made by reacting a solution of ferric chloride with hydrated lime to form crude ferrous hydroxide, which is refined into a pigment.
> *Antimony oxide;* in the preparation of antimony oxide lime is employed to neutralize acidity derived from the sulfide in the crude antimony ore.
> *Zinc oxide;* lime is used to form the intermediate, zinc hydroxide, which is subsequently calcined to zinc oxide.

Many of the proprietary water soluble *masonry paints* and "alleged" basement wall-waterproofing agents contain hydrated lime (10 to 35% by weight of the mix) along with white portland cement and minor additives. Such paints are packaged as dry powders in different colors for use by painters or homeowners. Many "do-it-yourself" enthusiasts simply mix white portland cement and hydrated lime (equal parts by volume of each) with water instead of purchasing the proprietary mixtures.[68]

Other proprietary water-soluble paints for interior decorating and inexpensive coatings like casein paints, Kalsomine, contain hydrated lime in varying amounts. Again, there are "home-made" counterparts of such paints in many "whitewash" preparations that have existed since antiquity—plain lime and water; lime, salt, and water; lime and skimmed milk; and more complicated formulas of lime, trisodium phosphate, casein, and formaldehyde.[68]

Lastly, lime is commonly used in *varnish* manufacture. It serves to neutralize the acid in the oleo-resin and assists in the clarification and hardening of the varnish.

Utility Plants. Commercial coal-derived producer gases are often purified by passing the gases through a milk-of-lime scrubbing tower in which H_2S, CS_2, and CO_2 are absorbed.

Rubber. Only a small amount of lime is used in rubber manufacture. However, quicklime has been used as a desiccant, in removing excess moisture from rubber during processing. Hydrated lime is used by a few rubber companies as an accelerator along with other chemicals in increasing the speed of rubber vulcanization.

Textiles. Little (if any) lime is employed in textile-finishing plants in the United States at present, but at one time it was commonly employed as the cooking agent in "kier boiling" and other functions in which caustic soda is almost universally employed. Lime still prevails in some European textile operations.

Dehydration. There are numerous, small, sporadic applications for quicklime as a dehydration agent in such processes as the purification and concentration of alcohol, ketones, chlorinated hydrocarbons, etc. Its former use of dehumidification of damp areas has been replaced with modern desiccants, like silica gel, or mechanical equipment.

AGRICULTURAL USES

Soil Liming. One of the oldest (and at one time most important) uses of burned lime products is in agricultural liming[69] in which its function is similar to the use of agricultural limestone, previously described on page 103 (namely, as a plant nutrient and neutralizer of acid soils). Actually it was burned lime rather than stone that was first used in agricultural liming. It greatly preceded the development of chemical fertilizers. The first recorded application of this material in agriculture is before the Christian era. Liming was commonly practiced in sixteenth- and seventeenth-century England. Benjamin Franklin in his prolific writings urged farmers to use it for more productive crops and soil conservation. As a result, liming was practiced to some extent in Colonial America. Quicklime remained predominant throughout the nineteenth century in the United States. In fact, until 1916 on an equivalent CaO basis more lime, in the form of burned lime products, was applied to soils in the United States than limestone or any other calcareous material. The consumption of burned lime in agriculture reached its peak in 1914 when 684,000 tons were consumed. In 1963 consumption was only 180,000 tons. What happened to curb this use?

In the 1915–1925 era more state agronomists in the Agricultural Experiment Stations directly and through the states' County Agents began promoting the concept of agricultural liming. Previously only a few states, like Pennsylvania, New Jersey, and Virginia, were active in preaching its virtues to the farmers. Long-range experiments had proven conclusively the acute needs for it and the desirability of periodic liming programs on most farms in the eastern half of the United States. At the same time these agronomists began promoting the use of fertilizers (nitrogen, phosphorus, and potassium) with great vigor. Many of these leading agriculturists concluded that their promotion of liming would be more fruitful with the farmers if they stressed the use of limestone over lime. They reasoned that ground burnt quicklime was two to three times more costly than ground limestone, that hydrated lime was even more costly than quicklime, and that on an equivalent basis limestone could be applied by farmers at a total cost of about one-half less than burned lime, even allowing for the fact that about 1.75 times more limestone would have to be applied than lime for the same equivalent neutralizing value. In other words, their liming recommendations were geared to the farmers' pocketbook. The immediate result was largely a drastic increase for limestone—with lime tonnage declining slightly. Accelerating this trend, of course, was the fact that limestone was much more widely produced and offered on an average freight savings and faster delivery service. At this time the large agricultural sections of the Middle West began liming on a large scale for the first time. The new users in these Middle Western states employed 100% limestone—undoubtedly due to the influence of their state agronomists. In these areas the agronomists not only recommend limestone but relatively coarse sizes, such as limestone meal and No. 10 mesh material. These lower-cost, coarse fractions of stone increased the cost disparity against lime still further.

In spite of this promotion many farmers in the Middle Atlantic states, which were the earliest users of lime in the United States, still clung to ground burned lime or hydrated lime.[70] They had used it for years, liked the results better than limestone, and continued to use it—as many do to this day. The famous Amish farmers in southeastern Pennsylvania have long been advocates of lime. This preference stems from improved soils tilth and faster action that lime provides. Soil acidity can be corrected in a week or less, whereas pulverized limestone requires much more time—weeks, months—depending upon how fine it is ground. The reaction of limestone meal may require a year because of its coarseness. Eastern truck farms that have a high potential dollar volume per acre desire to correct their soil acidity quickly, if a soil test in the spring reveals a low pH of 4 to 5. A slow reactive limestone can cause the loss of 20 to 25% of their first crop. Also in certain areas of the Middle Atlantic states farmers will strive for (and frequently obtain) three crops of some vegetables, like string beans, in a single season. As a consequence, these farmers prefer either ground burnt lime or hydrated lime because of its greater efficiency, which they reason, is the lowest *net* cost because of greater productivity that outweighs the higher material cost. Burned lime acts quickly, like fertilizer; its results are more visible than limestone. See Table 5-11 on p. 111 for the relative soil-neutralizing powers of these liming materials.

Another reason for the continued use of burned lime is that in some Eastern Seaboard areas, soils are often deficient in magnesium. This element can be most efficiently restored with dolomitic lime. Dolomitic limestone, unless extremely finely ground, reacts too slowly with the soil.

When the U.S. Department of Agriculture began subsidizing the purchases of liming materials in the 1930's under the AAA, burned lime was adversely affected again. Many of the state soil-conservation departments that administered this federal program would only reimburse the farmer for a small fraction of his burned lime purchases, basing the subsidy on the price of limestone. Thus, a farmer buying limestone at $2/ton on a 50% subsidy basis would be refunded $1/ton; with ground burnt lime at $10/ton he would receive a subsidy of only $1/ton, plus about $.70/ton for its additional neutralizing value

(Table 5-11, p. 111) or a total of $1.70/ton or only about 17% of his purchase price in contrast to 50%. There were state variances in these subsidy practices, but the above was quite typical. Naturally this further discouraged the use of burned lime.

The quality of burned lime used for soil treatment is probably the least exacting of all lime uses, since as previously discussed on p. 112 virtually any calcareous material can be used. The hydrated limes are usually the most unrefined types in which a minimum of milling is employed to remove core, silica, and impurities. Companies that actively supply this market need not be selective in their quarrying, and, as a consequence, the available lime in both the quicklime and hydrated lime is quite low—about 85% and 70%, respectively, on an average. In fact, before the decline in burnt lime sales in the 1920's, there were about 100 small, primitive lime plants in the United States that produced a small tonnage of quicklime for agricultural liming. Often these producers were farmers who consumed their own production and sold the balance to neighboring farms. The quality of their stone was often poor, and the quicklime was low in available lime due to improper burning in crude "field pot" kilns. In 1964 there were only a very few of such producers still in existence—in Pennsylvania. However, in most states allowance is made for this purchase of lime or limestone on an equivalent available CaO content, which also establishes extra credit for the inherent higher neutralizing power of MgO over CaO. Thus, a quicklime of 90% available CaO would receive a credit of 6% in bid prices over one with 85% available.

Liming in Europe somewhat paralleled the United States. All European countries used burned lime products largely until about 1925–1930. Then pulverized limestone began to grow. However, in 1964 the proportion of burned lime to limestone in all European countries exceeded the United States. In Germany, there actually has been very little loss to limestone; in 1960 consumption of burned lime totaled 556,000 tons as against 617,000 tons of limestone. In England burned lime has lost much ground since World War II, with federal subsidies favoring limestone. Germany's apparent resistance to the encroachment of limestone is largely due to the fact that limestone is not too plentiful, and most limestone producers also burn some lime. (In the United States limestone operations outnumber lime plants by 20 to 1.)

Home Use. Many thousands of bags of hydrated lime, although the total tonnage is of small consequence, are employed annually on lawns, vegetable, and flower gardens of homeowners.[71] Its application is similar to farm use except its scale of use is so relatively minute. (Magnitude of 50 to 200 lb. as against 50 to 200 tons for farm use per season.) Limestone, of course, is commonly used by the homeowner too. However, the ratio of lime to limestone is much higher than on farms. This is probably due to the fact that it can be purchased more easily than limestone. Furthermore, the homeowner has other uses for it around the home for whitewashing, for making masonry paints, for mortar, and even for making watermelon pickles. For the homeowner market it is packaged in 10 and 25 lb. paper bags instead of the standard commercial package of 50 lb.

For further information on the chemical and physical effect on soils, soil test evaluation, methods and rates of application for different soils, spreading equipment, and varying liming requirements of different crops, refer to pp. 103–113 on the use of agricultural limestone.

Miscellaneous Uses

Compost. A small amount of lime is regularly used each year by farmers and home gardeners in composting leaves, weeds, grass cuttings, etc.[71] This foliage is accumulated in a large pile. Decomposition of the foliage can be greatly accelerated by adding mixed fertilizers and lime. By providing an alkaline medium, the nitrogenous fertilizers decompose the foliage much faster into a humus, top-soil type of material, for enrichening farm and garden soils. Best compost procedure is to add a high nitrogen mixed fertilizer and hydrated lime to *separate* layers of the compost heap at the rate of about 10 lbs. of additives to 100 lb. of dry organic matter. Water is then added and top of pile is covered with soil to prevent the escape of ammonia gas. Hydrated lime is preferred; limestone reacts too slowly.

Chicken Litter. Floors of poultry houses are covered with a loose litter material of straw, hay, or peat moss. Litters that become wet, soggy, and matted tend to harbor coccidiosis and parasitic poultry diseases. Many poultry farmers distribute hydrated lime on their litters at the first sign of moist matting. The hydrated lime provides some degree of germicidal protection against parasites and disease. It also helps dry up the litter and restores it to a loose, fluffy condition. Liming fresh litters also prolongs their life. The rate of treatment is about 1 lb. of hydrated lime to every 3 to 5 ft.^{2} of litter area.

Additionally, it is frequently scattered on the dropping boards under poultry roosts. It tends to reduce objectionable odors and infestations of flies as well as preserve the nitrogen content of the manure, which is valuable to the farmer as a rich organic fertilizer.

Disposal of Animal Bodies. Dead animals, carcasses, hides, and other objectionable organic matter are easily decomposed with quicklime. Such material is placed in a pit or pile. Pebbles or lumps of quicklime are spread liberally over the refuse and water is added in small increments. The considerable heat of hydration resulting causes a rapid decomposition of this organic matter, including the bones. Lime application is repeated until decomposition is completed.

Fertilizer Manufacture. In the manufacture of mixed fertilizers, a small amount of dolomitic hydrated lime (20 to 40 lb./ton of fertilizer) is added at the mixing plant by many manufacturers on a standard basis.[72] There are several reasons for employing this additive: it tends to make the mixed fertilizer more free flowing (less "caking"); it reduces "bag rotting" by neutralizing minute traces of free acid still present in the superphosphate; and it contributes a valuable plant nutrient—magnesium—to the mix. High calcium hydrate is not used; it causes a loss of nitrogen.

Liming Ponds. In recent years there has been a growing trend in the United States to rehabilitate stagnant ponds and boggy, turbid lakes, in which aquatic life is poor or dormant due to heavy algae growth, decayed vegetation, and acid water, with fertilizer and liming applications. This has been practiced in Germany for many years. Hydrated lime alone or mixtures of hydrate and pulverized limestone are spread on the surface of the lake from a motor boat to increase the pH to the desired level. (Figure 11-6 shows lime being pumped into a lake as a slurry.) In addition to neutralizing the acid water, it, in particular, is credited with helping to remove turbidity from the water, permitting sunlight and oxygen to penetrate into the water. Such treatment is only performed after meticulous laboratory tests with the water, since an overdosage could be deleterious as well as expensive. Consequently, there is no standard rate of application for lime or fertilizers; it varies greatly depending upon the analysis of the water. A small lake in Wisconsin was successfully treated with 0.77 lb./1000 ft.^{3}, which elevated the pH of the water from 5.4 to 7.2; 3500 lb. of hydrated lime was necessary for this treatment.[73] Most of this liming has been done in Wisconsin, New Jersey, Ohio, Florida, and Michigan.

There have been documented cases where fish life had ceased to

Fig. 11-6. Pumping lime into small acid bog lake in northern Michigan. (Courtesy, R. C. Ball, Michigan State College.)

exist before treatment and where the water was not only murky and stagnant, but odiferous as well. After a series of treatments for two years, fish life became abundant and the waters regained remarkable clarity and freshness. Quicklime is not regarded as safe to use for this purpose.

Starfish Control. In certain areas of the United States, notably Long Island Sound, there have been periodic infestations of starfish to such an extent that oyster beds have been greatly decimated. To save oysters from their nemesis, the starfish, the Fish and Wildlife Service of the U. S. Government engaged in research to develop a method of eradicating starfish from oyster-bed areas without killing the oysters and other desirable marine life.[74] They developed a method of spreading ground or granular quicklime on the surface of the ocean from a motor boat in oyster-bed areas. If one particle of lime falls on the starfish, a lesion rapidly develops causing the demise of the starfish. The lime particles have, of course, no effect on the oyster should they fall on its hard shell, which is almost pure calcium carbonate. The rate of application commonly employed is 125 to 150 lb. per acre of water, which is a rather expensive treatment for oyster fishermen, since the oyster beds cultivated by them often cover hundreds of acres. Of course, much of the lime is necessarily wasted, since countless particles fail to land on the starfish. To alleviate the oyster men's plight, there have been several bills introduced into Congress which would provide the Fish and Wildlife Service with funds of up to $1 million for eradication of starfish using this method. None of these bills have been enacted to date.

For this purpose a hard-burned, slowly reactive granular quicklime of about No. 10 mesh is preferred. Hydrated lime is not used because of its minute particle size, and it is quicklime's heat of hydration that causes the lesion to develop on the starfish.

CONSTRUCTION

STRUCTURAL USES

History. As a building material lime is one of the oldest—probably only antedated by stone and mud. Antiquarians speculate that its use may date back to the caveman era, at least before recorded history. Its earliest documented use was about 4000 B.C., when it was used in Egypt for plastering in pyramids. Archaeologists have excavated two-coat lime plaster with hair as a backing utilized for mural painting in the Palace of Ruossos in Crete in 1500 B.C. It is mentioned on several occasions in the Old Testament of the Bible. Pliny writes how it was prepared by the Greeks in building the Temples of Apollo and Elis in 450 B.C. The Wall of China was largely laid with lime mortar. However, it was the Romans who really developed its structural uses. Shortly before the Christian era Vitruvius, a military engineer under Julius Caesar, even wrote specifications on lime for use as mortar in unit masonry (stone and clay brick), interior and exterior plastering, and road construction. The famous Appian Way contained lime in three of the four layers of this 36 in. depth road that has endured for 2000 years. Proof that the Romans introduced it in their occupation of England is seen in numerous old Roman ruins.

However, the Romans' greatest contribution involving lime was the discovery that when quicklime was mixed with volcanic ash and aggregate it would harden under water. This hydraulic characteristic led to the development of Roman cement or hydraulic lime made from lime and pozzolans, the latter deriving its name from Pozzuoli, a Roman city near Naples, where extensive deposits of volcanic ash existed. Thus, the Romans actually used this mixture as a mortar and to produce slabs, pavements, piers, and other concretelike masses. With the fall of the Roman Empire much of this Roman technology was lost to mankind, and with it construction suffered accordingly in the Dark Ages.

With the Middle Ages and Rennaissance period lime or lime-pozzolans again became widely used. In fact, until the nineteenth century in Europe and about 1890 in the United States when portland cement was developed, the standard mortar for masonry in civilized countries was either a pure lime or hydraulic lime or lime-pozzolan mortar.[75] In eighteenth- and nineteenth-century Europe the latter two mortar types were more widely employed than in the United States. The oldest brick buildings at Harvard University's Massachusetts Hall, dating back to 1730, Independence Hall in Philadelphia (1734), and numerous other historical structures were laid in straight (pure or "fat") lime mortars and many are still in surprisingly good states of preservation today. In view of lime's rich, venerable heritage in construction, it is understandable why it is still regarded as largely a building material even in 1965. However, between 1895 and 1910 its preeminence as *the* mortar and plastering material was seriously challenged, first by portland cement and shortly after by gypsum.

Possibly one of the most significant and exciting technological developments was the advent of the soon-to-be ubiquitous material known as portland cement. Certainly its introduction did more to revolutionize the construction industry than any other factor. It immediately captured the interest and imagination of architects and builders that this gray powder could be mixed with sand, graded aggregate, and water and produce in a few days materials of rocklike strength—much harder than the old Roman cements and lime-pozzolan mixtures. In contrast, pure lime mortars only develop strength very slowly through carbonation of the lime from CO_2 in the atmosphere. Final strength today is about 35 to 40 times less for lime than cement. Furthermore, this small budding cement industry had the foresight to organize and develop uniform standards of quality, something that the lime or lime-pozzolan industry had to that time never achieved. In a sense, it was possibly easier for the portland cement industry to accomplish this standardization since the product was a heterogeneous mixture or blending of limestone and siliceous materials in precise proportions in contrast to pure or hydraulic limes that were largely derived from one given stone deposit. Any cement producer that was unable to meet the specification developed by the portland cement industry through its association could not label its product, *portland* cement. In time most aspiring cement manufacturers saw the wisdom of complying with these early standards (and later modifications). The result was amazing industry solidarity, which led to the formation of a strong association that pursued vigorously research, development, and promotion.

Masonry Mortar

In spite of the fact that lime on an average had produced eminently satisfactory mortars and had centuries of durability experience, architects soon increasingly specified portland cement for mortars, apparently on the premise that the hardest and strongest mortars were best. Cement's strength and speed in setting had been amply demonstrated in the many types of poured concrete structures that were rapidly becoming the vogue in construction. Strength became the keynote. The first transition was toward mixtures of cement and lime in varying volumetric proportions ranging widely from 1:3 to 3:1 cement and lime, respectively. During this period even most of the strongest adherents of cement recognized the need of a plasticizer, like lime, to produce a workable mortar. Cement is harsh and contributes no plasticity to the sanded mortar. In spite of many conflicting opinions and theories among builders as to which specified mix produced the optimum-quality mortar, the trend, abetted by the momentum of this new exciting industry and its potent promotion, was for increasing proportions of cement. Thus, serious competitive inroads were made into one of lime's main "bread and butter" markets. Probably the economic repercussion on the lime industry would have been most severe if simultaneous to this loss of market, its many chemical uses had not commenced to develop. Also with the rapid population and industrial growth in America at this time, construction boomed with unprecedented activity, thereby easing the impact.

But the mortar strength complex among builders, nurtured by the cement industry, continued to grow until in the 1915–1930 era many builders were using either straight cement or mortars with only 10 to 25% lime as an admix to the cement by volume. Shortly after this change in mortar practice, there was an epidemic of leaky masonry in many cities. A new class of specialized building artisan was created to meet the problem called "waterproofing contractors." They cut out the masonry joints from leaky walls and tuckpointed the walls with fresh mortar. This problem became sufficiently critical that several independent research organizations investigated the causes. Usually they discovered that where leaks occurred, either very high cement or 100% cement mortars were employed. Granted the mortar was extremely hard, but it was also brittle, and due to the inherent tendency of cement to shrink, separation cracking had occurred between mortar and the masonry unit at the interface—not in the mortar itself. Driving rains in time penetrated these cracks. Masonry work with these mortars was studied during construction. It was discovered that the mortars lacked plasticity; they worked lean and harshly under the trowel and did not spread easily. As a result, joints were often not completely filled; there was not intimate contact between the mortar and masonry unit; oftentimes the mortar stiffened ("pancaked") prematurely due to the rapid suction or removal of moisture by the absorptive brick or concrete block from the mortar—all of which contributed to *poor bond* at the mortar interface and subsequent cracking and leaks.

These research findings caused the pendulum to swing back to lime in the 1930's and the high-strength complex on mortars was subdued. Mortars of 1:1:6 to 1:2:9 (cement, lime, and sand, respectively) by volume generally prevailed during this period. Such mortars had more than ample strength, high plasticity, were quick-setting due to the cement content, and provided watertight walls with strong bond.

However, lime's resurgence was short-lived. The cement industry introduced a new cementitious material for masonry—proprietary mixtures, called "masonry cements," "patent mortars," or "mortar mixes." While these mixtures varied in components and proportions, the prevailing products were largely mixtures of 50 to 60% portland cement, 40 to 50% pulverized limestone and a fractional percent of air-entraining agents or stearates. The latter organic chemical additives were used in lieu of lime to provide plasticity to the mortar that was not obtained with the former high portland cement mortars. By diluting the portland cement with pulverized limestone, the compressive strength of the resulting mortar was greatly reduced from straight cement mortars, with their objectionably high brittle strengths. Obviously, too, from the cement producer's standpoint, there was great economy in using the low-cost pulverized limestone and air-entraining agent combination. Lime through necessity would have to be purchased in bulk from the lime manufacturer, whereas most cement companies obtained their limestone from their own quarrying operations. A few of these patent mortars contained lime, instead of limestone, but usually these were blast-furnace slag-lime mixtures, with or without some portland cement. An alleged virtue of these prepared mortars was greater simplicity in that the cementitious materials were provided properly blended in *one* bag, in contrast to lime-cement mortars that are proportioned from *two* bags into the mortar mixer. The validity of this argument is questionable, since in large construction jobs entire bags of cement and lime, clearly labeled, are emptied into the mortar mixer along with the sand and water. Thus, 1:1:6 and 1:2:9 cement-lime mortars vied competitively with these proprietary masonry cements (1:3 mixtures of masonry cement and sand by volume).

This competition subsided during World War II years when construction activity ebbed, but resumed again in the postwar construction boom. Due largely to much greater concentrated promotional effort, the masonry cements steadily gained on lime-cement mortars during this period. A number of portland cement shortages during this time aided masonry cement because of the difficulty of securing portland cement to mix with lime on the job. (It is significant that there never was a shortage of the highly profitable cement specialty, masonry cement.)

In certain areas where the lime industry promoted masons' lime vigorously, masonry cements either did not gain or even lost markets to lime, but these instances were the minority. From the standpoint of mortar quality and economy, lime-cement mortar was not deficient (on an average they were superior) to masonry cements—only in sales effort.

Trends in composition of mortar materials, however, appear to be somewhat fluid. Since about 1958 there has been an increasing tendency for cement companies to add hydrated lime to their masonry cements, most of which products formerly contained no lime. In order for many masonry cements to develop high plasticities equivalent to lime-cement mortars, increasingly larger amounts of air-entraining agents were added by the cement manufacturers to the point that the resulting mortars contained 15 to 25% air. Although research has clearly proven that entrained air improves the weatherability of mortars, other research has shown that bond strength of mortar suffers as the air content increases, particularly when air contents exceed 15%. Consequently, some cement companies started altering their mortar composition by reducing the air-entraining agent so as to produce a 15 to 18% maximum range of air in the mortar. The resulting loss in plasticity was restored by substituting 5 to 12% hydrated lime (in place of the pulverized limestone). Thus, many cement companies are now purchasing bulk hydrated lime. In spite of this improvement, it is doubtful if this growth trend for masonry cements will continue. There will always be lime-cement mortars used, particularly in large projects.

Mix Adjustments. As previously mentioned there is no standard type or composition of mortar in the United States. Note the wide range of mortar types recognized by ASTM Specification C-270 in Table 11-4. Some masons even add hydrated lime to their masonry cements. Also the same contractor that uses lime-cement mortars frequently adjusts his proportion of lime to cement to meet such variables in job conditions as temperature, type of masonry unit employed, quality of sand, load-bearing requirements of wall, earthquake or wind conditions, and above or below grade.[76]

In *chilly weather* or where sudden drops in temperature are imminent (below 40°F) contractors frequently increase the cement content of their mortars to develop high early strength—cement sets much faster than lime. Thus, the mortar mix might be changed from a 1:1:6 to a 1:$\frac{1}{2}$:$4\frac{1}{2}$ (cement, lime, and sand) or from a 1:2:9 to a 1:1:6. Conversely, in extremely *hot weather* contractors frequently adjust

Table 11-4. Specifications for mortar for unit masonry[a]

Standards	Property specifications		Proportion specifications		
Mortar types	Flow after suction minimum percent	Minimum ave. comp. strength psi twenty-eight days	Parts by volume portland cement	Parts by volume of hydrated lime or lime putty	Aggregate, measured in damp loose condition
K	70	75	1	over $2\frac{1}{2}$ to 4	Not less than $2\frac{1}{4}$
O	70	350	1	over $1\frac{1}{4}$ to $2\frac{1}{2}$	and not more
N	70	750	1	over $\frac{1}{2}$ to $1\frac{1}{4}$	than 3 times
S	70	1800	1	over $\frac{1}{4}$ to $\frac{1}{2}$	the sum of volumes of the
M	70	2500	1	$\frac{1}{4}$	cement and lime used.

[a] Represents composite summary of proportion and property specifications contained in ASTM specification C 270 on mortar. Specification provides for the use of Types N and S hydrated lime and lime putty from quicklime without restriction.

their mix by increasing the lime content from 1:1:6 to 1:2:9 or 1:2:9 to 1:3:12, etc. This is done to reduce the chance of cement setting too fast and to increase the water retentivity of the mortar so as to resist the increasingly high rate of suction imposed by hot, absorptive masonry units.

The degree of *porosity* and *texture* peculiar to the various types of masonry units influences changes in lime and cement proportions.[77] Clay brick varies greatly in moisture-absorptive qualities. The greater the absorption of the more porous brick, generally the greater the use and need for lime. With the dense, more impervious brick the trend is to increase the cement content. The former unit demands high water retentivity in mortar for good bond; the latter requires less water retentivity.[78] The same principle applies to a lesser degree with concrete block and brick, although these units have a rather high uniform rate of suction. With these concrete units one school of thought contends that the mortar's compressive strength should never exceed the strength of the unit itself. Thus, a $1:\frac{1}{4}:3\frac{1}{2}$ (high cement) mortar with a compressive strength of 3500 psi should not be used with a lightweight cinder block of only 1500 psi. A mortar of 1:1:6 to 1:2:9 (higher in lime) would be more compatible with this lower-strength unit and would eliminate the possibility of stresses in mortar causing the unit itself to crack vertically. Stone used for masonry is usually quite dense, and high water retentive mortars (high in lime) are generally not preferred. Yet standard practice to lay glass block, which is completely nonabsorptive, is a 1:1:6 mortar. One theory is that stronger, higher cement mortars of $1:\frac{1}{2}:4\frac{1}{2}$ to 1:1:6 should be used for the weaker cavity-type concrete block and clay tile. Sand-lime brick and cellular concrete products (Ytong, etc.) are generally laid in 1:1:6 and 1:2:9 mortar. Virtually every type of masonry unit can be laid in lime-cement mortars, except gypsum block used for interior partition walls. This requires a sanded gypsum mortar or plaster.

The *gradation of sand* can affect mix proportions. While most specifications are based on a cementitious material (cement plus lime) to sand ratio of 1:3 by volume as an arbitrary empirical average, in practice frequent adjustments are necessary, depending on the gradation and workability of the sand. Some sand that is slightly loamy with high fines content actually lends slight plasticity to the mix. Because of this some contractors may reduce their lime content accordingly. Conversely, coarse, poorly graded sand is harsh and lends no plasticity to the mortar. With this latter type the lime content is often elevated to provide more plasticity and sand-carrying capacity. Generally, the higher the lime content, the more sand that can be used in the mortar, which, of course, contributes to economy, since the sand is only one-third to one-fourth as costly as the lime and cement combined. Tests with the same sands reveal a difference in sand-carrying capacity of 1:2–2½ for a 100% portland cement mortar to 1:3½–4½ for a 100% lime mortar. Therefore, depending upon the quality of sand used, a so-called 1:1:6 mortar might be adjusted to 1:1:5 or 1:1:7, or using the same sand a 1:½:3½ might be changed to a 1:2:9 proportion with nearly 30% more sand-carrying capacity. Some loamy sands tend to increase the compressive strength of high-lime mortars noticeably. This is believed due to the clay reacting chemically with the lime, forming in effect a mild pozzolanic reaction.

Compressive strength in mortar, of course, is more important in *load-bearing* walls than nonload-bearing ones. Then certain load-bearing walls may be subjected to unusually heavy loads. Most building codes recognize this by requiring greater amounts of cement in the mortar and will prohibit the use of high-lime mortars. However, the importance of compressive strength is usually grossly exaggerated.[80] Building technologists[76] calculate that a mortar with a compressive strength of only 80 psi can support a four-story building of solid brick masonry. Thus, a high-lime mortar of 1:2:9 made with type S hydrated lime that develops a compressive strength of 750 psi would provide a safety factor of 9.5.

In *earthquake* areas masonry is usually reinforced with steel to provide a monolithiclike strength and prevent the possibility of catastrophic cracking from earth tremors. Since maximum strength is desired, very little or no lime is used in such mortars. Usually the cement mortar is gauged with only 10 to 25% lime by volume to provide a little plasticity. A controversial trend is also to increase the cement content in hurricane areas where walls are subjected to high lateral pressure from heavy winds. Yet, paradoxically, builders of high industrial masonry chimneys desire a relatively weak but rich mortar of 1:2:5 (C, L, and S) since they desire a more flexible mortar characteristic of high-lime content. Such chimneys that are 375 ft. in height are known to sway as much as 2 ft. in winds of 80 mi./hr. High cement mortars would be too rigid and brittle for this purpose. Resilience is desired.

In masonry constructed at or *below grade* level, higher cement content mortars are specified because of cement's hydraulic characteristic and greater resistance to frost and moisture. Mortars of 1:¼:3 to 1:1:6 are only used.

The attributes of lime in mortar are summarized as follows:

1. Plasticity and workability
2. High water retentivity
3. High sand-carrying capacity
4. More flexibility under stress
5. Bond strength
6. Ease of retempering
7. Less efflorescence
8. Lighter-colored mortar
9. Autogenous healing

Forms of Lime Used. Until 1905 all lime used for structural purposes was *lump quicklime* that was slaked on the job in metal troughs or in pits or trenches dug into the earth. After slaking, the resulting putty was often aged or cured for days or weeks with additional water added as needed. Aging increases the plasticity of the lime putty. However, such practices are rarely employed now in most modern countries. Most of the American masons lime used (about 62%) is in hydrated form, delivered to the job in paper bags. There are three exceptions to this—the use of bagged pulverized quicklime that is slaked into a putty for use on the job; lime putty or ready mixed lime mortar delivered to the job in drums or trucks, like ready-mixed concrete, from central mixing plants; and bulk hydrate or quicklime used or sold for compounding proprietary mortars or masonry cements (about 24% of the total).

For sheer economy *pulverized quicklime,* slaked on the job, would be the lowest-cost building lime in high putty yield and material cost. Because it is pulverized, slaking proceeds much more rapidly than the old lump type. In spite of this, it has lost favor because of the inconvenience of slaking and the danger of burns to workers. In 1964 in the U.S. barely 3% of the structural lime used is pulverized quicklime; it enjoys greater usage in Germany of about 23%.

For greatest simplicity, exceeding the one-bag masonry cement, *ready-mixed lime mortar* is the easiest to use, on large projects in particular. It is shipped already mixed as a 1:3 lime-sand mortar—or gauged with portland cement in 1:2:9 or 1:1:6 proportions with a retarder from the commercial mixing plant. In the latter case it arrives at the job site already for the masons to use. More often this ready-mixed mortar is purchased as a 1:3 mortar and gauged with cement on the job, as needed. In this way and unlike lime-cement mortars, it can be used for many days without setting by adding water when necessary, and retempering. Because of factory-controlled proportioning and large mixing equipment, this is the most uniform type of mortar available. Its extent of use in the United States was estimated in 1964 at 10% of all masons lime.

Before 1940, in the United States all building *hydrated limes* had to be soaked in large metal troughs, for twelve to twenty-four hours to develop plasticity. As such, the lime was used as a putty on the job. Then *autoclaved hydrated lime* was introduced. This type of lime develops high plasticity without soaking so that it can be fed dry into the mortar mixer, like cement or gypsum. ASTM in Specification C 207-49 distinguishes between normal and special masons' hydrated lime by designating the quick plastic hydrate as Type S and the other hydrate that requires soaking as Type N. Table 11-5 presents a comparison between Types N and S based on the

Table 11-5. Specifications for Masons Hydrated Lime[a]

1. Uses. Masons Hydrated Lime is used principally as a constituent of mortars and concrete. It may be used in stucco and base coat plaster.

2. Scope. This specification covers two types of hydrate, N and S. Type N covers the type usually referred to as Masons Hydrated Lime. Type S covers the product characterized principally by its ability to develop high early plasticity and high water retentivity.

3. Chemical Composition. Both types of Masons Hydrated Lime shall conform to the following chemical composition calculated to the basis as indicated below:

	Type N	Type S
Calcium and magnesium oxides (minimum nonvolatile basis)	95%	95%
Carbon dioxide (maximum as received basis)		
If sample is taken at point of manufacture	5%	5%
If sample is taken at any other point	7%	7%
Unhydrated oxides (maximum as received basis)	No Requirement	8%

4. Fineness. Both types of Masons Hydrated Lime shall leave a residue of not more than 0.5% on a No. 30 sieve.

5. Soundness. Both types of Masons Hydrated Lime shall show no popping or disintegration when tested according to the method prescribed.

6. Plasticity. Type S Masons Hydrated Lime shall have a plasticity figure of not less than 200 when tested within fifteen minutes after mixing with water. Type N Masons Hydrated Lime shall not be required to meet this test.

7. Water Retentivity. Type N Masons Hydrated Lime shall have a water retentivity figure of not less than 75 when tested as either a mortar made from the dry hydrate or from the putty from that hydrate which has been soaked from sixteen to twenty-four hours. Type S Masons Hydrated Lime shall have a water retentivity of not less than 85 when tested as a mortar made from the dry hydrate.

[a] Represents composite summary of ASTM specification C 207-49. Tests referred to are provided by ASTM specifications C 25 on Chemical Analysis and C 110 on Physical Tests.

ASTM specification. The Type S hydrate commands a premium price and has grown rapidly in trade acceptance, until in 1964 nearly 95% of the hydrated lime used for masonry and 60% used in plaster was Type S.

Although both high calcium and dolomitic hydrated limes can be used for structural purposes, easily dolomitic hydrate is the most widely used in the United States, with over 80% of the total. This is believed to be due to two factors: (1) dolomitic limestone, particularly in northwestern Ohio and Nevada, seems to possess a unique, inherent crystalline structure from which hydrated lime of the highest plasticity and hoddability can be most readily produced; (2) high calcium lime producers generally have concentrated on chemical lime uses and except for a few companies have done little research to develop plastic building limes. There are a few high calcium producers that make good type S hydrates but for plastering, even more than masonry, contractors generally prefer the workability of dolomitic. This same marked preference for dolomitic lime over high calcium for building uses does not occur in other foreign countries; both types are used about equally with no apparent preference. Data from English testing of both limes indicate that dolomitic limes develop early strength much slower than high calcium, although within four to six months their strengths usually equal or exceed the high calcium. This is contrary to United States test results in which dolomitic exhibits superior strength. (Table 11-6). German test results dovetail with those of the United States.

Hydraulic Limes. Unlike Europe, hydraulic limes have enjoyed relatively scant usage in the United States. As of 1964 there was only one producer. The types, chemical compositions, and properties of the various European hydraulic limes[75] are described in Chapter 8. Although these products are mixed in 1:2–3 volumetric proportions with sand as mortars for masonry, stucco and plaster, there has been considerable application of the highly hydraulic types in place of portland cement for certain concrete work, like foundations, footings, and in the construction of underwater piers in connection with harbor construction. Subjected to continual water-saturated conditions, these mortars peculiarly gain more in strength than when they are exposed to air. Furthermore, until recent years they established a superior performance record in resisting the chemical attack of salt water than portland cement in seacoast areas. However, since its strength has always been of a much lower magnitude than cement and since it has had slower setting qualities, it has seldom been used for most

Table 11-6. Comparison of mortars of different proportions and constituents

Type of lime or cement	By volume			Approximate average compressive[a] strength in psi (twenty-eight days)	H$_2$O retentivity[a] %
	Cement	Lime	Sand		
Type N hydrate	—	1	3	75	85–95
Quicklime putty	—	1	3	60	92–96
Type S hydrate	1	2	9	750	88–95
Type N hydrate	1	2	9	500	75–90
High calcium quicklime putty	1	2	9	425	85–95
Type S hydrate	1	2	7$\frac{1}{2}$	1000	88–95
Type S hydrate	1	1	6	1650	85–90
Type N hydrate	1	1	6	1000	75–87
High calcium quicklime putty	1	1	6	825	83–90
Type S hydrate	1	$\frac{1}{2}$	4$\frac{1}{2}$	2400	77–82
Type S hydrate	1	$\frac{1}{4}$	3$\frac{1}{2}$	3300	70–75
100% portland cement	1	—	3	4100	50–60
Type I masonry cement	1	—	3	800	80–90
Type II masonry cement	1	—	3	1400	78–90
Weakly hydraulic lime	—	1	3	190	80–87[b]
Moderately hydraulic lime	—	1	3	375	75–82[b]
Eminently hydraulic lime (Roman lime)	—	1	3	875	67–75[b]

[a] Based on test procedures outlined in ASTM specification C 91, modified by specification C 270. There is a wide range in strength values by as much as 20 to 40% plus or minus the approximate average values shown above. The greatest variance is found in masonry cement due to the widely divergent kinds and/or proportions of components that constitute these proprietary materials. Other contributing factors are: differences in the analyses and uniformity of limes and cements; quality of sand; and some degree of laboratory error inherent in the tests that precludes exact reproducibility.

The variance in values for water retentivity is much less.

These values are calculated from the composite results of such investigators as Voss, Staley, and Murray of M.I.T.; Palmer, Parsons, Wells, Fink, Trattner, and Fischburn of the National Bureau of Standards.

[b] Estimated by author.

normal concrete purposes. It is excluded from reinforced concrete work, and since the resulting concrete has poor resistance to abrasion, it is not employed for the surface of pavements and poured floor slabs. Table 11-7 covers mortar proportions.

However, even in Europe use of hydraulic limes have ebbed in favor of lime-cement mortars, made with pure lime, and masonry cements. Lack of uniformity in their performance (even from the same source), less compressive strength and slower setting qualities are probably the cardinal reasons for their decline. In the concrete field European portland cements have improved in quality and uniformity further reducing hydraulic limes' applications in below grade foundation projects. They also lack the plasticity inherent in most quicklime putties and type S hydrated limes. Their old reputation for superiority probably stems from their comparison with straight pure lime mortars. The latter cannot set as rapidly or develop as much strength as hydraulic limes, even the feebly hydraulic types. However, 1:1:6 and even 1:2:9 mortars made with Type S hydrated lime develop faster setting and greater ultimate strengths than even most of the so-called eminently hydraulic limes. Probably to compensate for some of these deficiencies a few hydraulic lime producers are reported to have added portland cement to their products. In so doing, their product should be classed as a proprietary masonry cement.

There is still considerable hydraulic lime used in France and Germany in particular, and doubtless there always will be to some extent. However, France, which has been the largest and most famous pro-

Table 11-7. Recommended proportions of mortars and concretes made with highly hydraulic lime from Teil, France[a]

| Application | By volume | | |
	Hydraulic lime	Sand	Stone aggregate
Mortar for use in salt water	1	2	—
Fresh water	1	$2\frac{1}{2}$	—
Air	1	3	—
Concrete for use in salt water	2	4	3
Fresh water	1	$2\frac{1}{2}$	2

[a] Reported by E. C. Eckels, "Cements, Limes, and Plasters," p. 188.

ducer with a reported two million tons production before World War II has plummeted to nearly one-half that amount. At present, hydraulic limes account for about 28% of West Germany's annual building lime shipments. The weakly hydraulic limes of England known as "blue lias" and "Dorking greystone" have similarly lost ground.

The ASTM recognizes hydraulic lime in a use specification C 141-61, "for structural purposes." Some of the most pertinent provisions of this specification are enumerated as follows:

1. Compressive strength—minimums of 175 psi in seven days and 350 psi at twenty-eight days as provided by specified test.
2. Soundness test—maximum 1% expansion in autoclave as determined by ASTM test procedure (C151).
3. Fineness—maximum of 0.5% retained on no. 30 mesh sieve; maximum of 10% on no. 200 mesh sieve.
4. Time of Set—not more than two hours for initial set, with final set within forty-eight hours as provided by test.
5. Chemical composition:

	Minimum	*Maximum*
MgO and CaO	60	70
SiO_2	16	26
Iron and Al_2O_3	—	12
CO_2	—	5

Lime-pozzolans. The building-lime picture is not complete without considering another hydraulic mortar classification in which lime plays a role. This involves lime-pozzolan mixtures. Unlike cement, lime, or most hydraulic limes that are produced by calcining, hydrating, or refining limestone or blended mixtures of limestone and siliceous material, this group is made solely by compounding and milling two or more materials into a homogeneous material. Lime-pozzolans are of two basic types:

1. Mixtures of lime and *natural* pozzolans, like volcanic ash found in Italy, southeastern France, Azores, and Canary Islands; the trass of the German Rhineland; santorin from Greek volcanic islands.
2. Mixtures of lime and *artificial* pozzolans, like blast-furnace slag, fly ash from utility plants, and burnt clay.

Such combinations comprise proprietary mortar mixes, which have rather similar properties and uses to highly hydraulic limes. Portland cement is added to some of these mixtures transforming these products into masonry cements. In America only the artificial pozzolans have been utilized, largely granulated blast-furnace slag, but in recent years fly ash is being employed by some companies. Some of these slags contain such low iron contents that they have become almost the standard cementitious material for *nonstaining mortar* in laying marble, Indiana limestone, and expensive cut-stone masonry. Some former lime-slag producers in the United States now make only slag-based portland cement. India and other countries in Asia have mixed burnt clay (or pulverized brick "bats"), called "surkhi," with lime for mortar and even concrete work in varying proportions, depending on the purpose, both from central plants and at job site. Europe has utilized both natural pozzolans and blast-furnace slags. Some hydraulic lime manufacturers have added these pozzolans to their limes to increase its hydraulicity and strength. Lime-pozzolan ratios vary greatly from 1:4 to 1:1 lime and pozzolan, respectively, with the lime admixed in dry, hydrated form. While some of these products are of good, uniform quality, many, like hydraulic lime, lack *uniformity.* This is due largely to the difficulty of obtaining pozzolans of consistant analysis, rendering quality control difficult or almost prohibitively costly.

There are two ASTM specifications on the use of lime with pozzolans. These are C432-59T ("Pozzolans for Use With Lime") and C379-56T ("Fly Ash for Use as a Pozzolanic Material with Lime"). The chemical constituents of pozzolans have one characteristic in common—sizable percentages of silica and secondarily alumina.

Pure Lime in Concrete. At one time in the United States (between 1900 and 1930) there was usage of hydrated lime (nonhydraulic types) as an admixture in poured concrete, usually at or below grade line for foundations, dams, sewers, tunnels, reservoirs, bridge footings and abutments, highway pavements, grain silos, and stadiums. Apparently foreign countries never used anything but hydraulic lime for these purposes. Lime additions ranged between 5 and 12% by volume of the portland cement content; seldom was the lime increment used as a substitution for cement.

The principal advantages for this admixture claimed by adherents were improved watertightness and impermeability due to densification from filling voids in concrete; improved workability, facilitating the pouring and placement of concrete; minimized segregation of aggregates ("honeycombing"); permitting use of lower water-cement ratios; provides smoother surfaces. One of the largest projects using lime was the Wilson Dam at Mussell Shoals, Alabama, in the early 1920's in which 25 lb. of hydrated lime was specified to be added to each yd.[3] of concrete (total lime consumption about 15,000 tons). Other large projects were the University of Minnesota football stadium and Wacker bridge in Chicago, and others of comparable importance.

However, this use has largely disappeared, owing to the increased strength and finer grinding of portland cements in the past thirty years and the introduction of air-entrained portland cement with its superior workability characteristic. At its peak in the early 1920's, nearly 100,000 tons of lime hydrate were consumed annually for this purpose. As of 1964 only a few thousand tons at the most were used in poured concrete.

Cement from Quicklime. There is a popular misconception that burned lime is used in making portland cement. Granted, there is much CaO in portland cement, most of which is combined and including some free lime, but this is all derived directly from limestone with calcination occurring in the cement kiln. So generally lime is not directly used in making portland cement. There is, however, an interesting exception reported in addition to the German use of waste quicklime fines for portland cement manufacture (p. 267).

One large Japanese cement producer does prepare its raw-material mix (cement flour) from quicklime instead of limestone. The lime is produced captively in vertical kilns at the cement plant site and is interground with silica and alumina materials before the mix is sintered into clinker form. Advantages claimed by the Japanese for using quicklime instead of limestone are:

1. There is a net saving in fuel; 600 kcal/kg each is employed to burn the lime and sinter the cement flour mix or 1200 kcal/kg total. This is less than the 1400 to 1600 kcal/kg of fuel required for conventional dry-process cement sintering.

2. Lower operation costs are effected. Since quicklime is softer than limestone, there is less grinding cost and higher labor productivity. There is little or no loss on ignition, so that much greater capacities can be obtained from the same rotary kiln.

3. Gradual formation of uniform crystals improves cement quality.

Some of these alleged advantages are plausible except for claim of 600 kcal/kg thermal consumption for lime burning. This is obviously impossible, being below the theoretical minimum heat necessary for dissociation of the stone (p. 136).

Plaster and Stucco

Along with the introduction of portland cement in masonry mortar about 1890 in the United States was the use of the same material in exterior plaster, also known as "stucco." About the same time gypsum was promoted vigorously in the interior plaster field, and both of these materials encroached on the plastering uses of structural lime. Again it was the quick setting and greater strength of cement and gypsum that dislodged lime from its traditionally dominant position in the plastering field. By 1920, in *interior plaster*, gypsum had largely captured the base coat market with its "hardwall" plaster promotional campaign. Consequently, the two plaster undercoats, known as the "scratch" and "brown" representing about $\frac{1}{2}$ to $\frac{5}{8}$ in. combined thickness became standard gypsum sanded mixtures. Only for a small amount of specialized plastering was lime still used as base-coat plaster, like auditorium walls where it possesses better acoustical qualities than gypsum in cases where a smooth plaster coat is desired. Such lime base-coat plaster comprises the following average composition:

> *Applied on metal lath*
> *Scratch coat.* By volume one part of lime putty to two parts of plastering sand and $7\frac{1}{2}$ lb. of fiber or hair per yd.³ of mortar, gauged with $\frac{1}{3}$ part of portland cement or keene's cement.
> *Brown coat.* By volume one part of lime putty to three parts of plastering sand and $3\frac{1}{2}$ lb. of fiber or hair per yd.³ of mortar gauged with   to $\frac{1}{4}$ part of portland cement or keene's cement.

However, at present, virtually all of the lime used in interior plaster is for the third and final coat, known as the "finish coat," which is about $\frac{1}{8}$ in. thick. Gypsum is unable to provide the same degree of troweled, polished smoothness as lime. Consequently, lime finish coats applied by hand are standard in plastering.[83] For greater speed in application (less troweling) and obtaining a faster set, gypsum-gauging plaster (or keene's cement) is added to the lime in the following amounts:

> *Smooth Finish*
> By volume either four and one-half parts of putty or dry hydrate from finishing hydrated lime or four parts putty from pulverized quicklime to at least one part of gypsum gauging plaster. (For a harder finish coat, one to three parts lime to one part gypsum is used).

Lime is also used in sanded finish coats, called *sand-float finishes*, in the following approximate composition: one volume of lime putty to three volumes of sand gauged with keene's cement or gauging plaster. For hard-finish coats the gauging material is increased to equal parts with lime. Lime can also be employed for special textured and ornamental finish coats. ASA Standard for Gypsum Plastering, A42.1 contains full information on finish coat applications.[84]

Proprietary Plasters. A few of the best-known proprietary *acoustical plasters*—Kilnoise and Mute—are lime-based. This is applied to ceilings or walls on standard gypsum base-coat plaster or suitable lath material. After this light, fluffy plaster, which contains asbestos and an air-entraining agent, has hardened slightly, it is tamped with a special board with embedded nails that produces the optimum size and depth of hole for effective acoustics. Hydrated lime is one of the raw materials used in some of the new proprietary one-coat or thin-coat *veneer plasters* for application to special gypsum wallboard as a low-cost plaster. Actually, this material competes more with standard dry-wall construction (i.e., gypsum wallboard with taped and spackled joints) than conventional plaster.

Trend. Finishing lime, along with interior plaster per se, has lost considerable position in the United States since World War II due to competition from less costly dry-wall construction, in spite of being a much superior construction material from the standpoint of fire rating, sanitation, insulation, and acoustics. However, the skyrocketing wages of journeymen plasterers, always one of the highest in the building trades, along with no gain in productivity and poorer quality of workmanship have combined to price plaster out of much of the home building market. In addition, the number of plasterers has actually decreased, since the union for many years opposed apprentice training programs. Because of its high fire rating plastering is still employed in most public buildings.

Foreign Plaster. Plastering materials, proportions, and practices vary more than most other construction methods in different countries. For example, England uses lime mostly in the base coat, admixed with portland cement or gypsum, and rarely in the finish coat, which is composed of gypsum plaster—almost completely opposite from the American practice. In Germany a majority of plaster basecoats are prepared from mixtures of lime, cement and sand (1:1:6 to 1:2:9 mortars). The remainder are conventional gypsum basecoats. Similarly, the German finish coats are often lime-cement (in place of lime-gypsum) mixtures. Also unlike the United States, where most plaster is applied to gypsum lath or metal lath, the Germans usually apply their plaster (without lath) directly to the masonry wall. Most foreign plaster is not as smooth as standard American plaster. Textured or rough-cast plaster is more frequently employed.

Exterior Plaster. This is called "stucco" in the United States and some European countries, "rendering" in Great Britain, and simply "plastering" or "exterior plastering" in Germany and Scandinavia. From 1890 to 1920 stucco facaded exteriors on private homes and many public buildings were widespread in the United States, either applied to wood or metal lath. It was used often to renovate the exterior of old frame dwellings. During this period the universal stucco mix was similar to lime-cement mortars with rather wide ranges in the proportion of lime to cement. However, unlike masonry mortar, stucco mixes tend to contain slightly more sand, the theory being that the higher the sand content, the less cracking will result. Consequently, 1:4 and even 1:5 mixtures of combined lime and cement to sand are often used in addition to the conventional 1:3 mortar proportion. Gypsum is *never* employed in exterior plaster.

About 1920 a new type of stucco material was introduced in the United States called "oxychloride cement stucco," the "oxychloride" being a mixture of magnesium chloride and magnesium oxide. This quick-setting stucco soon replaced much of the lime-cement stucco, but within a few years serious expansion failures occurred with this new material. It proved to be completely unsound, causing severe cracking and spalling. The failures were so extensive and costly that many architects became disenchanted with *all* stuccoed buildings even though the older lime-cement stuccos had exhibited excellent durability. As a result, since about 1930, except for Florida, California, and parts of the Southwest, relatively little stuccoing has been done in most sections of the United States. In contrast, some European countries, where stucco has flourished for centuries, are using it even more than ever since World War II.[85] These countries, Scandinavia and Germany in particular, apparently did not experience the oxychloride cement stucco fiasco and its retarding effect. Its prevalence in Europe is simply based on economics—low-cost method of facing buildings combined with generally satisfactory durability experience. Its recent impetus has been triggered too by aesthetic considerations—improved pigments added to stucco finish coats to produce countless attractive pastel colors in intriguing combinations.

Stucco Mixes. Currently in the United States stucco mixes vary tremendously in composition and proportion. Some stucco base coats are prepared without lime from portland or masonry or "plastic" cement-sand mixtures. However, with the objective of influencing more uniformity in stucco practices, a new lime-cement stucco standard was prepared in 1960 by Committee A 42.5 of the American Standards Association, entitled "Standard Specifications for Lime-Cement Stucco." This specification provides essentially for 1:2:9 mortars (cement, lime, and sand) for the scratch and brown coats in which 3 to 6 lb. of hair or fiber is admixed with one yd.^{3} of the above mortar. Finish coats are one part portland cement to four parts hydrated lime or lime putty. Stucco is applied to $\frac{7}{8}$ in. grounds; 1:2:9–12 mixtures in sanded or "sand-float" finish coats prevail; or sand can be replaced by coarser aggregate in "pebble dash" or "pebble-crete" finishes, providing rougher textures.

Lime Requirements for Plaster and Stucco. Although lime requirements for plastering appear to be almost identical with masons' lime, there is a difference that is difficult to define. High plasticity is essential for plastering. Yet many highly plastic masons' limes cannot be used for plastering, at least not successfully. They are deficient in an indefinable characteristic which might be described as "hoddability" and which has never been measured in the laboratory. However, the plasterer (and apparently only the plasterer) can provide a swift appraisal on whether the lime is "hoddable" or not. His rejection of a lime is decisive and unequivocal. Consequently, plasterers—not laboratories—are the proving ground for testing prospective plastering or finishing limes. In view of this empirical situation, material specifications are not too meaningful. Yet there are two ASTM specifications on finishing hydrated limes, C6-49 for Type N and C206-49 for Type S (Table 11-8).

A Controversy. It was actually for plaster that type S hydrated lime was developed about 1940. In studying some expansion failures in the finish coat, the National Bureau of Standards concluded after prolonged research that delayed hydration of the unhydrated magnesium oxide in the normal (type N) dolomitic hydrated lime was the cause and in effect declared this type as unsound.[86] Based on these findings federal agencies soon amended their lime specifications for plastering to include only highly hydrated dolomitic lime or putties obtained from high calcium pulverized quicklime, the latter product which hydrates quickly and completely on slaking. Many architects and builders were similarly impressed with these research findings. As a result, Type S lime was developed to meet this revised requirement of no more than 8% unhydrated oxide content.

However, the Bureau also met with considerable opposition in having their hypothesis accepted. Many plasterers, on trying the new material, disliked its working qualities and wherever possible reverted to the use of their preferred Type N hydrate. Other building technologists and lime manufacturers attacked the Bureau's theory with the following arguments: that expansion failures only occur when Type N hydrate is improperly applied to solid backing (without furring out) and when it is applied to a brown coat that has been troweled smoothly. Scratch-scoring or roughening up the brown-coat surface before the white-coat application develops a strong bond and keying action sufficient to resist the acknowledged stress from delayed hydration of the magnesia. In contrast, a smoothly surfaced brown coat develops poor adhesion with the finish coat. There is no question that MgO in normal dolomitic hydrated lime may hydrate very slowly—even several years after it is applied to a wall. However, there is some evidence that the expansion force of this hydration is of a very low order of magnitude. Lastly it was pointed out to the Bureau that almost countless millions of square yards of finish coat had been applied for many years with Type N dolomitic hydrate with a very high percentage of success. In spite of these arguments, the Bureau was adamant, and in promoting acceptance they gained increasing adherents, resulting in changes in some building codes and specifications that eliminated Type N hydrate. Yet more than twenty years later, in 1964, an estimated 40% of all of the dolomitic finishing lime from northwest Ohio was still Type N. There are still many plasterers who insist that Type N produces a superior finish coat to Type S, at less cost, and these plasterers have to guarantee their work. Over 75% of the finishing lime in the United States originates from northwest Ohio. Yet many plasterers are completely reconciled to using Type S and profess to like it better than Type N. In contrast, Type S hydrate was introduced to masons, and it gained almost instantaneous acceptance. Of masons' hydrated lime that is consumed nationwide, about 95% is Type S. Lime's over-all position in structural lime would have degraded much more had it not been for the development of this improved lime, a purely American development.

Pitting and popping has always been a problem for finishing limes, although increasingly less in the last twenty-five years. The fineness requirements stipulated in the specifications for both Types S and N lime (Table 11-8) were designed to eliminate pits and pops on the surface of either interior or exterior plaster. As the name implies, these are unsightly small holes or pits that disfigure the plaster surface and are most commonly caused by coarse particles, +No. 30 mesh sizes, composed usually of silica or carbonate core (unburned lime)

Table 11-8. Specifications for finishing hydrated lime[a]

1. Uses. finishing hydrated lime is used primarily as the major constituent of the finish coat in plastering. It may be used, however, any place where masons hydrate would be used.

2. Scope. This specification covers two types of finishing hydrated lime, viz: the so-called normal type of hydrate which is produced by hydrating under atmospheric pressure and in which product a relatively large amount of the oxides are unhydrated. This product will be referred to as Type N. Also hydrate produced by any method in which relatively large amounts of the oxides are hydrated. This type will be referred to as Type S.

3. Chemical Composition. Both types of finishing hydrated lime shall conform to the following chemical composition calculated to the basis as indicated below:

	Type N	Type S
Calcium and magnesium oxides (minimum nonvolatile basis)	95%	95%
Carbon dioxide (maximum as received basis)		
If sample is taken at point of manufacture	5%	5%
If sample is taken at any other point	7%	7%
Unhydrated oxides (maximum as received basis)	No requirement	8%

4. Fineness. Both types of finishing hydrated lime shall conform to the following:

Maximum percent retained on a no. 30 sieve	0.5
Maximum percent retained on a no. 200 sieve	15.0

5. Pitting and Popping. Both types of finishing hydrated lime when gauged with plaster and tested in the steam bath in the manner outlined in the "Methods of Test" shall show no pits or pops.

6. Plasticity. Both types of finishing hydrated lime shall have a plasticity figure of not less than 200. Type N shall be tested after the putty has soaked for at least 16 hours but not longer than twenty-four hours. Type S may be tested any time after mixing. If it is tested after soaking less than eight hours and fails to pass, it shall be tested again after having soaked for sixteen hours but not longer than twenty-four hours. If it passes then, it shall be accepted.

[a] Represents a composite summary of ASTM specifications C 6-49 and C 206-49 on Types N and S limes, respectively. Tests referred to are provided by ASTM specifications C 25 on chemical analysis and C 110 on physical tests.

that are incompatible with finely milled hydrated lime. These unsound coarse particles exert pressure and pop from the plaster surface. Although this problem is greatly minimized, pits and pops still occasionally occur. It is believed that the above specification is not a sinecure against this problem. First, there is always the chance, even though admittedly rare, that some bags of finishing lime are produced that do not meet this fineness requirement. Also there are indications that minute traces of strontium oxide in the lime may cause this problem. Such an impurity of only a few parts per million could never be removed economically by the lime manufacturer. Also there are other causes of this problem that do not involve lime, such as iron or pyrites as an impurity in sand and organic or foreign matter that may be deposited in (or fall into) plaster and mixer troughs, including contaminated mixing water through plasterers' inadvertence. Thus, lime manufacturers have a potential liability to the contractor for repair work necessary in repointing pits and pops, including usually repainting.

Thus, an excellent finishing lime always qualifies as a good masons' lime, but a superior masons' lime does not always become a good finishing lime.

Road Construction

There is an old highway truism that no road is any stronger than its base. This is that part of the road unseen by the public, immediately underneath the bituminous or portland cement concrete pavement or wearing surface. As traffic counts and loads increase along with highway construction costs, engineers are necessarily becoming increasingly scientific in designing highways. More and more it is the base course and subgrade that is occupying their attention. Even highly designed thick pavements and rolled-stone base courses built on unstable clay subgrades are suspect. To correct such conditions, subbases are increasingly being constructed. As a result, road profiles on modern primary or Interstate highways frequently resemble Fig. 11-7, showing the various road layers. In stabilizing the in-place soils or borrow soils or aggregate, lime can be used in the subgrade, subbase, or base course for soil stabilization.[87]

SOIL STABILIZATION

Lime's reaction with soils is twofold. First, it agglomerates the fine clay particles into coarse, friable particles (silt and sand sizes) through base exchange with the calcium cation displacing sodium

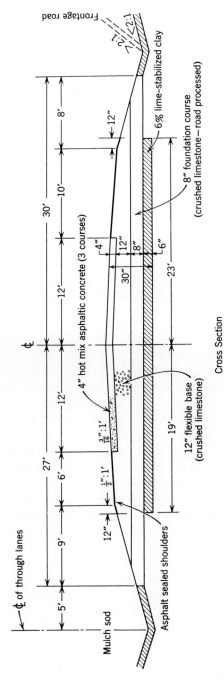

Fig. 11-7. Typical cross-section of U. S. Interstate #35, near Belton, Texas (between Waco and Austin), in which lime is used in stabilizing clay subgrade.

or hydrogen ions. Next, it provides a cementing (hardening) action in which the lime reacts chemically with available silica and some alumina in the raw soil or with pozzolan additives, forming complex calcium silicates and possibly aluminates (see p. 195).

In general, lime reacts readily with all clay soils, either the fine-grained clays or clay-gravel types. Such soils will range in Plasticity Index (P.I.) from 8 to 50+. Most soils with a P.I. lower than 8 are not as reactive with lime (although there are exceptions). In order for it to react with low-plastic to nonplastic sandy soils, generally a pozzolan (second additive) is needed. Fly ash is the most commonly employed pozzolan for this purpose in a patented process called Poz-O-Pac, although finely divided volcanic ash and burnt clay or shale can be effective.[88] Under certain conditions reactive raw clays have been employed as an additive to react with the lime.

Effect on Soil. In soil stabilization lime actually alters the physical characteristics of clay-bearing soils (in varying degrees), transforming these soils into more stable materials for improved road durability.[89] These physical changes are summarized as follows and illustrated in Table 11-9:

1. The Plasticity Index (or degree of plasticity of the soil) decreases sharply—as much as three or more fold.

2. The Plastic Limit generally increases and the Liquid Limit decreases. (The difference between these two values is the Plasticity Index.)

3. The soil binder (clay and fine silt sizes) content decreases substantially because of the agglomerating effect of lime.

4. The lineal shrinkage and swell drop markedly. Fine-grained clays without lime have greatest volume change in wetting and drying cycles as evidence of their instability.

5. Lime and water accelerate the disintegration (breaking up) of clay clods during pulverization construction step, resulting in coarser, more friable soils that can be manipulated more readily, thereby expediting construction.

6. Unconfined compressive strength increases considerably—in varying amounts, but as much as fortyfold.

7. Load-bearing values, as measured by California Bearing Ratio (CBR) and triaxial tests, increase substantially from two to ten times.

8. In swampy areas or with soils exceeding optimum moisture content, lime facilitates drying the soil by increasing both the Plastic Limit and optimum moisture content of the soil. This expedites construction under wet conditions.[90]

Table 11-9. Laboratory soils test data showing physical changes in clay soil with lime addition on interstate 35 (Texas)

Soil type	Raw soil					With lime added						Uncon. comp. strength (psi)		
	Tex. tri-ax. class	L.L.	P.I.	L.S.	S.R.	Lime	Tex.* tri-ax. class	L.L.	P.I.	L.S.	S.R.	Raw soil		With lime added
Del Rio clay	6.4	79.0	50	23.0	1.86	6%	2–3	61.8	27.7	13.1	1.52	8.3		4% 24.3 6% 42.8 8% 45.3
Darnoc clay	5.5	69.5	35	20.9	1.88	6%	1–2	40.8	9.2	5.2	1.44	5.0		4% 257.5 6% 219.8 8% 195.4
Crawford clay	5.1	56.4	25	17.7	1.86	$4\frac{1}{2}$%	1–2	43.1	10.4	5.7	1.41	10.6		3% 223.6 $4\frac{1}{2}$% 313.6 6% 412.1
Abilene clay	5.3	67.1	34	19.8	1.88	$4\frac{1}{2}$%	1–2	49.5	14.8	7.8	1.38	7.1		3% 175.9 $4\frac{1}{2}$% 242.9 6% 176.2
San Saba clay	5.4	64.7	30	20.5	1.92	3%	1–2	47.0	12.3	8.2	1.46	16.6		3% 79.9 $4\frac{1}{2}$% 105.6

Tex. tri-ax. class—Texas triaxial classification rating of soils for engineering purposes. The higher the number, the more unstable is the soil. Classes 1 to 2 qualify as high quality base material. Note greatly improved classification with lime addition.

L.L.—Liquid Limit
P.I.—Plasticity Index
L.S.—Lineal Shrinkage
S.R.—Shrinkage Ratio

[Last three vertical columns on right show gains in unconfined compressive strength with lime additions in contrast to raw (untreated) soil.]

9. A lime-stabilized subbase or base forms a water-resistant barrier by impeding the penetration of both surface and capillary moisture. Exposed but compacted lime-stabilized clay layers shed water readily during rains, thereby minimizing construction delays. This factor also provides some degree of protection against disruptive frost heaving.

Recommended Use. Lime stabilization can be divided into two main types—*subbase* (*or subgrade*) stabilization, involving fine-grained, cohesive soils containing little or no coarse material or road "metal," and *base* stabilization, involving plastic granular materials, such as clay-gravel, which contains less than 50% — No. 40 mesh soil. From ½ to 4% hydrated lime by weight of dry soil is required, depending on laboratory tests, design requirements, and engineering judgment for base stabilization; with the subbase, 3 to 6% hydrated lime is required. Either high calcium or dolomitic lime can be used. Slightly substandard limes can be utilized, but at higher percentages than commercial products, since their results are not as predictable. Quicklime is also at least as effective as hydrate but is rarely used in the United States because of the danger of severe burns to workers, on windy days in particular. Germany, however, employs pulverized

Fig. 11-8. Applying hydrated lime from 50 lb. bags for soil stabilization.

Fig. 11-9. Applying dry, bulk hydrated lime from truck tanker through blowing mechanism.

quicklime on most large subgrade projects. Better worker discipline and supervision enables them to do this safely and successfully. Lime can be employed dry or as a slurry composed of 30 to 33% lime solids. Slurry application eliminates dust nuisance, making it particularly suited for city streets and populated areas. Figures 11-8, 11-9, and 11-10 illustrate various methods of lime application.

Some lime stabilization specifications are listed below:

1. Texas Highway Department, "Special, Hydrated Lime for Soil Stabilization Purposes"
2. Texas Highway Department, "Bases"
3. Mississippi State Highway Department, Special Provision No. 292, "Lime Treated Subgrade"
4. American Public Works Association, Standard Specification L, "Lime Stabilization of Roads and Streets"
5. U.S. Engineers' Guide Specification for Military Construction

Fig. 11-10. Applying wet lime slurry from tank truck through spray bars—a dustless application method.

CE807.32, "Lime Stabilized Base Course, Subbase or Subgrade for Roads and Streets"

6. Poz-O-Pac Manual, Poz-O-Pac Company of America, Plymouth Meeting, Pa. (on lime-fly ash-soil-aggregate mixtures)—"Recommended Procedures"

Construction Procedures. For both base and subbase construction procedures are basically analagous.[87] For soil-in-place stabilization, scarifying to a depth of 6 in. and partial pulverization is the first step. Then the lime is applied evenly from bulk truck-drawn spreaders (in the case of slurry, from tank trucks) or from bags of hydrated lime spotted systematically on the road. Water is added up to the optimum moisture content of the soil, and lime, water, and soil is intimately mixed, preferably with rotary equipment. This mixture is then compacted to maximum practical or specified density, moist-cured for three to five days, and then a wearing course is applied. All *lime-stabilized roads must be surfaced*, except temporary or haul roads. Because of its poor abrasive resistance to traffic it does not produce a satisfactory unpaved road.

The above applies to road mixing. Mixtures of lime, water, soils, and aggregate or lime-pozzolan-aggregate mixtures are also centrally mixed from portable or permanent batching plants. This eliminates road mixing and only involves spreading the material on the road to the specified thickness and compacting.

With the subbase the only possible difference would be that following preliminary mixing of the lime, soil, and water, the soil is allowed to age or "mellow" for one to two days; then final mixing is performed. This delay helps to "rot" (break down) large, hard-clay clods in order to achieve more intimate mixing of lime and soil. Costs of lime stabilization range between \$.25 and \$.55/yd.^{2}/6 in. of compacted depth, exclusive of the wearing surface.

Scope of Use. This method of road construction has been employed on every type of road from farm-to-market to interstate freeways, including city streets and military roads. Its use has expanded to an increasing number of off-highway uses such as parking lots; airport runways, taxiways, and aprons; railroad beds; clay tennis courts and harness race tracks; and lastly, under footings and foundations in building construction. In all of these applications, lime is principally used to stabilize unstable clay-bearing soils. Before the advent of lime stabilization sticky clay was a perplexing, chronic problem for highway engineers. Usually it was excavated and wasted and replaced with oftentimes expensive select borrow material, or to minimize its inherent instability, extra thicknesses of granular base-course material were applied over the clay subgrade. Now, with lime, this heretofore undesirable clay is being used successfully as a road material, providing both economy and improved durability with less road maintenance. Lime competes with "soil cement" in stabilizing slightly to moderately plastic soils (P.I. of 10 to 20 and less) either alone or with lime-pozzolan mixtures. In the highly plastic soils (P.I. of 20 and above) lime has no competition among other soil-stabilizing additives, although a few states have used both lime and cement on the same project with the lime applied first and cement later after the lime has reduced the soil's plasticity.

Growth. From a practical standpoint this is lime's newest major market since lime stabilization really started in late 1945 with a few experimental roads built by the Texas Highway Department, the pioneers of this road-building method. It is true that the Romans used lime in roads, but their methods became a lost art. A few attempts to employ it in the 1920's and 1930's failed, due to lack of modern soil-testing methods and construction controls. The National Lime Association contributed materially to this road method in research and market development work from its inception in Texas. Figure 11-11, covering the growth of lime stabilization, shows that

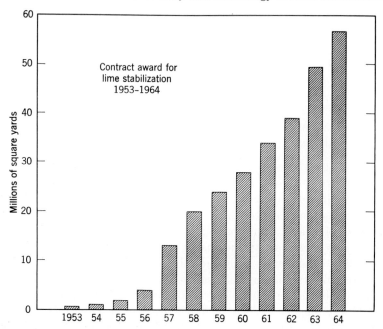

Fig. 11-11. Growth of lime stabilization. (Estimate by National Lime Association.)

this road method acquired no momentum until about 1955, ten years after the first Texas roads were built. By 1964 its total annual use in the United States had reached 55 million yds.²—a twenty-five fold increase over 1955. About 30,000 miles equivalent have been constructed from 1945 through 1963, including over 800 miles of interstate freeways in fourteen states. It has spread from Texas to forty other states and about thirty foreign countries, among which Germany, South Africa, Argentina, Australia, and Sweden are the largest. In spite of this rapid growth it is far from its total potential; steady further growth is anticipated for many years. Road stabilization per se was accorded increased stature and recognition in some of the research conclusions of the $26 million AASHO test road at Ottawa, Ill. in which about 6 in. of stabilized base was declared the equivalent of 12 in. of the best quality aggregate (rolled stone) base (1962). Lime consumption averages about 17 lb./yd.², or about 125 tons/mile equivalent of 20 ft. width and 6 in. depth.

However, on some of the highest-type roads, like interstate freeways, much more lime may be required. If it is applied across the full expanse of four or six lanes, plus road shoulders, and a portion of the median strip, a total width of 60 to 72 (or even more) ft. is possible. Additionally, lime may be used in multiple "lifts" of the road pavement, such as in the subgrade, subbase (with borrow soils), and/or base course, depths that may extend to 12 to 18 in. On one Florida interstate highway that was constructed on a saturated, hydraulic clay fill that traversed an old dried up lake bed, 24 in. of lime-stabilized thickness comprised the subbase. This was constructed in (4) 6 in. "layers." The clay subsoil that was used was so unstable, so saturated, and so deep that the engineers concluded this generous thickness was necessary to "bridge" across the mass of unstable soil in forming a "working table" on which to construct the road. In this highway about 2,000 tons of hydrate was used per mile of road. In spite of the tremendous expense involved, it was a less costly design than constructing an elevated highway across this lake bed. Generally, 6 in. is the maximum thickness of a road layer into which lime can be satisfactorily manipulated. Above this thickness it is applied in two or more "lifts," depending upon total thickness. Six inches is also the minimum recommended thickness for a lime-stabilized layer, particularly in lightly traveled roads where the 6 in. comprise the total base and where a subbase is not included.

Including consumption of waste by product lime of nearly 125,000 tons per year, total lime consumption for highway uses for 1964 was estimated by the National Lime Association at about 575,000 tons. One million tons/year of consumption would be a reasonable expectation for about 1970.

A few of the largest and most significant lime projects are recorded in Table 11-10.

Drill-lime Stabilization. A new stabilization technique employed in highway maintenance and involving hydrated lime was introduced by the Oklahoma Highway Department in the early 1960's. This is called "drill-lime stabilization"[91] and is referred to as "ionic diffusion" by some erudite soils chemists. Flexible pavements (asphaltic) that deteriorate due to base failure, caused by intrusions of unstable, plastic clay into the base course and/or sinking of road aggregate base into the clay subgrade during prolonged wet periods, are reconstructed by injection of hydrated lime through the road into the clay subgrade. There is evidence that the calcium ion of the lime migrates laterally and vertically through the road's soil substructure and by base exchange stabilizes the unstable clay in a manner some-

Table 11-10. Examples of important lime stabilization projects

Date	Location	Number and type of project	Pave-ment section	No. of yds.^{2}	Miles	% Lime used	Approxi-mate lime ton-nage[a]
1958	Bergstrom Air Base Austin, Texas	SAC runways	Subbase	1,000,000	—	4	13,500
1961–1962	Between Sioux City and Sioux Falls, S.D.	Interstate 1 29	Subbase	1,250,000	41	3–4	10,500
1956–1959	North and south of Waco, Texas	Interstate 1 35	Subbase	2,818,000	61.3	4–6	21,000
1957 and 1961–1962	West of Topeka, Kan.	Interstate 1 70	Subbase	800,000	23.5	4	8,200
1961–1962	East of Lake Charles, La.	Interstate 1 10	Subbase	1,250,000	25	4–6	15,000
1964–1965	Newport News to Williams-burg, Va.	Interstate 1 64	Subbase	500,000	—	—	10,000
1964	Dallas, Texas	Love Field Airport	Subbase	400,000	—	—	6,500

[a] Entire lime tonnage is commercial hydrated lime.

what similar to its more conventional road-mixing application, described in the previous section. In addition, each hole into which lime is injected appears to lend a pillarlike strength to the base and dries up a saturated subgrade, providing improved bearing value.

Technical evidence is still lacking on just how, at what rate, and to what extent these calcium ions migrate through an impervious clay. In fact, on some experimental projects subsequent tests revealed that no calcium ion diffusion had occurred, which might indicate a lack of effectiveness with certain types of particularly impervious clays. Yet even in such cases engineers have empirically observed that road disintegration and settlement have been arrested with this corrective treatment, and as a result of Oklahoma's success with many projects, its practice has spread to other states, on a trial-and-error basis. In some instances its use is often a desperation measure—the only possible low-cost solution to a perplexing construction problem.

Construction procedures of this new development are far from standardized. Generally, the following approximate procedure is pursued: 6 to 9 in. diameter holes are drilled through the distressed asphaltic road to a depth of 20 to 30 in. with a heavy duty portable power drill, mounted on the back of a truck (Fig. 11-12). Depth of penetration is usually at least 12 in. into the clay subgrade. Holes are either spaced in a random pattern through the distressed sections of pavement or at regular intervals (in event of complete road maintenance) at 5 to 6 ft centers. Road shoulders are also often similarly treated. Hydrated lime is then emptied dry into the holes from their shipping bag containers; water is then poured onto the lime, forming a paste or thick slurry, after cursory mixing with a stick; the wasted base material and soil removed by the drill is back-filled and tamped into the holes within 2 in. of the surface; then, after a bituminous "tack" coat is applied, the hole is plugged with standard asphalt "cold" mix. Excess excavated road material is bladed to the ditch, and the road can then be reopened to traffic. In four-lane highways single lanes can, thus, be maintained with a typical asphalt maintenance crew without closing the road to traffic. Within a few months

Fig. 11-12. Drill-lime stabilization method, showing drilling rig and holes filled with lime.

traffic action and some settlement cause the asphalt plugs to recede slightly below the road surface, so after maximum recession occurs in about six months, an asphalt-leveling course is applied over the treated area, largely camouflaging the holes and restoring a smooth surface.

It is conjectured that the kneading action and vibration of traffic stimulates lateral migration of the injected lime, with this influence accelerating as the traffic volume, velocity, and wheel loads increase. On one heavily traveled primary road in Oklahoma, lateral migration of the lime was measured up to 5 ft within five to six weeks after treatment. Other experimental methods of enhancing lime permeation are injection of lime slurry under pressure and injection with an electronic impulse to induce an electro-osmosis effect on lime diffusion. In the application just described about one-third to one-half of a bag of hydrate is used per hole (17.5 to 25 lb.), which at a cost of about $.01/lb. would be $.175 to $.25/hole for lime material cost. With a properly trained and equipped maintenance crew it requires only about one minute to drill a hole, and the other labor in filling the hole plus asphalt material cost adds up to a *total cost* of approximately $1/hole, or about $4100/mile for a roadway 20 ft. wide. This would correspond to a rate of application of about 40 to 55 tons of lime per mile of road, far less on an average in tonnage than conventional lime stabilization via road mixing.

Road embankments have also been stabilized with this same technique. Roadway cuts and fills that contain unstable clay have been a chronic problem in many areas due to slides and settlement, respectively. Such embankments have been consolidated and slides and settlement arrested with apparent permanence by deep vertical holes drilled through the cut or fill. In one treated fill some of the holes measured 30 ft. in depth, the maximum attainable with the particular drilling rig that was employed. In embankment holes lime consumption is much greater, one to ten bags (50 to 500 lb.) per hole, depending upon the diameter and depth of the hole. On one 20 ft. cut on one side of a road in Texas, 55 tons of lime were required.

This method also offers promise for stabilizing unstable soils for such miscellaneous off-highway purposes[92] as around and under *building foundations* where swelling subsoils cause differential settlement or heaving that will even crack masonry walls in extreme cases; it might also be used to resist erosion and slides of embankments along *waterways* and *irrigation ditches*. The lime stabilizes the moisture content of the subgrade so that less volume change occurs during severe alternate cycles of wetting and drying.

In all of these applications it is the *reaction of lime and clay* that improves stability, consolidation, and bearing strength of the soil mass, and in spite of rapid growth, stabilization is still in its infancy as far as ultimate development of many potential uses is concerned.

BITUMINOUS PAVING

In the period of 1910 to 1930 there was some application of hydrated lime in asphalt paving, notably in the Amiesite process, in which 1 to 2% hydrated lime by weight was admixed with asphalt concrete, both cold and hot mixes, and sheet asphalt. It was employed largely as an antistripping agent to prevent aggregate from raveling from the bitumen and as a void filler to provide the necessary percentage of fines for improved density and gradation. Several states recognized this application in their bituminous paving specifications. Then, rather unaccountably, its use practically disappeared. It was largely replaced by pulverized limestone and other lower-cost mineral dust fillers. Since its volume of usage was relatively low, it was never promoted by the lime industry. Early test data on hydrated lime as an additive tended to indicate some improvement in the quality of asphalt concrete, but not conclusively. Many materials engineers were, nevertheless, impressed with its performance and continued to provide for its use on an alternative basis in project specifications. But while its use was permissible, paving contractors at their option almost invariably selected the less costly mineral dust fillers. Meanwhile, European engineers, in England and Germany in particular, actually specified its use on some paving jobs, usually with tar instead of asphalt.

Recent Uses. This former use of lime[93] was revived about 1958 by the Colorado Highway Dept. In the laboratory they experimented with it, along with such mineral fillers as pulverized limestone, portland cement, and organic proprietary antistripping agents, with a test developed by the Bureau of Public Roads, called the Immersion-Compression Test.[94] This test, which measured asphalt concrete cylinders in compression, both dry after oven curing and wet after submersion in water, indicated that lime offered real efficacy in improving asphalt concrete (hot mixes) with certain asphalt-aggregate mixtures—in some cases with spectacular and very conclusive results. As a consequence, Colorado started specifying 1% by weight of hydrated lime (in a few cases 2%) gradually on some of their state hot-mix paving projects (Fig. 11-13). The field results were also encouraging, so they

Fig. 11-13. Lime being fed from bulk tank truck into feeding hopper at an asphalt "hot mix" plant; 1% lime additive.

increased its usage until by 1964 it was employed on over 50% of their state projects. Meanwhile, other states in the Far West and the Middle West, on the strength of Colorado's experience, began employing it. There is every indication that its use will continue to grow, since it is predicated on a realistic scientific test (Immersion-Compression).

Effect of Lime. This does not mean that hydrated lime improves *all* asphalt mixtures. With some aggregates other additives are either more effective or economical. As a generalization, it appears that lime is most efficacious with inferior quality siliceous acid aggregate (pit-run gravels and siliceous stone). It is conjectured that lime, an alkaline cation, produces a cationic reaction with the opposite charged, anionic acid aggregate. Because of this it appears to be not as effective with like-charged limestone or basic aggregate, although tests even with some of these latter aggregate have indicated some promise. For predictable results, laboratory testing is necessary.

Table 11-11. Asphalt concrete test results on Colorado project no. F-019-I

	Untreated aggregate	Pulv. lime-stone treated (3% wt.)	Hydrated lime treated (1% wt.)	State spec. limits
Asphalt grade	100/120	100/120	85/100	
% of asphalt	5.4	5.4	6.0	
Spec. gravity	2.26	2.30	2.27	
Wet strength psi	153	133	547	
Dry strength psi	315	350	375	
Stability W/D ratio	48/100	38/100	146/100	75/100
% Absorption by wt.	3.64	3.75	0.99	4.0
% Swell by vol.	0.56	0.82	0.03	

Immersion—compression test

	Untreated aggregate	Pulv. lime-stone treated (3% wt.)	Hydrated lime treated (1% wt.)
Asphalt grade	100/120	100/120	85/100
% of asphalt	5.4	5.9	6.0
Stabilometer	39	32	38
Cohesiometer	89	137	173
R^t value	92	91	96
Voids total mix %	6.0	—	5.0

Stabilometer test

Benefits claimed by engineers and contractors[95] are summarized as follows and shown in Table 11-11:

1. *Heat stable additive.* Hydrated lime does not decompose like some organic antistripping agents in the temperatures necessary to operate asphalt hot-mix plants.

2. *Increases asphalt content.* Generally lime additions permit an increase in the optimum amount of asphalt that can be used in the mix without a resultant loss in stability (2 to 10% more asphalt). The richer asphalt mix should provide increased waterproofing and improved durability.

3. *Void filler.* As a filler it reduces the total voids content by 1 to 3% without sacrificing stability, thereby helping to provide a denser, more impervious asphalt mix.

4. *Added strength and toughness.* The tensile strength is often increased by 25 to 100%, as determined by Cohesiometer tests. Hveem Stabilometer tests reveal a slight increase in stability of about 10%.

5. *High wet strength.* There is very little or no loss in compressive strength when lime-treated asphalt concrete has been saturated with

moisture. In fact, it is not uncommon for the wet strength to exceed the dry strength. This indicates a waterproofing and moisture-stabilizing effect as well as *antistripping* qualities.

6. *Less volume change.* With inferior aggregate it markedly reduces the percentage of swell.

7. *Less moisture absorption.* With inferior aggregates it absorbs up to 100% less water than untreated aggregates.

8. *Upgrades aggregate.* Thus, submarginal aggregates can be upgraded into satisfactory hot-mix material. Such inferior aggregate that often contains some clay and possesses slight plasticity (P.I. of 2 to 4) can be rendered nonplastic—the same function it displays in soil stabilization for road bases.

9. *Faster set.* Some contractors claim the resulting asphalt concrete sets faster after distribution, enabling them to roll it sooner; specified densities can be obtained with less rolling.

10. *Dries aggregate.* In cold-mix asphalt, lime assists in drying the aggregate, a prerequisite in this paving method.

11. *Low cost.* It is relatively low-cost, adding around $.20 to $.25/ton of asphalt concrete for a 1% application (or $.01/yd.^{2}/in. of pavement).

REFERENCES

1. *U. S. Bureau of Mines Minerals Yearbook* Chapters on "Lime" (annually for 1908–1963).
2. N. Knibbs, *Lime & Magnesia,* E. Benn, Ltd. (London), 1924.
3. O. Bowles, "The Lime Industry," *U. S. Bu. Mines I.C. 7651,* 43 pp. (Nov. 1952).
4. P. Hatmaker, "Lime," *Industrial Minerals & Rocks,* AIME pp. 395–426, 1st Ed. (1937).
5. N. Rockwood, "Lime" Chap. *Industrial Minerals & Rocks,* AIME, 2nd Ed. (1949), pp. 467–512.
6. R. Boynton and K. Gutschick, "Lime" *Industrial Minerals & Rocks,* AIME, 3rd Ed. (1960), pp. 497–519.
7. W. Wing, "Lime as a Chemical Raw Material," *Chem. Inds.,* 14 pp. (May and June 1942).
8. U. S. Bureau of Standards, "Lime—Its Properties & Uses," *Circ. 30,* 25 pp. (1920).
9. National Lime Association, *Chemical Lime Facts,* 42 pp., 2nd Ed. (1964).
10. R. Boynton, "Lime—an Industrial Chemical," *Chem. Eng.,* 5 pp. (July 1950).
11. C. Herty, Jr., "Burnt Lime & Raw Limestone in the Basic Open Hearth Process," I & E Chem. **19** (May 1927), p. 592.
12. C. Parker, *Metallurgy of Quality Steels,* Reinhold (New York, 1946), pp. 14–54.

13. Econ. Comm. for Europe, *Comparison of Steel-Making Processes,* United Nations (New York), 83 pp. (1962).

14. B. Kalling et al, "Desulphurization of Pig Iron with Solid Lime," *Blast Furn. & Steel Plant* (May 1957).

15. D. McBride, "Physical Chemistry of Oxygen Steel-Making," *J. Metals* **12** (1960), p. 531.

16. R. Pelhke, "Some Considerations on the Kinetics of Lime Solution in Basic Oxidizing Slag," *Proc. AIME Steel Conf.* (1964).

17. R. Limons et al, "Use of Burnt Lime as Partial Replacement for Dolomite Flux in Sinter Mixes," *Proc. AIME Iron-Making Conf.* (1964).

18. O. Nyquist, "Effects of Lime on Sintering," *Agglomeration,* Interscience (New York, 1962), pp. 809–858.

19. T. Miller, "The Importance of Lime in Wire Drawing," *Wire Prods.* **29** (Aug. 1954).

20. A. Schallis, "Dolomite—Base Refractories," *U. S. Bu. Mines I.C. 7227,* II pp. (1942).

21. J. Chesters, "Dolomite Refractories," *Iron Age* (Aug. 5, 1943), p. 48, and (Aug. 12, 1943), p. 86.

22. W. Trauffer, "The Non-Metallic Phases of Magnesium Mfgr.," *Pit & Quarry* (Mar. 1944), p. 71.

23. E. Schrier, "Silicothermic Magnesium Comes Back," *Chem. Eng.* (Apr. 1952), p. 148.

24. J. Grindrod, "Dolomite + Seawater = Refractory," *Pit & Quarry* (Sept. 1959), p. 102.

25. R. O'Meara et al, "Compendium on Limes in Hydrometallurgy & Flotation," *U. S. Bu. Mines I.C. 6423,* 54 pp. (1931).

26. A. Engel, "Simple Treatment Methods for Oxide Gold & Silver Ores," *U. S. Bu. Mines R.I. 4758* (1951).

27. O. Fraser, "Developments in Nickle," *I. & E. Chem.* **44** (May 1952).

28. J. Lee, "Making Alumina at Mobile," *Chem. & Met.* (Oct. 1940), p. 674.

29. W. Trauffer, "Bauxite & Alumina Output Soars," *Pit & Quarry,* (June 1944), p. 115.

30. F. Littman and H. Gaspari, "Causticization of Carbonate Solutions," *I. & E. Chem.* **48** (Mar. 1956), p. 408.

31. J. Olsen and O. Direnga, "Settling Rate of $CaCO_3$ in the Causticization of Soda Ash," *I. & E. Chem.* **33** (Feb. 1941), p. 204.

32. Anon., "Dicalcium Phosphate," *Chem. Eng.* (Nov. 1955), p. 370.

33. O. Wicken, "Magnesite & Related Minerals," *Industrial Minerals & Rocks,* AIME (1960), pp. 533–544.

34. H. Comstock, "Magnesium & Mg Compounds," *U. S. Bu. Mines I.C. 8201,* 128 pp. (1963).

35. R. Taylor, "Precipitated $CaCO_3$," *Chem. Inds.* (June 1947).

36. M. Riehl, *Water Supply & Treatment,* Nat. Lime Assn. (Washington), 220 pp., 9th Ed. (1962).

37. A. Janzig, *Water Wks. Eng.* (Nov. 1952), p. 1045.

38. D. Jones, "Use of Pulverized Quicklime for Water Softening," *Water & Sewage Wks.* (Dec. 1955), p. 503.

39. W. Rudolfs, *Principles of Sewage Treatment,* Nat. Lime Assn. (Washington), 130 pp., 4th Ed. (1955).

40. W. Parsons, *Chemical Treatment of Sewage & Trade Wastes,* Nat. Lime Assn. (Washington) (1965), 168 pp.
41. R. Hoak, C. Lewis, et al, *I. & E. Chem.* **36**, No. 3 (1944), p. 274; **37**, No. 6 (1945), p. 553; **39**, No. 2 (1947), p. 131; **40**, No. 11 (1948), p. 2062.
42. E. Druschel, *Metal Finishing* **58**, No. 3, p. 58, and **58**, No. 4 (1960), p. 66.
43. G. Barnes, *Proc. 3rd Purdue Ind Wastes Conf.* (1947), p. 179.
44. B. Dickerson and R. Brooks, "Neutralization of Acid Wastes," *I. & E. Chem.* **42** (1950), p. 599.
45. R. Landau and R. Rosen, "Fluorine Disposal," *I. & E. Chem.* **40**, No. 8 (Aug. 1948), p. 1389.
46. L. Johnson, *U. S. Bu. Mines I.C. 7382* (Nov. 1946).
47. S. Braley, "Acid Mine Drainage," *Mechanization* (Apr. 1955).
48. H. Jacobs, "Neutralization of Acid Wastes," *Sewage & Ind. Wastes* **23** (July 1951).
49. W. Rudolfs, "Acid Waste Treatment with Lime," *I. & E. Chem.* **35** (Feb. 1943), p. 227.
50. R. Hoak, "How to Buy & Use Lime as an Acid Neutralizing Agent," *Water & Sewage Wks.* (Dec. 1953).
51. W. Rudolfs, *Sew. Wks. J.* **15**, No. 1 (1943), p. 48.
52. W. Parsons and H. Heukelekian, *I. & E. Chem.* **46**, No. 7 (1954), p. 1503.
53. S. Faust and H. Orford, *Ind. Wastes* **2**, No. 2 (1957), p. 36.
54. D. McSwiney, "The Function of Lime in Glass," *Glass Ind.* **7**, No. 8 (Aug. 1926), p. 186).
55. "Calcium Silicate Bricks," *Sand Lime Brick Mfgrs Assn. Bull.* (London, 1959).
56. R. Valore, "Cellular Concretes," Parts 1 and 2, *J. Am. Conc. Inst.* **25** (May and June 1954).
57. E. Ahlstedt, "Lightweight Building Units of Autoclaved Lime & Siliceous Materials," *Rock Prod.* (June 1952), p. 106.
58. J. Selden, "Lime, Fly Ash, Silica, and Cement in Autoclaved Concrete Products," *Rock Prod.* (May 1954).
59. W. Shearon, "Cane Sugar Refining," *I. & E. Chem.* **43** (Mar. 1951).
60. B. McDill, "Beet Sugar Industry," *I. & E. Chem.* **39** (1947), p. 657.
61. J. Heid, *Food Inds.* **17** (1945), p. 1479.
62. C. Eaves and C. Lockart, *J. Hort. Sc.* **36** (Apr. 1961), p. 85.
63. V. Kalichevsky, "Sweetening & Desulphurization of Light Petroleum Products," *Pet. Ref.,* Part VI, p. 111; Part VII, p. 117 (Apr. and May 1951).
64. R. Hafsten and K. Watson, "Neutralizers & Inhibitors To-day," *Pet Ref.* (May 1955).
65. O. Van Dyke et al, "Chemicals Used in Red-Lime Drilling Muds," *I. & E. Chem.* **42** (Sept. 1950).
66. D. Smith, "A New Material for Deep Well Cementing," *J. Pet. Tech.* (Mar. 1956).
67. J. Etzel et al, *Ind. Wastes* **1**, No. 7 (1956), p. 237.
68. National Lime Association, "Whitewash & Cold Water Paints," *Bull. 304-F* (1956).
69. J. White, "Use of Burned Lime Products in Soil Improvement," *Pit & Quarry* (May 1947).
70. National Lime Association, *"100 Questions & Answers on Liming Land,"* 32 pp., 4th Ed. (1960).

71. National Lime Association, "Lime on Lawns & Flower Gardens," *Bull. 180-A* (1958).

72. R. Boynton, "Ca & Mg in Mixed Fertilizers," *Chemistry & Technology of Fertilizers,* Reinhold (New York, 1960), pp. 446–453.

73. A. Hasler, "Improving Conditions for Fish in Brown Bog-Lakes by Alkalization," *J. Wildlife Management* (Oct. 1951).

74. V. Loosenoff et al, "Use of Lime in Controlling Starfish," *U. S. Dept. of Int. Rpt. 2* (Fish & Wildlife Serv., 1942).

75. E. Eckels, *Cements, Limes, & Plasters,* John Wiley (New York, 1922), pp. 123–125, 172–199.

76. W. Voss, "Lime Characteristics & Their Effect on Construction," *ASTM Publ. 40* (1939), p. 103.

77. W. Voss, *Exterior Masonry Construction, Nat. Lime Assn. Bull. 324,* 70 pp. (1960).

78. L. Palmer et al, *U. S. Bu. Stands. J. of Res.* **6** (Mar. 1931), pp. 473–492.

79. L. Minnick, "Effect of Lime on Characteristics of Mortar in Masonry Construction," *Am. Cer. Soc. Bull.* 38, No. 5 (1959), p. 239.

80. T. Ritchie and J. Davison, "Cement-Lime Mortars," *Building Research* (Mar. and Apr. 1964).

81. National Lime Association, "Mortar Technical Notes," *Bulls. 1–4* (1964).

82. National Lime Association, *Bull. 322-A* (1965).

83. J. Diehl, *Manual of Lathing & Plastering,* Nat. Bu. Lath & Plast. (1960), pp. 103–122.

84. Am. Stands. Assn, "Standard Specification for Lime-Cement Stucco," *A 42.5* (1960).

85. R. Boynton, "European Stucco Practices," *Plast. Inds.* (Oct. 1960).

86. L. Wells et al, *U. S. Bu. Stands. J. Res.* 41 (1948), p. 179.

87. Am. Rd. Bldrs Assn, "Lime Stabilization Construction Manual," *Bull. 243* (1962).

88. L. Minnick and R. Williams, *Hwy Res. Bd Bull. 129* (1956).

89. C. McDowell, "Stabilization of Soils with Lime," *Hwy Res. Bd Bull. 231* (1959).

90. K. Gutschick, "Expedite Construction with Lime Stabilization," *Mod. Hwys* (June 1958).

91. Anon., "Subgrade Improved with Drill-Lime Stabilization," *Rural & Urban Rds* (Oct. 1963).

92. Anon., "Stabilization Technique for Treating Expansive Soils," *Bldg Const. Illust.* (Mar. 1960).

93. National Lime Association, "Lime in Asphalt Paving," *Bull. 325* (1961).

94. B. Brakey, "Colorado Upgrades Gravel Hot Mixes with Lime," *Rds & Streets* (Apr. 1961).

95. R. Boynton, "Stabilizing Aggregate for Bituminous Paving," *Tech. Bull. 249 Am. Rd. Bldrs Assn* (1961).

CHAPTER TWELVE

Economic Factors
of Lime and Limestone

In the United States lime and limestone, while very closely related, are two distinct industries, since nearly 95% of all limestone, exclusive of stone for cement, is produced by companies that do not make lime; in fact, limestone operations outnumber lime plants by about 20 to 1. But this situation does not prevail in all countries. For example, in Germany an opposite condition exists. Interpolating from 1963 German statistics, burned lime producers predominate, since as a group they produce for their own captive lime requirements and the open market more than an estimated 65% of *all* German limestone, exclusive of that used in cement manufacture. Even their stone for lime is nearly equal in tonnage to their total commercial shipments. This means that proportionately, German limestone consumption is much less than the United States and their lime consumption is much greater. The major reason for this is that limestone per se is much less abundant in Germany. Unlike the United States, other siliceous stone and sand and gravel greatly predominate over limestone as construction aggregate. (In 1963 limestone shipments for construction were only 9.46 million tons against 293 million for the United States.) The ratio of limestone to lime plants in Germany is only about 2 to 1. Therefore, German lime producers dominate limestone to such an extent that both products are regarded as *one* industry with *one* association embracing both. Limestone is more of an adjunct to lime and is far less in total value.

Lime Industry

Competition. Traditionally the lime industry has been intensely competitive (at times even savagely so)—more than most industries.

Just exactly *why* this situation has prevailed cannot be explained explicitly, but the following are some principal reasons:

1. Except in time of war there is usually considerable *excess capacity* available. As an example, it is estimated that the industry as a whole only operated at 65 to 70% of capacity in 1962–1963. Cost of production invariably increases as the percent of operation decreases. The resulting profit incentive, plus often the existence of unamortized new plant facilities, creates economic pressure on companies to expand sales during such "buyers' markets." The result oftentimes leads to price instability on a regional or even a national basis, including so-called price wars, in which prices on occasion have plummeted to or below the cost of production.

2. Insufficient lime storage facilities are also a contributing factor. Sudden depressed business in consuming industries causes cancellations of orders and inevitable "peaks and valleys" in shipments. When silo storage is full, there is a seeming reluctance by at least some manufacturers to retard the rate of production. All manufacturers, of course, have some storage facilities, but the cost of building storage is relatively quite expensive due to the relatively light weight (low bulk density) of the material. Hydrated lime only weighs about 35 lb./ft.^{3}, nearly one-third less than portland cement. Yet many manufacturers fail to realize that an increase in storage facilities, although seemingly costly, is also tantamount to an increase in capacity through leveling production rates—and at much less cost.

3. One product, quicklime, is quite perishable, and even when stored under the best conditions, it should be consumed at least a month or two after manufacture for maximum utility.

4. Unlike such chemicals as soda ash, chlorine, and sulfuric acid, which are almost completely standard in quality and analysis, there are many slight differences in limes, even though they will meet certain basic, broad specifications. With the former it is strictly price and service; with the latter *quality* is an added factor. There are high calcium and dolomitic limes that compete bitterly in certain joint markets. Limes of high reactivity and slow settling rates versus less reactive, faster settling products. Then there are available lime, magnesia, silica, iron, sulfur, phosphorus considerations versus corresponding products with lower impurity contents; rotary versus shaft kilns; varying particle or pebble sizes; etc. Complicating the situation further is the growing tendency among some companies to "tailor-make" lime to a consumer's particular (possibly peculiar) requirement or preference—and usually at a standard price. Then too, the con-

sumers' appraisal, even for the same use, is frequently contradictory. Thus, at equal prices a consumer oftentimes has a definite preference for one or more specific brands. Yet this same consumer may quickly switch to another less preferred brand if he can save $.50 or $1 a ton. So a valid purchasing consideration exists—quality, but it certainly intensifies competitive relationships.

Reciprocity is also involved but no more than exists with other large-volume industrial materials.

All of this results frequently *in low average profit margins* with generally a *low return on capital investment*. The exceptions to this among companies are few.

Prices. Lime has always been a low-priced commodity. Even during its rare instances of scarcity, there has been very little fluctuation in price. There have been no so-called bonanza years. Even in times of inflation prices are slow to rise, e.g., during the steady inflationary period from 1941 to 1963 when many products and services increased in price three to five times, lime only advanced 96%. More surprising still, 1963 prices were only 31% higher than the peak prewar year in 1920 and 120% above 1933, the depth of the great depression. Yet the capital cost of lime production facilities has increased at least 350% since 1941. The only explanation for this incongruity has been greater productivity through mechanization and the attrition of relentless competitive pressures. Under these circumstances plant amortization cost is lengthy—fifteen or twenty years on an average, longer since World War II.

The Bureau of Mines is the only source of price statistics. Figure 12-1 portrays price trends for total lime and building, chemical, and refractory limes from 1908 through 1963.

Average F.O.B. plant prices per ton on a bulk basis in 1963 for the following classifications were:

Total lime	$13.73
Agricultural lime	14.30
Construction lime	16.39
Chemical lime	12.79
Refractory lime	16.96

These prices do not include freight absorptions. Traditionally the industry has always sold on a basing point system in which those companies who desire to compete must equalize their higher freight rate with the lowest rate to a given destination, absorbing the difference. The result is that most producers will ship to some extent

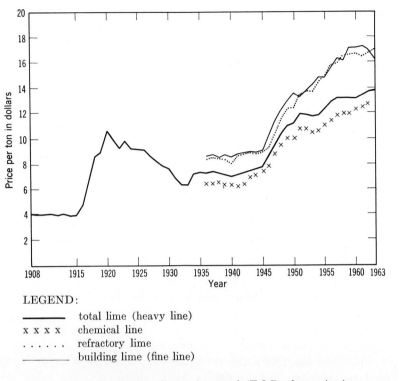

LEGEND:
—————— total lime (heavy line)
x x x x chemical line
. . . . . . refractory lime
——————— building lime (fine line)

Fig. 12-1. Average lime price trend (F.O.B. plant prices).

outside of their natural marketing area, where they do not establish the lowest rate. In contrast to cement and some chemicals, most carload shipments are directly from the plant, not distribution terminals, as basing points. There are only a few such distribution centers in the American lime industry. Naturally the lowest-cost producers are able to ship profitably a higher percentage of their products into freight absorption areas and/or ship further (or equalize more transportation cost). Only a few companies are geographically so isolated as to have little or no incursions from their competitors. Even these plants experience considerable competition on the fringes of their natural marketing areas. F.O.B. plant prices will vary somewhat by geographical area, but not within a producing district. Pacific Coast prices have been consistently the highest. Such plant prices are predicated on an area's production costs, reflecting principally differences in fuel and labor and study of freight rates. The maximum distances that a producer can economically ship vary greatly—300 to 400 miles would be average. However, 150 miles is the limit for a few plants, and a few can or have shipped up to 2,000 miles. Shipments from the Mississippi River to East and West Coast cities and from northwest Ohio to Miami are common occurrences. Rates are often erratic and even seemingly discriminatory, e.g., instances of rates for a 500 mile haul being greater than a 1000 mile movement are not unusual. Consequently, in this highly competitive business a vigilant and aggressive traffic department is a necessity. Transportation costs are often a very high percentage of the delivered price—on an average nearly 30%. In extreme cases freight has even exceeded the F.O.B. plant price.

Transportation. Railroads at one time were the universal carrier of lime. However, in the 1950's truck transport encroached considerably on the railroads, particularly on short hauls. This was caused by lower rates and superior and faster consumer service. In some cases trucks do more than just deliver lime; in soil stabilization of roads, for example, tank trucks deliver directly to the job site (not at the nearest rail siding) and for a slight extra cost are employed to distribute the bulk hydrated lime onto the road through spreaders attached to the truck's rear. Thus, materials handling by the contractor is eliminated and a valuable construction step is actually performed. In the 1960's the railroads started a campaign to recoup this large tonnage loss to trucks by establishing lower rates; use of larger cars that can transport 75 to 80 tons; six-car tandem loads; and faster service. It is difficult to predict the outcome of this increasingly intense competition among the carriers. Railroads are still the prime carrier of lime in tonnage.

Also in the late 1950's some water transport of lime occurred for the first time—2,000 to 3,000 ton barge loads on the Mississippi, Ohio, Missouri river waterways. Such transportation with its large payloads is, of course, by far the lowest of all transportation rates. However, the economics of reshipment inland by rail or truck from river terminals is still not proven.

Bulk versus Packages. The only package used for lime since about 1925–1930 is the paper bag, usually the multiwall, valve type. The large majority of hydrated lime produced is packaged in 50 lb. paper bags; some companies do not even have facilities for bulk shipment of hydrate. However, the reverse is true with quicklime. At least, 98% is shipped in bulk with the remainder in heavier-ply paper bags, weighing usually 80 lb. There is a trend toward increased shipment of hydrate in bulk due to its growing highway use. The favored rail car for shipment is the covered hopper, although box and container cars are commonly employed.

Before paper bags became the standard lime package, barrels and textile bags were employed. Many lime companies during this period operated their own cooperage. The barrel was even recognized in a federal statute of 1916, known as the "Standard Lime Barrel Act," which established 280 lb. and 180 lb., net weight, for large and small barrels.

The cost of packing hydrated lime in paper bags is variously estimated between $3 and $3.50/ton, including cost of bags. Undoubtedly this extra packaging cost has encouraged a consumer trend to bulk shipments. Improved bulk pneumatic materials handling equipment is another contributing factor.

Since about 1915 all lime in the United States has been sold by the *net short ton*. Before this the *bushel* was a common unit of measure.

World Trade. It is rather obvious that there would be relatively sparse world trade for such a low-cost commodity as lime. Furthermore, it is produced to some extent by nearly every nation in the world, large and small.

In spite of this, until 1961 United States lime exports had always exceeded imports. Since 1961 a sudden and increasing disparity between United States exports and imports has occurred, with imports reaching an all-time high in 1963 of 100,800 tons. Meanwhile, 1963 U. S. exports declined to an all-time low of 17,500 tons, about 0.2% of total shipments, most of which is hydrate. Before 1962, peak imports were about 40,000 to 50,000 tons. Total United States imports and exports since 1955 are contained in Table 12-1.

Table 12-1. U. S. world trade on lime

Year	Exports	Imports
	thousands of tons	
1955	82.4	69.6
1956	82.7	41.6
1957	65.2	49.7
1958	45.8	25.5
1959	52.8	35.4
1960	61.1	32.0
1961	30.0	36.6
1962	19.5	77.6
1963	17.5	100.8

Since 1946 over half of all annual lime exports have been to Canada, with the balance, averaging about 35%/year, to Central American nations—Costa Rica, Honduras, and Panama in particular. Meanwhile, 95% or more of United States imports are from Canada. Originally, until 1961, most of the Canadian exportation was into the Pacific Northwest, which for many years consumed more lime than it produced. Then a paradox occurred. In 1962–1963 Pacific Northwest imports declined sharply due to new domestic production, but a large new Canadian exportation was commenced into the Buffalo-Niagara Falls area, primarily for U.S. steel manufacture. The latter far exceeded the former loss, so that Canadian exports to the U.S. have pyramided rapidly. It is believed that these Canadian exports have nearly reached their peak and probably will not exceed 125,000 ton/year. Sporadically there has been a small trickle of lime imported from Mexico. The record year for U. S. exports was 1956, when tonnage reached 82,700 tons, nearly 1% of total shipments that year. In addition to reduced Canadian imports, a large Central American hydrated lime requirement for insecticide use has plummeted. So on lime the United States has become a deficit nation in foreign trade, and no marked change in this situation is anticipated in the foreseeable future.

Since most of the foreign trade involving the United States is with Canada, this is the only tariff that warrants citing. In 1964 the U. S. tariff on quicklime and hydrated lime was $.50 and $.60/ton, respectively. However, Canada's tariff is about four times as high as that of the United States—15% ad valorem (equivalent of about $2.25/ton for quicklime and $2.50/ton for hydrated lime.) In 1930 tariffs for both countries were nearly equal. United States duties were $2.40 and $2/ton for hydrated lime and quicklime, respectively, when lime prices were about 95% less than 1964. The Trade Agreements of 1936 and 1939 halved the above in a two-stage reduction. Then the U. S. duties were again cut in half at the 1948 Geneva Trade Conference. Meanwhile, there was no change in the Canadian tariff over the same period.

World trade is also inconsequential in most other countries. The most notable exceptions would be German and Belgium exports to the Netherlands of 230,000 and 238,000 tons, respectively, in 1963. Total German exports were 345,000 tons (or over 3% of total production) in 1963, which also included substantial shipments to Luxembourg. Belgium also exports to the French steel industry. There have been small, intermittent exchanges among Scandinavian countries. In 1963 Poland shipped 17,500 tons to the Netherlands.

Number, Size, Location of Plants. Lime has generally paralleled most other industries with a steady decline in number of plants but accompanied by a rapid increase in productive capacity. Thus, a plant of 25,000 tons annual capacity in 1908 was a large producer; in 1964 it is considered small. In 1909 there were 1,232 commercial plants as opposed to only 136 in 1964 (actually only 105 of real commercial significance). Figure 12-2 shows the decline in number of commercial plants over the years.

The size of plants, classified by established shipment or production records in 1963 comprising both commercial and captive plants, is depicted in Table 12-2. In 1963 only 19 of the plants canvassed (208) made 49% of the total lime produced. However, since there are many multiplant companies, the actual size of certain individual companies' total production and sales is considerably larger than indicated in Table 12-2. Of the commercial companies the most operations by

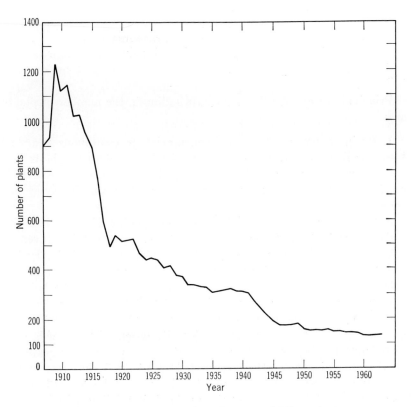

Fig. 12-2. Decline in number of commercial lime plants.

Table 12-2. Relative size of U. S. lime plants in 1963[a]

| | | 1963 | |
| | | Production | |
Annual production (Short tons)	Plants	Short tons	Percent of total
Less than 10,000	67	298,238	2
10,000 to less than 25,000	37	610,918	4
25,000 to less than 50,000	39	1,423,364	10
50,000 to less than 100,000	25	1,828,411	13
100,000 to less than 200,000	21	3,211,690	22
200,000 and over	19	7,149,636	49
Total	208	14,521,000	100

[a] U. S. Bureau of Mines.

one company is seven; one company has six; three have four plants; and seven operate two plants. However, no company is nation-wide in scope.

Principal reasons for this steady dwindling in the number of plants are:

1. Business closures through bankruptcies or continuous years of losses or little or no profits, are common. Some of these plants had enjoyed profits in the past, but due to high production costs, lack of finances to mechanize, mismanagement, they simply could not compete.

2. Quality problems abound. The same quality of lime that was satisfactory in 1910 (even in 1930) is often unacceptable (even at submarket prices) in the 1960's. The chemical process industry has steadily demanded and readily procured purer, more completely calcined, and more reactive limes over the years. Even the building trades' requirements are more stringent. Thus, faced with a declining market, these producers were forced to cease operations or commence a new plant with a new stone deposit. Often disillusioned, they abandoned the business altogether. However, many simply remained in the stone business. It is estimated as of 1963 that nearly 200 limestone producers (or their antecedents) in the United States formerly operated lime plants at or near their existing quarries.

3. Exhaustion of the limestone deposit or its high-quality vein

frequently occurred. In some cases, through inadequate or misinterpreted analysis and core drilling, the plants were poorly conceived in proving the extent of the deposit, thus causing their premature demise, in some instances unamortized. Exploration for new economical stone deposits is most difficult, costly, and uncertain (see p. 34 on limestone exploration).

4. Loss of principal market (bread-and-butter customer) is not unusual due to new technological changes in consumer industries, revised architectural practices, and even acts of God. Examples would be closure of an old established steel plant due to shifting markets for steel or marginal production costs; a chemical plant revamped to include some improved, new nonlime-consuming process. (In 1933 the earthquake in California triggered a sweeping overhaul of West Coast building codes that proved most damaging to masonry per se, and in addition, downgraded lime in mortar.) Many of these former plants placed too much reliance on one customer or market. When it was lost, they lacked the profit incentive or sales ability to recoup. Lime markets historically have been fluid, unpredictable, and even hazardous (Fig. 11-2).

5. Loss of strategic freight rate position to a new plant that operates in a more favorable marketing location.

6. Abrupt, soaring increase in fuel costs has made operations unprofitable and untenable. At one time plants in the north, located long distances from coal or oil, burned wood. A combination of forest conservation, scarcity, and greatly increased wages to lumberjacks made wood prohibitive in cost. There was no economical substitute.

7. Adverse reciprocity has caused loss of key accounts.

There are many vicissitudes in the lime business.

Figure 12-3 reveals the location of all commercial lime plants of significance. Open-market lime is produced in thirty-three of the fifty states. There is production in Hawaii but none in Alaska. Some states simply do not possess limestone deposits of sufficient economical size and/or purity to justify plants. For the past sixty years the largest lime-producing areas have been the Berkshire Mountains of Massachusetts; southeastern and central Pennsylvania; northwest Ohio; Wisconsin; the Appalachians of Tennessee and Virginia; Shelby County, Alabama (near Birmingham); central Texas; the southern half of Missouri; and southern Nevada. Until 1940 there were a few other formerly large producing sections that now are very small or nonexistent. These are: coastal Maine (near Rockport), Vermont, and central Indiana. Table 12-3 shows apparent lime consumption by

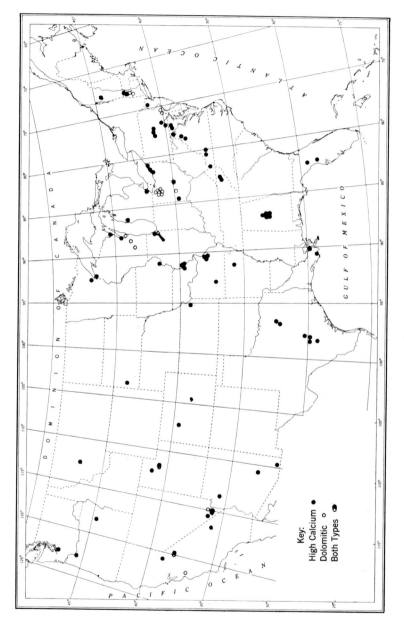

Fig. 12-3. Location of commercial lime plants in the United States (1965).

Table 12-3. Apparent total consumption of lime by states[a] (short tons)

1960

State	Quicklime	Hydrated lime	Total
Alabama	257,262	75,830	333,092
Alaska	—	231	231
Arizona	135,570	25,259	160,829
Arkansas	63,027	9,333	72,360
California	409,590	97,941	507,531
Colorado	17,471	11,526	28,997
Connecticut	46,096	25,107	71,203
Delaware	35,783	9,587	45,370
District of Columbia	—	8,635	8,635
Florida	248,866	57,096	305,962
Georgia	75,139	22,623	97,762
Hawaii	—	(1)	(1)
Idaho	2,315	2,649	4,964
Illinois	373,461	108,728	482,189
Indiana	528,847	62,530	591,377
Iowa	80,381	18,130	98,511
Kansas	37,896	15,478	53,374
Kentucky	503,653	18,094	521,747
Louisiana	617,640	35,668	653,308
Maine	48,526	10,404	58,930
Maryland	168,902	26,586	195,488
Massachusetts	35,098	46,731	81,829
Michigan	806,390	595,599	1,401,989
Minnesota	95,372	19,910	115,282
Mississippi	39,750	10,578	50,328
Missouri	110,703	53,255	163,958
Montana	78,054	5,989	84,043
Nebraska	10,499	11,556	22,055
Nevada	356	36,546	36,902
New Hampshire	4,861	5,465	10,326
New Jersey	33,254	71,258	104,512
New Mexico	4,869	50,886	55,755
New York	1,271,977	124,103	1,396,080
North Carolina	102,007	27,174	129,181
North Dakota	7,357	1,532	8,889
Ohio	1,671,747	117,163	1,788,910
Oklahoma	38,397	12,680	51,077
Oregon	49,454	8,975	58,429

Table 12-3. (*Continued*)

		1960	
State	Quicklime	Hydrated lime	Total
Pennsylvania	1,028,769	238,498	1,267,267
Rhode Island	8,707	6,342	15,049
South Carolina	12,790	7,711	20,501
South Dakota	8,429	1,058	9,487
Tennessee	47,036	27,019	74,055
Texas	441,058	400,968	842,026
Utah	86,672	23,762	110,434
Vermont	—	1,825	1,825
Virginia	272,516	37,199	309,715
Washington	19,007	11,536	30,543
West Virginia	166,218	20,578	186,796
Wisconsin	94,981	55,464	150,445
Wyoming	502	4,017	4,519
Total	10,197,255	2,676,812	12,874,067

[a] U. S. Bureau of Mines.

states for 1960, including most captive lime. In order of size the following states were the largest open-market producers in 1963:

1. Ohio
2. Missouri
3. Pennsylvania
4. Virginia
5. Alabama
6. Texas

Including captive production, Ohio is still by far the largest producer—as well as consumer.

Profits. There is no reliable information on profits among lime companies. Financial statements are either not available or meaningless. The reason is that lime companies are almost invariably either small, closely held (or even family-owned) corporations or divisions of large, highly diversified corporations. With the former, such information is private; with the latter, published information is profuse but only for the large corporation *as a whole* and not for its divisions and subsidiaries. The trend has been away from small, family-owned companies to large corporations since World War II, with many mergers and plant acquisitions resulting. Fifteen of the commercial producers are listed on the N. Y. Stock Exchange in 1965; there were only four in 1946.

World Production. Many industries are able to estimate world production precisely; however, this is impossible for lime. The United States, West Germany, Canada, and the Union of South Africa are the principal large-producing nations that systematically conduct a detailed canvas on annual lime production. Even the United Kingdom makes only a periodical "estimate." Nevertheless, it is possible to hazard an educated guess on approximate world production from reports by United States consulates and commercial attachés; estimates by large foreign producers and associations; occasional reports from "iron curtain" countries that may be based on bonafide surveys. Red China is the real imponderable. Insofar as the author can determine, Table 12-4 on world lime production is the most complete survey of this type ever attempted.

Surprisingly, *Russia* is easily the *largest producer* in tonnage in the world—19 million tons per year. (This has been confirmed by the U. S. Bureau of Mines and the German Lime Association.) The United States is second, followed by West Germany. But note how proportionately large production behind the iron curtain is. With Russia's enormous production it is larger than the total for Western Europe.

But in per capita consumption (based on 1960 population) Russia is only seventh, and the United States is tied with Canada for eighth, as revealed below. Belgium easily leads all nations with a staggering 499 lb. per capita, with Czechoslovakia a strong second and West Germany is third. No allowance in these figures is made for exports or imports.

lb. lime per capita

1. Belgium	499	6. France	183
2. Czechoslovakia	394	7. Russia	182
3. West Germany	328	8. United States	152
4. Sweden	295	9. Canada	152
5. Austria	221	10. Poland	146

Reports indicate that lime production is increasing in Poland, Hungary, Yugoslavia, Czechoslovakia, and Italy since 1960. The only explanation why the U. S. lags behind the other countries in proportionate usage is the much higher rates of use in steel manufacture, structural lime in building construction, far more extensive manufacture of sand-lime brick, and greater proportionate application of acetylene (via calcium carbide).

Table 12-4. 1962 World lime production thousands of short tons

North America

Canada	1,380
Mexico	850
Nicaragua	28
United States	13,752
Other Central Am.[a]	50
West Indies and Bermuda[b]	90
Total	16,150

Africa

Algeria	25
Egypt	150
Kenya	18
Libya	19
Morrocco	50
Tunisia	139
Uganda	18
Union of South Africa	725
Other African[c]	145
Total	1,289

South America

Argentina	800
Bolivia	25
Brazil	1,400
Chile	60
Columbia	60
Ecuador	20
Peru	88
Paraguay	16
Uruguay	40
Venezuela	49
Total	2,558

Asia

China (red)	750
India	600
Israel	50
Japan	1,000
Kuwait	60
Pakistan	100
Taiwan	50
Thailand	75
Turkey	125
Other Asian[d]	200
Total	3,010

Europe (free)

Austria	775
Belgium	2,250
Denmark	165
Eire	100
Finland	243
France	3,900
Italy	1,750
Greece	350
Luxembourg	50
Malta	53
Netherlands	75
Norway	200
Portugal	50
Spain	250
Sweden	1,110
Switzerland	215
United Kingdom	2,450
West Germany	8,723
Total	22,409

Oceania

Australia	300
Indonesia	150
New Zealand	60
Phillipines	50
Total	560

Europe (Iron Curtain)

Bulgaria	450
Czechoslovakia	2,600
East Germany	1,200
Hungary	675
Poland	2,175
Rumania	250
U.S.S.R.	19,000
Yugoslavia	800
Total	27,150
European Total	49,559

[a] Known production in Costa Rica, Guatemala, Honduras, Panama, and Salvador.

[b] Known production in Bahamas, Barbados, Bermuda, Cuba, Dominican Republic, Dutch West Indies, Haiti, Jamaica, and Trinidad.

[c] Known production in British East Africa, Cape Verde Island, Congo, Ethiopia, Ghana, Liberia, Nigeria, North Rhodesia, Sierre Leone, South Rhodesia, Southwest Africa, Sudan, and Tanganyika.

[d] Known production in Burma, Cambodia, Ceylon, Iran, Iraq, Korea, Laos, Lebanon, Malaya, Ryuku Island, Syria, and Viet Nam.

Table 12-5. West German lime production in 1963; size of plants.

Plants classified Annual production long tons	Number of plants	Total production %
0–5,000	28	0.6
5,000–10,000	15	1.2
10,000–60,000	44	14.5
60,000–120,000	10	10.4
120,000–500,000	12	34.6
Over 500,000	2	38.7
Totals	111	100

The West German lime industry in their miraculous industrial recovery following World War II merits attention. (They produced less than 1 million tons in 1946.) Table 12-5 presents pertinent statistics on the relative size of their producing plants in 1963. Included in this tabulation are two plants in the Ruhr that are the largest in the world in production records and capacity. An end-use breakdown for West Germany is given as follows for 1963.

Use category	Thousands of short tons
Total chemical and industrial	4,476[a]
Sand-lime brick manufacture	1,513
Building (structural) lime	2,037
Agricultural lime	351
Export	345
Total	8,722

[a] Includes 2,319 fluxing lime and 760 dead-burned dolomite for steel and 1,425 for the chemical industry.

Captive Lime. According to the U. S. Bureau of Mines there was 5.632 million tons of captive lime production in 1963. (See Table II-I, p. 342.)

The largest captive category is the alkali industry—over 3 million tons. It is mandatory for this industry and the beet sugar industry (586,000 tons) to burn their own lime, since they must have low-cost carbon dioxide for their processes to be economical. CO_2 is obtained as a co-product with lime. Thus, over two-thirds of U. S. captive production is inevitable.

The next classification in size is nonferrous metals (0.6 million tons). This includes primarily large copper companies in the West that locate their lime plants at their copper ore beneficiation mills and smelters and also manufacturers of alumina and magnesia principally in Arkansas, Texas, and Louisiana. Usually these captive plants were built because lime supplies were uncertain and/or freight costs too high. Many of these Western plants are quite old, pioneering industries. At the time these operations were built, the then Western lime industry was a fledgling and was probably apathetic about these new large requirements. So integration was deemed necessary.

Another large category is calcium carbide (368,000 tons). The remainder is small and scattered. Steel only makes 7.5% (71,000 tons) of its fluxing and 6% of its refractory lime requirements.

An estimated 250,000 tons of this total is actually lime used by the open-market producers in their manufacture of other products, like proprietary masonry mortars, plasters, precipitated calcium carbonate, other chemicals, water treatment, etc.

Generally speaking (and excepting alkali and beet sugar), the difficulty of correlating an adequate supply of high-grade limestone at low cost with existing consuming plant locations makes on-site captive plants an unprofitable investment—unless possibly the prospective captive operation is located a long distance from commercial lime sources. Many former captive plants have been closed down and even dismantled as evidence of this. Furthermore, the cyclical nature of many industries, like steel, alumina, and others, generally makes captive operations imprudent. Since there is only one use for this product in their processes, operating kilns must be retarded or shut down often. This resulting lack of production continuity increases fuel and relining costs considerably. Under the most favorable circumstances captive kilns rarely yield more than a scant profit on capital investment—2 to 3% is typical. (See p. 280 on the vital role that limestone kiln feed occupies in lime production costs and p. 34 on the speculative exploration of high grade, economical limestone deposits.) However, the threat of captive plants among large consumers does exert a retarding influence on lime prices.

Until 1953 the Bureau of Mines collected very little information on captive lime. Then gradually they commenced adding these plants to their annual canvas, until by 1961 their survey was virtually complete. As a result, their figures give the erroneous impression that captive production has grown rapidly. If captive figures were available for many years back, it is the author's opinion that such figures would confirm that captive lime has even lost position to commercial production on a percentage—not tonnage—basis. It is the lime recovery installations, in paper predominately and most recently in the carbide and water field, that have wedged inroads into commercial tonnage. Since these latter companies are utilizing a waste calcium carbonate sludge of negative or at best zero value, because of a disposal and space problem, the economics are completely different from the commercial and captive producers, who make virgin lime from limestone.

Captive production exists in other countries—in some to a much greater extent than others. It is relatively quite low in West Germany, proportionately less than the United States, but quite high in France and Sweden where many steel and chemical companies produce their own requirements.

Research. As in the case of limestone, there has been inadequate research conducted on lime over the years—a pittance as compared to the progressive cement and chemical industries. In defense of the industry, profit margins are often so slender that there are only limited funds available for fundamental, applied, and market research. Yet it is obvious that there must be many research opportunities available for this low cost, versatile material. It is also axiomatic that expanded research activities could, if intelligently and vigorously pursued, cure most of the ills of this industry, such as excess capacity, which leads to cut-throat competition and low profits. Developing new uses and expanding existing markets through improved applications would stimulate its growth. However, even during periods of moderate to good profits, the industry does not increase research and tends to be complacent. This failing is generally recognized and even deplored, but little is done to rectify it. One possibly encouraging sign is the increased laboratory testing on quality control since World War II, probably motivated more by defensiveness as a result of intense competition than for progressive reasons. Many companies that had no laboratories previously, now have them; a few have modern, impressive facilities.

Research has been divided between a small minority of the more progressive companies and the National Lime Association. The latter has sponsored most of the fundamental research at M.I.T., National Bureau of Standards, and the University of Illinois. Results of some of this association research are included in Chapters 6 and 7. In applied research the association has initiated fellowships on soils stabilization, asphalt paving, masonry mortars, autoclaved concrete products, steel fluxing, acid neutralization, trade-waste treatment, and agricultural liming. A few of these research projects, notably in highways, has been pursued by the National Lime Association in market development and promotion with marked success.

Generally, individual companies have expended more research funds on the production phase of the business—new improved kilns or equipment, patented hydration processes, specialized preheaters and coolers for greater fuel efficiency, and other production methods designed to improve the quality and uniformity of their products. As a result, technically the industry has progressed more in production than in the use and application of its products. With respect to the latter it is again the National Lime Association that has initiated most of the applied research on uses, but in total effort the industry as a whole has been backward. In effect, the industry has largely relied on its many consuming industries in the chemical-process field to develop the hundreds of uses of lime, and the construction uses were inherited from antiquity—namely, the Egyptians. Of markets of consequence the *sole exception* is the association's pioneering research in road stabilization.

In addition, except for a few companies there is no concerted market research being pursued. The substantial losses in some markets could have been minimized or even possibly averted had there been more activity exerted on market research and development and creative selling instead of resorting to unimaginative commodity selling that still prevails throughout 90% of the industry. Lime's principal competition from the cement and chemical industries, which are renowned for their market development successes, may force a change in the industry's traditional, but now archaic, research and sales approach. If there is no change, then future over-all market prospects for lime are bleak.

This same research lethargy has existed in most other nations, with the one exception of Germany. This nation, through its large and strong association has resisted the inexorable competitive pressures of cement, gypsum, and limestone more successfully than any other country. It is estimated that proportionately Germany spends ten times as much as the United States on research and promotion. That is why, as an example, it consumes about seven times as much structural lime per capita as the United States and are easily the world's leader in this field. Of course, cartelization has aided Germany in organizing its potent and united industry program. The West German industry maintains through its association, *Bundesverband der Deutschen Kalkindustrie*, an impressive laboratory in Koln. Since World War II some excellent fundamental research on lime has been pursued by Dr. Rune Hedin, Cement and Concrete Research Institute of Technology, Stockholm, Sweden. Since 1955 England has also made an encouraging contribution through CLAIRA (Chalk Lime and

Allied Industries Research Association) at Welynn, Herts., near London. In addition to lime they research sand-lime brick, ready-mixed lime-based mortars, and whiting. This organization is financed on a joint basis by the government and the industries involved.

To stimulate greater interest in lime research, an American (and internationally known) lime plant engineering consultant, Victor J. Azbe, generously established in 1959 a perpetuating fund through an irrevocable trust for an annual award of $1,000 for the best technical paper on lime manufacture or fundamental research on lime. The National Lime Association is delegated to administer this annual contest, appoint judges, and present the award at its annual fall Operating Meeting. The four winners and their prize-winning articles to date are:

1960 T. C. Miller, "A Study of the Reaction between Calcium Oxide and Water."

1961 Dr. Rune Hedin, "Structural Processes in the Dissociation of Calcium Carbonate."

1962 Dr. Rune Hedin, "Processes of Diffusion, Solution, and Crystallization in System $Ca(OH)_2 \cdot H_2O$."

1964 (Co-winners). Dr. R. P. Mayer and Dr. R. A. Stowe, "Physical Characterization of Limestone and Lime."

T. C. Miller, "Improving Calcium Hydroxide for Better Calcium Hypochlorite Production."

(Rules for the Azbe contest can be obtained from the National Lime Association, Washington 16, D. C.)

Wartime Stimulus. As evidence of the strategic importance of lime, invariably during wartime the demand for it surges sharply upward. During both World Wars I and II annual lime shipments soared to all-time record heights. Even the relatively confined Korean War stimulated a brisk demand for lime.

This is caused largely by the burgeoning requirements of steel, nonferrous metals (aluminum, magnesium, and copper), explosives, and chemicals. These expanded requirements dwarf lime shipment losses for structural lime, which invariably declines abruptly as a result of federal wartime restrictions on building construction. During both World War I and II short-lived but acute *lime shortages* periodically occurred, and the commercial industry was severely taxed to satisfy the requirements of war industry and essential civilian use, like water treatment.

During World War II the lime supply problem was uniquely solved. As a result of construction restrictions the portland cement industry was operating at less than 50% of capacity. As many as twelve of the idle rotary kilns in six cement plants in New York, Michigan, Wisconsin, Missouri, Tennessee, and Texas were converted to the manufacture of peddle quicklime. These plants shipped to ordnance (smokeless powder), steel, alumina, and chemical plants, thus balancing the precarious supply and demand picture. Conversion of these cement plants also conserved critical materials by obviating plant expansions among existing lime producers, which would have otherwise been necessary. A certificate of necessity for plant expansion was issued to only one producer of chemical lime—a single large rotary kiln. Consequently, the commercial lime industry, unlike other essential industries that expanded greatly, emerged from the war with very little change in the industry's capacity. (The cement kilns were reconverted to portland cement.) There was, however, some expansion of captive lime facilities, notably magnesia and alumina. Paradoxically, in the Korean War certificates of necessity were granted to about eight plants for increased capacity of chemical lime and dead burned dolomite.

In spite of the "apparent" periodic shortages that stimulated some panic buying of lime, neither the War Production Board nor its predecessor in World War I had to allocate lime to industry. Voluntarily, the commercial industry was able to satisfy the urgent, essential requirements. Issuance of WPB directives was necessary in some instances to ensure the supply to a war requirement. At several critical stages the over-all supply problem necessitated several Lime Industry Advisory Committee meetings to be called by the WPB in Washington.

Probably the most persistent problem in wartime is *manpower*—sufficient competent labor to operate plants at capacity. Owing to conscription and the lure of higher paying wages in war plants, many workers were lost by the lime industry. With "frozen" wage scales and product prices, lime companies could not retaliate and entice other workers into their employ with remunerative inducements. Furthermore, with most lime plants located in rural areas near their quarry site, their wages have been traditionally less than the urban areas. Since replacements were insufficient to compensate for these losses in manpower, most plants were compelled to extend working hours and mechanize their operations ingenuously to alleviate their deficiency in manpower. In spite of this, some producers sporadically still lost production due solely to the labor shortage; a few small producers were forced to close their plants. Finally in late 1944 and 1945 when the labor shortage became most acute, high priorities, equal to steel, were accorded chemical lime producers to forestall some possibly serious losses in production. This enabled the War Manpower Commission to provide some relief through federal labor recruitment. To alleviate the chronic labor problem a few plants resorted to use of imported Mexican labor and prisoners of war.

A second problem in World War II was rigid price control on lime. Regardless of federal control over prices and wages, lime manufacturing costs increased, in some instances considerably, due to lower labor productivity, more overtime pay, dislocations in normal sources for fuel and supplies. A few small producers, who had always been dependent on wood as their fuel, were unable to secure it and were obliged to import coal from long distances, often increasing their fuel costs by 100 to 200%. Indirectly this problem adversely affected the lime supply. The higher costs precluded the operation of some submarginal stand-by kilns that otherwise might have been employed to ease the supply. Even with the robust demand net profits were not very high during wartime. The industry was fortunate in being made exempt from federal renegotiation.

Immediately following the war, the ceiling on wages was removed. Thereafter followed a selective decontrol of commodity prices that exerted much greater pressure on lime profits and "frozen" prices. Finally price relief was granted by the Office of Price Administration in June, 1946, nine months after the cessation of hostilities.

Wartime conditions exerted relatively little effect on the limestone industry. The tremendous reduction in the demand for aggregate stemming from construction controls more than counteracted the increased demand for chemical and metallurgical grades of limestone, so there were no real shortages.

Limestone Industry

Competitive Factors. It is apparent from Table 5-13 on p. 117 that limestone encounters tremendous competition from many other types of aggregate engaged in construction activity. As a consequence, in this field interindustry competition is as intense as intraindustry, since most construction specifications do not dictate the particular type of aggregate that can be used. Often any of several to seven or eight different kinds of aggregate are permissible for use. However, in the chemical, metallurgical, and agricultural fields, competition is strictly intraindustry, since contrary to construction, usually there is no possible substitute for this extremely basic material.

When chemical and/or physical requirements are severely stringent, potential sources of supply are narrowed greatly to the extent that in some areas, usually isolated locations, little or no competition exists. But in the industrial Northern, Middle Western, and Eastern states intense competition usually abounds in all markets, regardless of consumer specifications.

There are *two prime reasons* that account for these highly competitive conditions, both of which involve *by-product stone:*

1. Sale of *reprocessed stone spalls.* Necessarily in most stone-processing operations there is an inevitable accumulation of large tonnages of stone sizes and gradations that cannot be used captively or sold to major markets, at least without further processing and/or classification. Such piles of stone, called "spalls," may accumulate to the extent that they constitute a space problem for the manufacturer. The cost of producing such waste stone has long since been absorbed. So with an impelling need to dispose of these piles, they may possess *zero* or even negative value. With some ingenuity exercised in market research and development and in reprocessing techniques, it is often possible to obtain new salable products from such wastes by judicious grinding and screening at low cost—often at costs appreciably lower than other producers manufacturing the same gradation as a *prime* product. The upshot is that the market becomes sated, prices become badly depressed, and the prime producer loses both volume and profit. Such by-product stone sources may be only temporary, disappearing in a year or two after the stone spalls are exhausted. But the unpredictable availability of such by-products and the sporadic losses it effectuates is a serious problem to the prime manufacturer. The constant threat of such competition serves as a depressant on stone prices.

2. *By-products from captive stone producers.* A large captive stone producer, as typified by lime, steel, alumina, alkali, and other industries, processes stone primarily for its own use, i.e., as kiln feed, flux stone, raw material, etc. The cost of producing this stone is absorbed by the end product. Therefore, the inevitable plus and minus stone sizes that cannot be consumed are regarded as waste by-product stone. Any monetary return that can be derived from the sale of such stone is simply applied to reduce the over-all cost of the end-product operation. The subsequent processing entailed in converting such stone to marketable products enables the captive plant to sell at a price below the prime stone producer. Too often such stone is not merchandised; as a matter of quick expediency, it is sold or "dumped" at unreasonably low prices. Subsequently prices of such by-products may be raised to a reasonable level, but its accentuating

effect upon competition is clearly evident. In many areas by-product stone determines price in many markets that the prime producer is obliged to meet.

Thus, the prime stone producers' markets can be quite precarious in the face of such formidable competition, except possibly with extremely large producers. But even the largest prime producers are small compared to an integrated steel or chemical company. A somewhat analogous situation can also prevail between large and small independent stone companies.

There has always been an adundance of limestone per se. So the industry has never enjoyed the stimulus of a so-called seller's market.

Prices. Crushed, broken, and ground limestone prices vary tremendously, much more so than lime, depending upon size, gradation, and purity. In 1962 prices literally ranged from about $1/ton to $30/ton, the latter being the most refined type of limestone whiting.

The following average plant price/ton values on limestone for different markets, reflecting different sizes and gradations, were calculated, from the Bureau of Mines 1962 statistics on total tonnages and values.

Prices per short ton

Cement manufacture	$1.11	Limestone sand	$1.80
Rip rap	1.20	Asphalt filler	2.15
Concrete & road stone	1.32	Glass manufacture	3.20
Flux stone	1.42	Coal mine dusting	4.18
Agricultural liming	1.70	Mineral food	5.42
Lime manufacture	1.73	Shell poultry grits	8.33
		Limestone whiting	11.50

(See Chapter 5 on the above uses in which requirements are described.)

Except for the highest-cost limestone products, notably whiting and mineral food grades, shipments are of much shorter distance than with lime—100 to 150 mile maximum range, and in the heavily populated areas of the East, maximum shipping range is even less, about seventy-five miles. There is some equalization of freight, but at the extremely low plant prices prevailing on most tonnage, there cannot be as much freight absorption as with lime, chemicals, and other more expensive materials.

The pattern of limestone prices over the years parallels the graphic trend of lime in Fig. 12-1. Limestone prices are also inflated less than most commodities.

Table 12-6. Crushed stone sold or used in the United States, by methods
of transportation

Method of transportation	1962		1963	
	Thousand short tons	Percent of total	Thousand short tons	Percent of total
Commercial:				
Truck	388,023	59	416,864	61
Rail	83,221	13	82,578	12
Waterway	53,200	8	51,348	7
Unspecified	63,935	10	67,634	10
Total commercial	588,379	90	618,424	90
Government-and-contractor:				
Truck[a]	65,846	10	67,326	10
Grand total	654,225	100	685,750	100

[a] Entire output of Government-and-contractor operations assumed to be moved by truck.

Transportation. Unlike lime, truck transport greatly predominates in the shipment of limestone, and has for many years. The only statistics applicable to transportation pertain to *all* types of crushed and broken stone, as shown in Table 12-6, but limestone is believed to be similar to percents shown—about 68 to 70% by truck transport.

Most of the water transport originates from about five large plants in northern and upper peninsula Michigan through the Great Lakes to lake-port cities.

Packaging. Only an inconsequential amount of all limestone is packaged, estimated at less than 0.25%.

World Trade. Similar to lime there is relatively little foreign trade on limestone, virtually all of which is between the United States and Canada.

In 1962 United States exports of crushed and broken limestone to Canada totalled 576,000 short tons, whereas imports from Canada amounted to 466,000 tons. Total 1962 exports to all destinations totaled 621,000 tons. United States imports of British whiting were 11,663 tons. Future U. S. imports from Canada will be somewhat higher due to importation of Canadian stone for newly established lime and cement plants in the Pacific Northwest.

The American duty on crushed and broken limestone imports has been reduced from an original amount of $1/ton to $.20/ton as of 1964. There is no Canadian duty on United States limestone exports.

Similarly, throughout the world there is relatively little world trade on limestone, since it occurs in some form in virtually every nation. Probably the Netherlands has the greatest proportionate deficiency of it.

Number, Size, Location of Plants. The number of limestone operations in the United States is not computed by the Bureau of Mines, since their only applicable figures pertain to *total* crushed and broken stone plants. Since there are about 3,000 such plants in existence (Table 12-7), it is estimated that at least 2,350 of these are limestone producers, including marble and shell operations.

Table 12-7 also reveals ranges in size of operations based on annual production of crushed and broken stone for *all* types of stone aggregate. Since limestone represents about 70% of the total United States crushed and broken stone, 65 to 75% of the figures in Table 12-7 would represent a reasonable approximation of limestone's participation in these tonnage brackets. (The higher percentage, 75%, applies to the largest tonnage categories and the lowest percent, 65%, for the smallest ranges.)

Figures 2-2 and 2-3, which reveal the cartographic location of high calcium and dolomitic limestone deposits, are also a reliable general guide on locations of commercial quarries, mines, and stone-processing plants and serve to depict those areas where the industry is most concentrated. In 1962 the Bureau of Mines recorded production from forty-six of the fifty states. Of the four nonlimestone-producing states—Delaware, Louisiana, New Hampshire, and North Dakota— one of these, Louisana, is a large supplier of oyster and clam shells from tidal waters and river estuaries. So, in effect, forty-seven of the fifty states produce limestone. The largest producing states with their annual production in 1962 are:

Millions of tons

1.	Pennsylvania	41.5	7.	Texas	25.7
2.	Illinois	41.3	8.	Tennessee	24.0
3.	Ohio	33.9	9.	New York	23.8
4.	Michigan	28.2	10.	Iowa	21.6
5.	Missouri	27.9	11.	Kentucky	19.5
6.	Florida	25.9	12.	Indiana	18.1

Taxes. The United States limestone industry, along with most mineral industries, has enjoyed the benefit of percentage depletion

Table 12-7. Number and production of crushed stone plants by size of operation

Annual production (short tons)	1961				1962			
	Number of plants	Production Thousand short tons	Percent of total	Cumulative total, thousand short tons	Number of plants	Production Thousand short tons	Percent of total	Cumulative total, thousand short tons
Less than 25,000	[a]813	[a]6,670	1.2	6,670	889	7,462	1.3	7,462
25,000 to 50,000	323	11,586	2.1	18,256	292	10,805	1.8	18,267
50,000 to 75,000	270	16,536	3.0	34,792	241	14,828	2.5	33,095
75,000 to 100,000	186	15,922	2.9	50,714	193	16,899	2.9	49,994
100,000 to 200,000	[a]496	[a]70,088	12.9	120,802	499	71,421	12.1	121,415
200,000 to 300,000	248	60,715	11.1	181,517	248	60,554	10.3	181,969
300,000 to 400,000	[a]173	[a]59,765	11.0	241,282	187	65,654	11.1	247,623
400,000 to 500,000	113	50,604	9.3	291,886	137	61,046	10.4	308,669
500,000 to 600,000	81	43,944	8.1	335,830	94	50,556	8.6	359,225
600,000 to 700,000	[a]45	[a]29,245	5.4	365,075	50	32,325	5.5	391,550
700,000 to 800,000	33	24,622	4.5	389,697	36	26,900	4.6	418,450
800,000 to 900,000	20	17,009	3.1	406,706	21	17,627	3.0	436,077
900,000 tons and over	[a]86	[a]138,787	25.4	545,493	92	152,302	25.9	588,379
Total	[a]2,887	[a]545,493	100.0	545,493	2,979	588,379	100.0	588,379

[a] Revised figure.

in connection with their federal income tax since 1951. The original Congressional amendment on percentage depletion to the Internal Revenue Code stipulated 15% percentage depletion for chemical and metallurgical grades of limestone, 10% for dolomite, 10% for calcium and magnesium carbonates. These provisions were later amended so that all limestone or dolomite used for construction purposes is given a 5% rate; shell, regardless of end use, is accorded 5%; and all other uses 15%, including stone for lime and cement manufacture, regardless of their end use.

The cut-off point in computing percentage depletion has been the subject of considerable controversy and litigation. In 1963 an Internal Revenue Ruling No. 62-5 established a limit on the "grinding" costs that are allowable based on fineness of the product and limited by the Congressional restriction disallowing "fine pulverization." This permitted grinding down to a maximum of 100% passing a No. 20 sieve and 95% passing a No. 45 sieve. Producers of material as prime products that are more finely ground than the above must negotiate a cut-off point, representing a proportion of their processing costs with their Internal Revenue agent.

Profits. The profit picture is the same as lime. Closely held corporations do not publish financial statements, and the lime and limestone activities of the large integrated corporations are well concealed and unidentifiable in their consolidated financial reports.

However, there is one element affecting profits on limestone that is different from lime. Due to fortuitous circumstances, occasionally a stone company will receive a "windfall." Usually this occurs as a result of a large construction project—a dam, interstate freeway, federal works project, etc.—being constructed adjacent to or in very close proximity to the stone plant. Huge tonnages of aggregate are required with just a minimum haul and are sold at higher plant prices, which reflect the protection afforded by haulage costs from the nearest (but fairly distant) competitors. Sales costs are at a minimum. There have been instances of relatively small, unprofitable companies being quickly transformed into larger, prosperous concerns due to *one* windfall of this type. Otherwise, it is a grueling, undramatic, highly competitive type of business with close profit margins.

All conditions being equal, the most prosperous stone producers are those who exercise the greatest resourcefulness in developing methods for selling and reprocessing their *odd* stone sizes and gradations (spalls) at the maximum return. There is a decided difference manifest among stone producers in their sales orientation and ingenuity in stone merchandising. Some may sell their whole output regularly, but at weak, substandard prices that sacrifice profits; others sell much more advantageously. It may be possible to add $.50 per ton in processing costs that will yield $1 to $2 per ton more in price.

Captive Production. There are no statistics that delineate captive stone from commercial (open-market) stone production, but, although sizable in tonnage, it is less in percentage than with lime. Of the 488 million tons production in 1963, an estimated 125 million tons were captive, of which portland cement accounts for about 70%. Stone for lime manufacture would be easily the second largest captive use.

There are no evident trends on captive versus commercial production; about the same percentage has been captive for many years. Since 1950 a few large concrete products companies have acquired their own stone plants and quarries, some of which are limestone, to supply their own aggregate requirements, but statistically the amount involved is insignificant in the national picture. Otherwise many of the factors previously discussed on captive lime are also applicable to stone.

CHAPTER THIRTEEN

Analytical Testing
of Limestone and Lime

Methods of test and chemical and physical analysis of limestones and limes are increasingly important to the consumer as a means of evaluating the specific (type or grade of) product required for optimum performance among the myriad uses in which these products are consumed. Such analyses are considerably more involved (with more alternative methods) than most other basic chemicals and materials, owing to their wider range in quality and chemical and physical characteristics. For example, sulfur is generally more uniform in its chemical characteristics than limestone; even granite, basalt, and trap rock tend to be more uniform in their physical properties than limestone. Certainly sulfuric acid, soda ash, chlorine, and most basic chemicals are more uniform than quick and hydrated limes per se in both chemical and physical properties. Therefore, the consumer generally enjoys a much wider range of qualities from which to select, thus leading to many individual specifications or tolerances for the producer to meet. This explains the producer's increasing reliance on quality control testing in manufacture and processing to ensure conformity with these requirements.

Many of these tests and analytical methods are controversial, and often there is disagreement between buyer and seller on the tests that are the most accurate, realistic, and equitable to adopt. Furthermore, there is appreciable inconsistency among similar classes of consumers on the tolerances they demand. For example, one steel company may be emphatic that 0.03% sulfur is the maximum amount permissible in quicklime that is consumed for fluxing steel; yet another steel company making the same type of steel may be satisfied with 0.06% sulfur. As a result, a lime or limestone producer who desires to sell to a wide variety of consumers must necessarily develop a systematic program of quality control testing. Delivery refusals of substandard material by the consumer is most costly to him.

Probably the greatest influence toward quality standardization of lime and limestone in the United States and Canada are the specifications promulgated by the American Society of Testing Materials. Many consumers may ignore these ASTM materials specifications and methods of tests, since adoption of these standards is strictly voluntary. Yet these specifications do comprise a guide or basis on which a consumer can mold his own individual specification by modifications. These standards are jointly prepared by the consumers, producers, and general interest members, so that presumably these standards should be objective and unbiased—equitable to all concerned. Two ASTM committees are involved: Committee C-7 on Lime has prepared lime materials specifications for numerous uses, sampling procedures, and methods of test for chemical and physical analyses for both lime and limestone; Committee D-4 on Road and Paving Materials has written many tests and specifications applicable to limestone and other mineral aggregates for all phases of construction in which the physical characteristics are the controlling factor. There are also federal specifications and test procedures, but these are usually almost a duplication of ASTM. Some foreign countries have similar specifications, but these are generally instigated by their governments.

Some of these ASTM tests are too laborious and time-consuming for quality-control testing by producers and consumers except as a periodic check or when maximum accuracy is needed in event of a dispute between buyer and seller. Because of this such detailed tests are regarded as "referee" tests. For quality-control testing often special rapid short tests are contrived and are widely employed, since they provide reasonably or sufficiently accurate results for routine testing. Some of these short tests along with other supplementary investigative procedures will be reviewed briefly along with the more important ASTM standards.

PHYSICAL TESTING OF LIMESTONE

ASTM specifications applicable to mineral aggregates, concrete, and bituminous highway materials are enumerated as follows by number and name:

D 1663-59T.   Asphalt Paving Mixtures, Hot-Mixed, Hot-Laid.
D 692-59T.   Bituminous Paving Mixtures, Coarse Aggregate for.
D 1073-59T.   Bituminous Paving Mixtures, Fine Aggregate for.

D   242-57T.   Bituminous Paving Mixtures, Mineral Filler for.
D   693- 54.   Bituminous Macadam Base and Surface Courses of Pavements, Crushed Stone and Crushed Slag for.
D 1139-61T.   Bituminous Surface Treatments, Single or Multiple, Crushed Stone, Crushed Slag, and Gravel for.
D 1863-61T.   Built-Up Roofs, Mineral Aggregate for Use on.
C   33-61T.   Concrete Aggregates.
C   35- 59.   Gypsum Plaster, Inorganic Aggregates for Use in.
D   694-57T.   Macadam Base Courses, Dry-Bound or Water-Bound, Crushed Stone, Crushed Slag, and Crushed Gravel for.
C   404- 61.   Masonry Grout, Aggregates for.
C   144-54T.   Masonry Mortar, Aggregates for.
E   11-61T.   Sieves for Testing Purposes.
D 1241-55T.   Soil-Aggregate Subbase, Base, and Surface Courses, Materials for.
D   448- 54.   Standard Sizes of Coarse Aggregate for Highway Construction.

Methods of Test for:

C   131- 55.   Abrasion of Coarse Aggregate by Use of the Los Angeles Machine.
D   289- 55.   Abrasion of Graded Coarse Aggregate by Use of the Deval Machine.
D   2- 33.   Abrasion of Rock by Use of the Deval Machine.
D 1411-56T.   Chlorides, Water Soluble, Present as Admixes in Graded Aggregate Road Mixes.
C   142-55T.   Clay Lumps in Natural Aggregates.
D 1664-59T.   Coating and Stripping of Bitumen-Aggregate Mixtures.
D 1865-61T.   Hardness of Mineral Aggregate for Use on Built-Up Roofs.
C   123-57T.   Lightweight Pieces in Aggregate.
C   117-61T.   Materials Finer than #200 Sieve in Mineral Aggregates by Washing.
D 1864-61T.   Moisture in Mineral Aggregate for Use on Built-Up Roofs.
C   87-58T.   Mortar-Making Properties of Fine Aggregate, Measuring.
C   40- 60.   Organic Impurities in Sands for Concrete.
C   227-61T.   Potential Alkali Reactivity of Cement-Aggregate Combinations (Mortar Bar Method).

Chemistry and Technology of Lime and Limestone

C   289-61T.   Potential Reactivity of Aggregates (Chemical Method).
D   75- 59.   Sampling Stone, Slag, Gravel, Sand, and Stone Block for Use as Highway Materials.
C   136-61T.   Sieve or Screen Analysis of Fine and Coarse Aggregates.
D   546- 55.   Sieve Analysis of Mineral Filler.
C   235-57T.   Scratch Hardness of Coarse Aggregate Particles.
C   88-61T.   Soundness of Aggregates by Use of Sodium Sulfate or Magnesium Sulfate.
C   127- 59.   Specific Gravity and Absorption of Coarse Aggregate.
C   128- 59.   Specific Gravity and Absorption of Fine Aggregate.
C   70- 47.   Surface Moisture in Fine Aggregate.
D   3- 18.   Toughness of Rock.
C   29- 60.   Unit Weight of Aggregate.
C   30- 37.   Voids in Aggregate for Concrete.
C   342-61T.   Volume Change, Potential, of Cement-Aggregate Combinations.

Sampling. More than any other factor the accuracy of tests on limestone aggregate is contingent upon how representative the sample is. In testing graded aggregate the principal pitfall is segregation of coarse and fine fractions, obscuring their true gradation. Considerable care is necessary to minimize this inherent tendency of stone samples to segregate, and sampling procedures of ASTM D 75-59 for stone and other aggregate is recommended. This includes sampling of stone from ledges or quarries; from outcroppings of field stone and boulders; at the crushing plant after final blending and screening; at point of delivery. The minimum weight of the sample depends upon its maximum particle size, as shown in Table 13-1.

Hardness, Toughness. Tests employed to determine hardness, toughness and resistance to abrasion and impact of all types of highway aggregate are described briefly as follows:

The *Los Angeles Abrasion Test* (ASTM C 131-55) is a severe accelerated test for measuring the abrasive resistance of limestone. It involves testing different specified weights and gradations of stone in the Los Angeles machine, which comprises an enclosed, hollow steel cylinder into which an abrasive charge of a prescribed number of steel spheres (adjusted for the different stone gradation classifications) are tumbled together by a rotating mechanism. After the specified number of revolutions at 30 to 33 rpm, the sample is screened

Table 13-1. Minimum weights of stone samples

Nominal maximum size of particles passing sieve no.	Minimum weight of field samples lb.	Minimum weight of sample for laboratory test[a] g.
Fine aggregate		
10	10	100
4	10	500
Coarse aggregate		
$\frac{3}{8}$-in.	10	1,000
$\frac{1}{2}$	20	2,500
$\frac{3}{4}$	30	5,000
1	50	10,000
$1\frac{1}{2}$	70	15,000
2	90	20,000
$2\frac{1}{2}$	100	25,000
3	125	30,000
$3\frac{1}{2}$	150	35,000

[a] This is obtained from field sample by mixing and quartering or other means to assure a representative gradation and portion.

to determine the abrasive loss in weight of the stone and from this is calculated the *percent of wear* of the stone. Specifications for maximum percent of wear generally range between 30 and 60%, contingent upon quality of stone available and stringency of requirement.

The *Deval Test* (ASTM D 289-55 and D 2-33) is another accelerated abrasion resistance test. It consists of a small cast-iron cylinder that is inclined at 30° with the horizontal axis, around which it is rotated at 30 rpm; 5 kg. of stone, consisting of 50 cubical-shaped fractions, is introduced into the machine and subjected to 10,000 revolutions. The loss in weight due to attrition is calculated by sieving through a No. 12 mesh screen, and the *percentage of wear* is then determined. The Deval test may also be modified into an *impact test* by introducing steel shot with the stone. This accentuates abrasion resulting in a higher percentage of wear.

The *Page Impact Test* (ASTM D 3-18) measures the toughness of rock by determining the resistance to impact. In a specially designed machine a small cylinder of stone, cored from the rock to be tested, is placed upon an anvil and upon the stone specimen is laid a 1 kg. plunger. A 2 kg. hammer is then dropped upon the plunger from successively increasing heights of 1 cm. until the core fractures. The final height in centimeters from which the hammer falls determines the toughness of the rock. There are modifications of this hammer test.

The *Dorry Hardness Test* is still another accelerated abrasive resistance test. A small core of stone, specially machined for the test, is placed in firm contact against a revolving steel disc upon which 30 to 40 mesh quartz sand is fed. After 1,000 revolutions the loss in weight of the stone core is calculated. The coefficient of hardness is equal to $20 - \dfrac{\text{weight of loss}}{3}$.

Compressive strength tests with hydraulic presses in which crushing strength of the stone in psi is measured is rarely employed, since most stone is regarded as amply strong in compression for highway uses. Furthermore, it is a costly test since stone blocks or cylinders must be precisely machined so that both sides are completely level and smooth.

Of the above tests the Los Angeles Abrasion Test is the most widely utilized. It is the most severe test but is also regarded as the most realistic in simulating the abrasive action of traffic on highways.

Specific Gravity. There are various calculations on specific gravity of stone that will reveal the degree of density, porosity, and absorption and provide an indirect indication of the hardness and weatherability of the stone. Two types of specific gravity are determined ASTM C 127 128-59):

Bulk specific gravity of the stone is defined as the ratio of weight in air of a given volume of a permeable material, including both its permeable and impermeable voids, in relation to weight of an equal volume of distilled water.

Apparent specific gravity is the ratio of weight in air of a given volume of the impermeable portion of a permeable material (the solid rock) in relation to weight in air of the same volume of distilled water. In effect this is the same as the bulk specific gravity minus the volume of permeable voids.

These values are obtained from the so-called *wire-basket test.* A 5 kg. stone sample is oven-dried to constant weight (A), immersed in water for twenty-four hours, then removed and surface-dried by hand with a towel and weighed (B). The sample is then placed in a wire basket, and its weight in water (C) is measured. From these values the following calculations can be made:

$$\text{Bulk specific gravity} = \frac{A}{B - C}$$

$$\text{Apparent specific gravity} = \frac{A}{A - C}$$

$$\text{Percentage of absorption} = \frac{B - A}{A} \times 100$$

A *quick test* for specific gravity can be measured with the Beckman Air Comparison Pycnometer. This has been employed successfully with hydrated lime and whiting.

The *unit weight* of a fine, coarse, or mixed aggregate is governed by ASTM C 29-60, which prescribes three alternative methods of weighing the aggregate in a specified size cylinder that is dependent on the maximum size of the aggregate. These three procedures that comprise rodding or jigging or shoveling techniques are also contingent upon the maximum size of aggregate that is present and serve to consolidate the aggregate to the maximum extent. The *net* weight of the consolidated aggregate is determined. The *unit* weight is then calculated by the following formula:

$$\text{Unit weight} = \text{net weight of aggregate} \times \frac{62.355 \text{ lb.}^a}{\text{net wt.}} \text{ of water in}$$

cylinder

Thus, in classifying the bulk density of stone gradations, reference may be made, for example, to "loose" weight or "rodded" (compacted) weight, etc.

Voids in aggregate (ASTM C 30-37) are determined by the following formula:

$$\% \text{ of voids} = \frac{(\text{sp. gr.} \times 62.355) - \text{wt.}}{\text{sp. gr.} \times 62.355} \times 100$$

sp. gr. = bulk specific gravity (defined above)

62.355 = weight in lb. of 1 ft.^{3} of water at 62°F

wt. = weight in lb./ft.^{3} of the aggregate, as determined by ASTM C 29-60

Size and Gradation. The physical sizes and gradations of stone are determined by screening the material through standard round or square hole wire cloth or plate sieves (ASTM E 11-61). A method of

---

[a] Weight of 1 ft.^{3} of water at 62°F

sieving by oven-drying the aggregate for plus No. 200 mesh sizes is stipulated in ASTM C 136-61. Both the percents passing or retained on the different sieve sizes are recorded.

For measuring size of mixed aggregates finer than the No. 200 sieve, a procedure is outlined in ASTM C 117-61. This consists of *washing a sample* of the aggregate in a prescribed manner. The decanted wash water containing suspended and dissolved materials is passed through a No. 200 sieve. This might consist of clay, stone dust, and water-soluble extraneous salts. The loss in weight from the wash treatment is calculated as percent by weight of the original sample. This is the percentage of material finer than the No. 200 sieve.

A *dry method* of sieving mineral fillers, most of which pass a No. 200 sieve, is described in ASTM D 546-55.

The constancy of gradation is expressed by the *fineness modulus* of an aggregate. It is an empirical number determined by adding the cumulative percentages retained on, say, the No.s 100, 50, 30, 16, 8, 4, and $\frac{3}{8}$ in., $\frac{3}{4}$ in., $1\frac{1}{2}$ in., and 3 in. square-opening sieves and dividing this sum by 100. This provides a figure of the average size of a given gradation. However, different gradations may possess approximately the same fineness modulus. This factor may be specified to control the uniformity of graded aggregate for portland cement concrete, and specifications may require that the variation in the fineness modulus shall not exceed 0.2.

See Table 5-14 on p. 119 for standard aggregate gradations for various construction uses.

Soundness. Associated with the degree of absorption of a rock is soundness, since generally the more porous the stone, the more moisture it will absorb, which accentuates its susceptibility to weathering. If the moisture freezes, a disruptive, expansive pressure is exerted by the ice lenses which induce the aggregate to spall, defoliate, or disintegrate. Aggregates vary greatly in their resistance to the deleterious effect of freezing and thawing cycles. Also visible predictability of an aggregate's soundness is speculative since the size, shape, and strength of the stone's pore structure, which cannot be observed, influence the severity of disruption through frost action. Therefore, accelerated laboratory tests have been contrived to evaluate soundness and weatherability. The most widely employed soundness tests are the *sodium sulfate or magnesium sulfate* tests (ASTM C 88-61).

This consists of immersing the aggregate samples in a saturated solution of either sodium or magnesium sulfate at 70°F for eighteen hours. The sample is then removed, oven-dried at 215 to 220°F, and then cooled to 70°F. This cycle of immersion and hot-drying is repeated usually five times (or a specified number of cycles) and the percent of loss is determined. Usually the limiting percent is 10 to 20.

For stone used in trickling filters of sewage plants an extremely severe requirement of twenty cycles is often specified with only 10% loss. Only relatively few sources of limestone can meet such a requirement. Use of $Mg(SO_4)_2$ is more disruptive than $NaSO_4$. This test is regarded as unreasonably severe by many engineers in spite of widespread application, and other special freeze-thaw tests have been introduced. However, no test yet advanced appears to be generally acceptable and realistically indicative.

The presence of an *excess of alkali* (Na_2O and K_2O) in aggregate may create a deleterious effect on portland cement concrete, particularly if the cement or mixing water is also high in such alkalis. In order to test the soundness of a combination of aggregates, cement, and mixing water, the so-called mortar bar test, an *alkali reactivity* test with cement-aggregate mixtures, is employed (ASTM C 227-6IT). This is an involved, detailed test with all conditions, equipment, and materials meticulously prescribed.

Briefly, mortar bars are made from one part portland cement to two and one-fourth parts graded aggregate by weight. After molding, they are stored in a moist closet for twenty-four hour; then they are removed from the mold. The length of the bar is measured and returned to the mold in the damp closet. Daily or monthly recordings of lineal changes (usually expansions) are made for twenty-eight days, one year, or any specified length of time. At conclusion total expansion is calculated along with observable physical changes in the bar, such as warping, cracking, surface mottling, etc.

This "alleged" test resembles more of a research project—very complicated, costly to conduct, and lacking in reproducibility.

Another soundness test (ASTM C 289-61T) is a chemical method of determining the potential reactivity of aggregate with alkalis in portland cement. It is indicated by the extent of reaction during twenty-four hours at 80°C between IN NaOH solution and the aggregate that has been crushed and screened to pass a No. 50 sieve and retained on a No. 100 sieve. This is also a complicated test, but is more realistic and simpler than the mortar bar soundness test.

Stripping in Bitumen–Aggregate Mixtures. Aggregate varies in its ability to retain a bituminous film in the presence of water. The bitumen–aggregate combinations that possess poor retention cause the aggregate to strip or ravel from the pavement. ASTM D 1664-59T describes coating and static immersion procedures to determine if stripping occurs or the degree of stripping prevalent. This test is only applicable to cut-back and semisolid asphalts and tars. Much of this test is dependent on visual observation, so it is only generally and not quantitatively indicative. There are many modifications of this test as well as other stripping and bituminous durability tests, one of the most important of which is the Immersion-Compression test developed by the U. S. Bureau of Public Roads (p. 423).

Although the reaction of aggregates with bitumens cannot be predicted with certainty, as a generalization the acidic types of rocks are more prone to strip than basic types, like limestone and slag. Acidic stone or aggregate are classed as *hydrophilic*, since they possess a stronger affinity for water than asphalt; the basic aggregates are called *hydrophobic*, since they possess better adhesive qualities for bitumens, even under wet conditions.

Petrography. When the texture, color, grain size and pattern, or geologic origin of the limestone is of importance, polished thin sections of the stone in question are subjected to microscopic examination. Such petrographic analysis is more related to research information than testing, since no limits can be set, except as pertaining to reflectivity. Otherwise, the information obtained is utilized arbitrarily or as "engineering judgment."

Other *miscellaneous tests* for use as construction aggregate include:

1. Determination of the percent of *flat and elongated fractions* in a graded aggregate, since cubical shapes of stone are most desired. This is controlled by specifying a maximum allowable percentage for odd shapes.

2. Determination of percent of *lightweight pieces* and *soft particles* in coarse aggregate that would prove to be unsound.

3. Determination of *fire-resistant* qualities of aggregate.

4. The extent of *clay lumps* in natural aggregate. (This is more applicable to sand and gravel.)

5. The tendency of limestone to *polish* under simulated action of traffic. There is no ASTM test, but several states have established arbitrary criteria to eliminate aggregate that contribute to slippery pavements in bituminous surfaced roads.

Limestone: Use Specifications

There are twenty-one ASTM use or aggregate specifications for many construction purposes, which were enumerated on p. 460. There are countless modifications of these specifications written by many diverse consumer groups, such as state highway departments, public works department, U. S. Engineers, U. S. Navy, Bureau of Reclamation, consulting engineers, architects, etc. Generally one authoritative group, American Association of State Highway Officials (AASHO), endorses the ASTM standards. Since most of the salient features and tolerances of these use specifications have been reviewed in Chapter 5 on p. 120, the gradation requirements for most construction purposes contained in Table 5-14, p. 119, and methods of tests on which they are predicated have just been described, very little will be added to avoid needless repetition.

The following requirements for aggregates, not previously cited, have

Table 13-2. Typical Los Angeles abrasion test requirements

Use	Maximum % wt. loss
Coarse aggregate for bituminous paving mixtures:	
Surface courses	40
Base courses under paving surface	50[a]
Crushed stone for bituminous macadam base and surface:	
Surface courses	40
Base courses	50[a]
Crushed stone for bituminous surface treatments	
Single or multiple	40
Coarse aggregate for concrete	50[a]
Crushed stone for dry-bound or water-bound macadam base courses	50[a]
Aggregates for soil-aggregate subbase, base and surface courses	50[a]
Railroad ballast	40

[a] Other standards may permit these values to extend to 60%.

Table 13-3. Typical soundness test requirements

Use	Maximum % loss, $NaSO_4$	Maximum % loss, $Mg(SO_4)_2$
Coarse aggregate for bituminous paving mixtures	12	18
Fine aggregate for bituminous paving mixtures	15	20
Fine aggregate for concrete	10	15
Coarse aggregate for concrete	12	18
Crushed stone for dry-bound or water-bound macadam base courses	20	30
Aggregate for cement grout	10	15
Aggregate for cement mortar	10	15

Table 13-4. Fineness modulus requirements

Use	Maximum variation
Fine aggregate for bituminous paving mixtures	±0.25
Fine aggregate for concrete	0.20
Aggregate for cement grout	0.20
Aggregate for cement mortar	0.20

Table 13-5. Gradation requirements of mineral fillers for bituminous paving mixtures[a]

Sieve no.	% Passing
30 (595μ)	100
50 (297μ)	95 to 100
100 (149μ)	90 to 100
200 (74μ)	70 to 100

[a] May be composed of any clean rock or mineral dust, including portland cement, hydrated lime, asbestos flour, etc.

Table 13-6. Percentage tolerances on deleterious substances in aggregate

Deleterious substance	Bituminous surface	Fine aggregate concrete	Coarse aggregate concrete
Clay lumps	0.1	1.0	0.25
Soft particles	2.5		5.0
Chert:	1.0		
Severe exposure			1.0
Mild exposure			5.0
Lightweight material (specific gravity of 2.0 or less)	1.0		
Flat or elongated pieces retained:			
Aggregate sizes, no.s 5 and 6[a]	10.0		
Aggregate sizes, no.s 7 and 8[a]	10.0		
Material finer than no. 200 sieve:			1.0
Concrete subject to abrasion		3.0	
All other concrete		5.0	
Coal and lignite:			
Where surface appearance is:			
Important		1.0	0.5
All other concrete		3.0	1.0

[a] Enumerated in Table 5-14, p. 119.

been consolidated largely from the above ASTM specifications into tables:

Table 13-2. Los Angeles Abrasion Test.
Table 13-3. Sodium and Magnesium Sulfate Soundness Tests.
Table 13-4. Fineness Modulus.
Table 13-5. Gradation of Mineral Fillers for Bituminous Paving.
Table 13-6. Typical Percentage Tolerances on Deleterious Substances in Aggregates.

Ranges in gradation requirements for chemical, metallurgical, and other uses of limestone are contained in Chapter 5.

Reflectivity. The degree of whiteness is an extremely important determination in limestone, chalk, marble, and shell whiting and precipitated calcium carbonate, although there is a dearth of uniformity in the many individual consumer requirements. Reflectivity measurement can be obtained with the G. E. Spectrophotometer at 550 wavelength (similar to green tristimulus filter) or equivalent equipment. It is also applicable to hydrated lime.

Limestone: Chemical Analyses

Analytical chemical test methods on the composition of limestone, including its impurities, is adequately contained in ASTM C 25-58 and which also contains identical methods for quicklime and hydrated limes. Methods that are equally applicable to both limestone and lime, such as quantitative determinations of impurities, will be reviewed later in the section under Lime: Chemical Analyses. Those tests directly applicable to limestone, such as loss on ignition, CO_2 content, free water, and organic content, will be described briefly as follows:

Loss on Ignition. 0.5 or 1.0 g. of limestone is placed in a weighed platinum crucible, covered with a lid, and then heated gradually in an electric muffle furnace up to about 1000°C. The sample is maintained at this temperature until a constant weight is obtained. This means that all volatiles have been expelled. The difference between the original weight of the sample and the final weight represents the loss on ignition.

Mechanical Moisture Content. 1 g. of limestone is placed in a flat-bottomed weighing bottle and is heated uncovered in a ventilated drying oven at 120°C for two hours. After heating, the bottle should be quickly sealed, cooled in a desiccator, and then unsealed just before weighing. The loss in weight represents only mechanical or hygroscopic moisture, no other volatile.

Carbon Dioxide Content. There are two standard methods of determining CO_2 content. The *first* involves boiling 0.5 g. of limestone with dilute HCl (1:1) in a small Erlenmeyer flask attached to a condenser from above. The CO_2 gas evolved from this reaction then passes through a drying and decontaminating system of $CaCl_2$ and anhydrous $CuSO_4$ to absorption tubes filled with soda-lime, which entraps the CO_2. $CaCl_2$ desiccates final traces of water. Air, free of CO_2, is introduced gently throughout the system during the test and at a greater rate after the test. When CO_2 is being absorbed, the absorption tubes heat up readily. When these tubes cool, they are disconnected and allowed to stand in the balance case until two weights, taken thirty minutes apart, agree within 0.5 mg. Soda-lime used in this test must be porous.

The *second* (alternative) method is quite similar to the first except that a specially designed Midvale bulb, packed alternatively with glass wool, P_2O_5, and ascarite is used to absorb the CO_2 that is evolved. This bulb is detached and weighed with a second Midvale bulb as counterpoise. Free moisture is desiccated and acid volatiles are recovered separately so that a true CO_2 content is obtained by difference.

Organic Content. Determination of the organic content of limestone is extremely difficult to measure with accuracy and no simple, completely accurate method has been advanced. (ASTM makes no provision for this test.) It is usually calculated by determining the free carbon content. One method consists of liberating the CO_2 by boiling the stone in acid and passing the volatiles, including the free carbon, into a mixture of sulfuric and chromic acids that oxidize the organic matter. In other tests the volatiles are boiled off and filtered through asbestos, which collects the carbon for subsequent organic combustion of the free carbon. At best most organic values are simply close quantitative estimates.

Physical Tests of Lime

There is only one ASTM specification (C 110-58) embracing physical tests on burned lime products, but it is an omnibus type of specification since it includes tests on residues of quicklime and hydrated lime, standard consistency of putty, plasticity measurements, soundness of hydrated lime, popping and pitting, water retention, slaking rate, and settling rate. Other analytical methods not recognized by ASTM will be included whenever appropriate, but before describing these tests a review of sampling procedures on lime and limestone (ASTM C 50-57) is propitious.

Sampling. The importance of agreement between producer and consumer on *where* the sampling will take place—at producer's plant or at destination—is emphasized. Since a producer can be victimized by the carrier in careless handling, improper protection, and delayed shipment that vitiate quality, from his standpoint sampling at *his* plant is preferable. Stress is placed on expeditious sampling so as to minimize exposure of the lime to air. (With limestone there is no problem in this particular.) Samples should be taken in *triplicate* and immediately sealed in air-tight, moistureproof containers. Each sample shall weigh at least 5 lb., provided no screen size is specified. If a screen size is specified, then larger samples shall be taken as follows:

Size, in.	Minimum sample, lb.
$\frac{3}{4}$	5
1	10
$1\frac{1}{2}$	30
2	75

One sample should be immediately delivered to the consignee; the second sample is for the consignor, if requested; the third is retained with the seal unbroken for a possible independent test.

In order to obtain a representative sample of lump or granular material in bulk, 50 lb. are sampled for each unit of thirty tons of bulk material, even though 5 lb. may only be required for the test. When lime is stored in piles or carloads, care should be exercised not to select material from the top or bottom. Ten shovelfuls are recommended to be taken from different parts of the pile or carload, at least 1 ft. below the surface. The 5 lb. samples are obtained by mixing and quartering all of the shovelfuls. When sampling is effected at producer's plant from conveyors or bins, a similar volume of material should be taken at regular intervals instead of all at once.

With packaged material at least 1% of the packages should be sampled with at least 1 lb. per bag being removed. Total sampling should then be mixed and quartered to provide the (3) 5 lb. samples.

With powdered material in bulk, a sampling tube is specified that will remove a core at least 1 in. in diameter and is sufficiently long to penetrate into the interior—not the bottom or top. A minimum of

50 lb. shall be taken from each 30 ton unit, mixed and quartered into the 5 lb. triplicate samples.

At the laboratory lump or granular material to be tested should be ground to pass a No. 100 sieve.

In case of *rejection* the manufacturer must be notified within one week after completion of tests with the cause of rejection stated. Each of the contracting parties may make claim for a retest within one week of the original test report. Expense of retesting must be borne by the party demanding it. Should the parties still be in disagreement, the third sample of material should be delivered unopened to a mutually satisfactory referee laboratory for test, and their findings are binding upon both parties.

Residue. The insoluble matter (core and impurities) in quicklime is determined by weighing the residue and calculating its percent from the weight of the original sample, which is at least 5 lb. Lump material should be first crushed so that it passes a 1 in. sieve; pulverized material is tested as received. The sample is then slaked carefully into a putty of maximum yield and is allowed to stand for one hour. The putty is then washed through a No. 20 sieve by a stream of water at moderate pressure. Washing is continued until the residue remaining visually appears to be nothing but coarse sand-like particles. The residue is then dried at 212 to 215°F.

With *hydrated lime* a 100 g. sample is placed on a No. 30 sieve which is nested above a No. 200 sieve. The material is washed through the meshes by a stream of water. No material must be washed over the sides of either screen. Washing is continued until both sieves are clear of hydrate, but in no case longer than thirty minutes. Residues from both screens are then dried to a constant weight in an atmosphere free of CO_2 at a temperature between 212 and 248°F. The percentage residue on each sieve is calculated, based on the original weight of the sample, and reported separately or cumulatively.

Standard Consistency of Putty. 300 g. of hydrated lime is mixed with sufficient water to form a thick putty and stirred to assure intimate mixing. Type N hydrate putty is stored, covered with a wet cloth for sixteen to twenty-four hours before testing; Type S hydrate putty can be tested immediately after preparation.

The equipment employed is a modified Vicat apparatus. Putty is added flush to the top of a nonabsorbent mold, 4 cm. in height, 7 cm. at base, and 6 cm. at top, which rests on a glass base. A plunger is lowered in contact to the putty surface and an initial reading is taken. The plunger is released for thirty seconds and another reading is taken. Standard putty consistency is obtained when a penetration of 20 mm. ±5 mm. is obtained in thirty seconds. If putty is substandard in consistency, the sample is returned to original putty, more water is added, mixed for two to three minutes, and a retest is made as above. If penetration then exceeds standard, the sample should be discarded and a new one prepared.

Plasticity Determination. A special apparatus known as the Emley Plasticimeter (Fig. 13-1) is used in this test. A similar type mold as employed in the standard consistency test is lubricated with water, placed on a porcelain or disposable plaster-base plate, and filled flush to the top with a lime putty of standard consistency. The mold is then removed carefully and vertically without distorting the mound of putty. The base plate with putty is on a turntable. A motor rotates the turntable at one revolution in six minutes, forty seconds and rises a thirteenth of an inch per revolution. Before rotation is commenced, a disc from above is placed in contact with the paste. This is intended to simulate the action of a trowel applying lime putty to an

Fig. 13-1. Emley Plasticimeter apparatus.

absorbent wall base. Timing is critical. The turntable should not be started in motion until exactly 120 seconds after the putty is first added to the mold. Thus, the following times must be precisely recorded:

Starting time
Starting motor (and rotation)
Record scale reading *each* minute until completion of test

The test is complete when:

1. The scale reaches 100 or
2. The scale value falls (reading less than previous one) or
3. The scale reading remains constant for three consecutive readings or the specimen has visibly ruptured or broken bond with the base plate.

The plasticity value is calculated from the following equation:

$$P = \sqrt{F^2 + (10T)^2}$$

where:
P = plasticity figure
F = scale reading at end of test
T = time in minutes of total test from time putty was first placed in mold

Essential to some degree of reproducibility in this test is a base plate with the proper degree of absorption that has been standardized at not less than 40 g. of moisture absorption when the plate is immersed in water at room temperature for twenty-four hours. Meticulous care of base plates is necessary—cleaning thoroughly, drying, etc, to maintain proper degree of porosity.

OTHER PLASTICITY TESTS. The Emley test above is the only test recognized by ASTM. However, even adherents of it readily acknowledge its shortcomings, particularly lack of reproducibility caused principally by base plates that lack uniformity in suction as well as the varying skills of operators. Yet in spite of such criticism it still is the most satisfactory method thus far advanced and certainly is informative on the rheological properties and workability of limes. However, this method has only been adopted by the United States and Canada.

The United Kingdom employs a nonabsorbent *flow table* for measuring workability. The number of times the flow table is bumped is recorded in flattening a truncated cone of putty of standard consistency to three times its original diameter. On this test the British specify a minimum of 10 and 13 for hydrated lime and quicklime putty, respectively—values that would be regarded as extremely low in the United States, probably substandard.

A so-called *Carson blotter test* will also indicate in a crude way the degree of plasticity. A glob of putty of standard consistency is placed on a filter pad or blotter and is spread over the surface with a spatula or putty knife. The number of strokes required before the putty loses its moisture and "balls up" under the spatula is an approximate measure of plasticity, although obviously of completely unpredictable reproducibility.

Water Retention (and Flow Values). An indirect measure of plasticity and workability of lime in sanded mortars is the water retention test, since frequently lime that possesses a high water retentive capacity has a correspondingly high Emley plasticity value (and vice versa) with limes of low water retentive capacity.

500 g. of lime and 1500 g. of standard Ottawa sand are intermixed with a measured amount of water in a nonabsorbent bowl. Hydrated lime is first wetted in the water before the sand is added or if lime putty is employed, it is first converted to a slurry before sand is gradually added, amid stirring and kneading action with a gloved hand.

The apparatus initially used for this test is a flow table, mounted on a vertical shaft in such a manner that it can be raised and dropped a fixed height of $\frac{1}{2}$ in. by means of a rotating cam. Table surface is of nonabsorbent, noncorrodible metal and should be so calibrated that it slowly revolves on each drop (1 revolution for every 25 drops). A nonabsorbent flow mold, 2 in. in height, 4 in. in inside diameter at the base, and 2.75 in. inside diameter at the top, is placed on the center of the flow table and filled, flush to the top, with the above sanded lime mortar. The mortar is pressed into the mold with finger tips to assure complete filling without voids, and the mold is carefully removed. The table is then dropped twenty-five times in fifteen seconds. The *mortar flow* is the resulting increase in diameter of the flattening mortar mass, expressed as percent of original diameter. This standardizing value is an essential prerequisite to the water retention test itself. If the flow is less than 100%, more water is added to the mortar until on retesting the flow measures 100 to 115%. The rejected mortar is returned to the original bowl and remixed for 135 seconds. If the flow exceeds 115%, the sample should be discarded and a new batch made.

The water retention apparatus consists essentially of a water aspirator controlled by a mercury column relief and connected by way of a three-way stopcock to a funnel upon which rests a perforated dish made of nonabsorbent material. A specified grade of filter paper is then fitted flatly and tightly to the bottom of the dish and moistened.

Immediately after the flow test described above, the remaining mortar on the flow table is returned to the original mortar bowl and remixed for thirty seconds. It is then rapidly distributed uniformly, without compaction, over the sheet of dampened filter paper in the perforated dish, flush to the top of the dish by use of a straightedge. The dish is returned to the funnel, and a vacuum of 2 in. is applied. After sixty seconds of suction of water from the mortar, the stopcock is turned off in order to expose the funnel to atmospheric pressure. The mortar is then removed from the dish with a spatula and returned to the mold on the flow table, where it is filled completely with gloved finger compaction flush to the top. A second flow value is then obtained in the same manner as previously described by dropping the flow table; this value is known as the *flow after suction*. The whole test (or what appears to be a series of tests) should be conducted as rapidly as possible and should never exceed thirty minutes from start to finish.

The water retention value is calculated as follows:

$$\text{Water retention, as } \% = \frac{A}{B} \times 100$$

where: A = flow after suction and B = flow immediately after mixing (or initial flow).

A high value or a percent of 85 to 95 indicates that a lime possesses high water retentive capacity by resisting the artificially imposed suction, described above. A value of 75 to 85% would indicate low to intermediate water retention. In contrast to lime, a portland cement mortar would only possess 50 to 60% water retention. This test is most indicative of all mortar tests, since water retention is an essential characteristic for use with absorbent masonry units. Otherwise, such porous units would quickly absorb moisture from a mortar of low water retentivity before it could harden properly and a poor bond at the mortar-unit interface would result. That is why it is an almost invariable requirement in all masonry mortar specifications.

Differences in the water retentivity capacity of mortars composed of different limes, cements, and proportions of both can be empirically observed by a crude *quick "blotter" test*. A glob of unsanded lime putty of standard consistency or a flow of 100 to 115 is placed on filter paper (150 lb./ream). The filter is immediately inverted and the number of seconds from its application to the appearance of moisture soaking through the paper is recorded. Contingent upon the test conditions, such as percent of flow, thickness of paper, and atmospheric humidity, four to thirty seconds might expire before visible moisture occurs.

Soundness. Unsoundness in limes, characterized by cracking, pits and pops, bulging or distorted surfaces of a freshly applied neat lime putty, gauged putty, or sanded putty, is caused primarily by presence of coarse particles, particularly +No. 30 mesh materials, and to a lesser extent +No. 200 mesh particles may be contributory. The presence of unhydrated oxides (generally MgO) in the lime is also a source of potential unsoundness due to the possibility of delayed hydration (see p. 407). Actually careful milling of the hydrate in an air separator at the point of manufacture to remove all coarse particles is the most effective sinecure against unsoundness, and in a similar manner removal of all course particles from quicklime putty by screening is likewise effective. If the gradation of a hydrated lime reveals a virtual absence of coarse particles, then from a practical standpoint there is no need for a soundness test, except possibly in the case of the controversial normal dolomitic hydrated lime. Yet before modern hydrate milling techniques were perfected, unsoundness from coarse particles was a chronic problem. The following test dates back many years to the era of relatively coarse hydrates, and although it is of doubtful value today, it is still recognized by ASTM. The meticulous detail and manual dexterity required in performing this test is almost ludicrous, at least from a reproducibility standpoint. Certainly a prohibitive amount of practice would be paramount for a laboratory technician to develop any competence whatsoever with this test.

It consists of mixing 20 g. of hydrate and 100 g. of standard Ottawa sand with sufficient water to produce a plastic mortar of rather dry consistency. The mortar is then spread evenly $\frac{1}{4}$ in. thick on a 4 in.^{2} glass plate. The resulting mortar pat is placed in an enclosed closet for twenty-four hours at room temperature. The pat is then soaked in water until a film of water stands unabsorbed on the specimen. If the pat cracks, the consistency of the mortar was originally too wet and a new pat must be made. (Trial and error is necessary to obtain the unspecified but optimum consistency peculiar to this test.)

When the correct consistency is obtained, 20 g. of the sample is mixed with enough water to form a thick cream, which is spread in a thin layer on the pat's surface and allowed to stand for fifteen minutes to prevent air bubbles from forming and is then troweled to an even surface, making this skim coat as thin as possible without exposing the sand through the surface. The pat is then returned to the same closet for twenty-four hours. If no crack or pops exist, it is then suspended in a vessel that is partially filled with cold water, in such a way that the water can boil without contacting the pat. The water is boiled gently for five hours, enveloping the pat in steam, and after twelve hours of cooling, the pat is removed and examined for distortion. Whether distortion occurs or not, the extent of disruption is left entirely to the often erratic judgment of the operator. Fortunately this test is seldom used anymore except as an alternate in event of faulty gradation.

POPPING AND PITTING. Another soundness test applicable to lime's use in plastering is another ancient test, bearing the unscientific title "popping and pitting." Presumably it reveals if there is likelihood of coarse particles in the lime defacing a plaster surface by popping or pitting.

It consists of mixing 100 g. of hydrated lime with water to form a putty of standard consistency and mixing the resultant putty with 25 g. of gauging plaster, which has been previously tested for assurance that it is totally sound. More water is added to maintain a workable consistency. A pat is prepared by troweling this mixture to a smooth finish, $\frac{1}{8}$ in. thick, on a glass plate, 6 × 8 in. The pat is allowed to stand overnight. Then it is placed on a rack in a steam bath so that it is not in contact with liquid water and is protected from condensation drippings falling from above. The steam bath is maintained at boiling for five hours; then the pat is removed and examined for pops and pits.

The British employ substantially this same test. A severe modification requires that all +No. 200 mesh residues be removed from the lime under test and then mixed with gauging plaster and steamed as above. Some pops and pits are almost certain to occur with this modified test.

Expansion. The ASTM lime committee has never adopted a test to measure the type of unsoundness characterized by expansion, usually caused by delayed hydration of the unhydrated oxides, largely because no accelerated test yet proposed is considered to be reasonably emulative of field conditions.

England, South Africa, and other countries employ the *Le Chatelier mold test* (or slight modifications of it) in measuring potential damaging expansion in lime. These molds are filled with a workable mortar composed of varying ratios of lime, cement, and standard sand, respectively, such as 1:1:6, 2:1:9, or 3:1:12. After being allowed to set for forty-eight hours, they are steamed for three hours and the extent of linear expansion is measured. One permissible limit of expansion is 10 mm.

Another expansion test, proposed by the U. S. Bureau of Standards, is more rigorous, involving an *autoclave*. It consists of preparing $1 \times 1 \times 10$ in. mortar bars of equal parts of hydrated lime and portland cement by volume, neat without sand, and storing in a moist closet for at least twenty hours. Then the bars are steamed in the autoclave for one hour at 216°C, equivalent to 295 psi; they are removed and measured linearly for the percent of expansion calculated from the original length of the bar. A 1% maximum expansion is proposed.

All Type N dolomitic hydrates fail this test badly since they exhibit expansion of 3 to 6%. The reproducibility of this test is suspect, since in a reported "round robin" testing program, several highly hydrated (Type S) dolomitic limes easily met this 1% limit, but all of them when mixed with a certain brand of portland cement exhibited considerable expansion, well above the limit. Tests on this particular cement revealed that it was sound, as determined by ASTM criteria. This test has also been bitterly resisted by the United States lime industry and declared unrealistic, since it does not remotely simulate field conditions. As a result, its adoption, except in isolated cases, is unlikely.

Sand Carrying Capacity. This characteristic of mortar materials has some relationship to plasticity and workability, although it has never been correlated and it may not be directly proportional in a quantitative way. However, as a generalization, the most plastic limes possess the greatest sand-carrying capacity, which means they can be mixed with larger volumetric proportions of sand and still provide adequate workability without being oversanded. Limes of poor to moderate plasticity have less sand-carrying capacity and portland cement averages 50 to 75% less capacity than a "fat," plastic lime, as determined by the *Voss Extrusion Energy* machine (also called *Plastometer*).

This apparatus provides the only realistic test for measuring this property. Although it is not included in the ASTM standards and rarely in any specification, a brief description of it is warranted.

It consists of extruding a mortar under confined conditions in a cylinder and measuring the pressure exerted in its extrusion. Unless the mortar, standardized at a flow of 100 to 110, extrudes completely and preferably easily, it is adjudged unworkable. Harsh, unworkable mortars consolidate quickly under initial pressure, and high pressures up to 200 psi are necessary for complete or only partial extrusion. In contrast only 3 to 10 psi is required for a properly proportioned (sanded) mortar for complete extrusion. In essence the machine operates on the principle that mortars must be readily amenable to deformation without "bleeding" or segregating. Thus, by trial-and-error testing it is possible to determine the optimum amount of sand that a given lime or cement (or combinations of both) can carry without sacrifice of workability. Generally limes will carry sand on a 1:3–4 ratio in contrast to 1:2–2.5 for portland cement.

Compressive Strength. Although pure lime is never employed in mortar for its mechanical strength, compressive strength requirements for lime and lime-cement mortars are frequently specified. With hydraulic limes compressive and tensile strengths would be of some importance, much more than with "fat" (pure) limes.

The criteria for compressive strength tests on mortars is largely ASTM C109-63, a specification promulgated by the ASTM cement committee.

It consists of molding 2 × 2 in. mortar cubes and then deviating from the above specification in the method of curing the cubes. Instead of under-water curing specified for cement mortars, curing is performed in a moist closet, or with cycles of wetting and drying, or in the case of straight or very high lime content mortars, just laboratory air. Cubes are then broken at seven and twenty-eight days with strengths reported in psi.

The *setting time* of "fat" lime mortars (without cement) is of no interest and is only applicable to hydraulic lime, cement-lime, or straight cement mortars. The same standard test is employed—a penetration test with the Gilmore needle and Vicat apparatus (ASTM C266-58T).

Settling Rate. ASTM has recognized the following settling rate test for many years:

Ten g. of lime hydrate is placed in a 100 ml. glass-stoppered, graduated cylinder and is wet down with 50 ml. of CO_2-free distilled water at room temperature. It is mixed thoroughly for two minutes, including inversion of the cylinder by hand. The contents are allowed to stand for thirty minutes, and then it is diluted to the 100 ml.

mark with more CO_2-free distilled water. After further mixing, contents are allowed to stand for twenty-four hours. Sedimentation height at bottom of meniscus is then visually taken at 0.25, 0.5, 0.75, 1, 2, 4, and 24 hour intervals. Or results can be reported as the time required for complete settling or to settle to half volume (50 ml. level), etc. (For sedimentation test results, see Tables 9-5 and 9-6, pp. 315–316.)

Slaking Rate. Probably the most equitable slaking rate test for quicklime is the current ASTM tentative standard, contained in C 110, Physical Tests for Lime, since allowances are made for different types and qualities of quicklimes. However, this and all slaking tests are invariably controversial.

The procedure consists of preparing the quicklime sample, −#6 sieve size, as rapidly as possible to prevent air slaking. Temperature and quantity of slaking water and amount of quicklime to be used is specified as follows:

	Type of lime	
	Dolomitic	High calcium
Temperature of water, °C	40	24
Quantity of water, ml.	400	380
Quantity of quicklime, g.	120	76

The temperature of 500 ml. of slaking water is adjusted to conditions above, ±0.5°C, and is added to the Dewar flask. The agitator propelled by a mechanical stirring device is started at 300 ±50 rpm, and the required amount of quicklime (above) is added without delay; the flask is then covered immediately. An initial temperature rise reading is taken at thirty seconds. Further temperature readings are recorded at the following specified intervals for different categories of lime:

Degree of lime reactivity[a]	*Reading interval*
High	30 sec.
Medium	1 min.
Low	5 min.

[a] Reactivity is defined: High reactivity is completely slaked within ten minutes; medium is completely reacted within ten to twenty minutes; low requires more than twenty minutes for complete reaction.

Readings are continued until temperature rise is less than 0.5°C in three consecutive readings. Total active slaking time will be computed from the first of these three readings.

This initial starting temperature is subtracted from the final temperature to obtain the *total* temperature rise. Before completion of slaking, temperature rise can be similarly computed for any earlier reading, like thirty seconds. Values are plotted on a curve.

There are many slaking rate tests, but most are analogous to the above. Many lime manufacturers check lime reactivity in a very cursory manner by simply mixing "approximately" a like amount of quicklime and water and visually observe the reaction. There is nothing learned quantitatively from such a crude, *quick test*, but it can be meaningful to a person experienced with lime. Any appreciable change in slaking rate will be quickly detected.

In Murray's research on lime reactivity (p. 143), he employed a temperature rise test. Since some of the experimental quicklimes that he tested were explosively reactive, he concluded that an ordinary mercury thermometer would react too slowly to record the immediate surge in temperature that occurs when quicklime contacts water. So he employed an iron-Constantan thermocouple with a Brown Electronic recorder that reacts instantly to any temperature change. Slaking was performed in a Waring Blender, modified by reducing its normal agitation speed to prevent the reaction from becoming too turbulent. He used a lime-water ratio of 1:7 by weight. Since preliminary slaking tests indicated a wide range in slaking rates, an empirical compromise point was selected as indicative of the rapidity of slaking. Temperature rise in five seconds was selected, and the reactivity coefficient was designated as ΔT_5. Yet he readily acknowledges that his test was inequitable for the extremely reactive limes in which slaking was actually completed in three to four seconds, so that a reading at five seconds made them appear to be slower than they actually were.

The American Water Works Association in a newly prepared, but unissued, standard on lime for water softening subscribes largely to the above ASTM test on slaking, except that no differentiation is made between different limes, probably since they specify only high calcium. Their proportions of water and quicklime are fixed at 400 ml. to 100 g., respectively, at 25°C. Quicklime is adjudged unsatisfactory and may be rejected if it does not produce a temperature rise of 10°C in three minutes and complete slaking within twenty minutes. A 40°C temperature rise or higher is the objective.

Gradation. Standard mechanical sieving, based on sieve sizes established in ASTM E 11-61 is virtually the only gradation test required for all lime and limestone, except whiting, which employs other methods of particle size determination.

Sieve sizes extend down to the No. 400 sieve (37μ), but the lime industry has generally standardized on the No. 325 sieve (44μ) as being the lowest practical sieving limit to employ for reasonable accuracy. The standard *dry sieving method* is satisfactory for quicklime and limestone but is totally unsatisfactory for hydrated lime. With the latter a *wet sieving* procedure, such as described on p. 474, is strongly recommended. Otherwise, on dry sieving hydrates agglomerate, even though they adhere loosely together, and will not pass through to the small sieve sizes, thereby distorting the gradation and making it appear to be coarser than it actually is. However, in wet sieving with a stream of water applied at low pressure, the agglomerates readily break up into countless microparticles, usually of subsieve sizes, and pass through the finest screens. With hydrates most of the gradation specifications are utilized to control or limit the coarse fractions (residues), such as a maximum of 0.5% retained on the No. 30 mesh and a maximum of 15% retained on the No. 200 mesh, etc. A few individual consumer's specifications are known to require 95 or 98% passing a No. 325 mesh, but this would represent the most stringent gradation requirement for hydrate.

For quality-control testing of No. 325 mesh limestone whiting, wet sieving is also practiced, generally adhering to the test procedure 1-61 developed by the Pulverized Limestone Institute, which, after wet screening, involves drying, mechanical agitation, and final weighing of the No. 325 mesh residue.

However, other evaluative tests are employed largely for research and general technical information that pertain to particle size and its distribution. These are employed more for determination of *subsieve* (micron sizes); the *shape* and *form* of the microparticles, whether they are cubical, platelike, needle-shaped, etc.; the range in surface areas for different particles and the average surface area for a representative volume or weight; petrographic analysis (thin sections); molecular structure; and even the porosity (pore size distribution) of these microparticles in angstrom units. The increasing recognition that *surface area* is of greater significance than particle size for many applications of limestone flour, whiting, and lime will probably lead ultimately to specifications predicated on a minimum surface area. (This method is now employed as standard practice for portland cement in ASTM specifications C 150-63, C 204-55, and C 115-58.) At least, one United States lime manufacturer uses the Blaine Air Permeability method that measures surface area in quality-control testing of hydrated lime and quicklime putty. Thus, in the future lime consumers may conceivably specify that hydrated lime must possess a minimum surface area of, say, 15,000 or 20,000 cm.^{2}/g.

The most commonly used micro-test methods are summarized as follows:

1. *Microscopy*. This may consist of an ordinary optical microscope, electron microscopy, microprojection, or photomicroscopy, in which tremendous magnification is achieved by the latter two methods by projecting the image or photomicrographic slides on screens. These methods are used for the estimation of particle size in the range of 75 to 0.2μ and are ideally suited for studying the shape and form of the particles, but are not as adaptable to surface area measurement as other methods. The electron microscope is much more versatile for minute submicron sizes than plain optical equipment.

2. *Sedimentation*. There are many individualized sedimentation tests, but most are predicated on the descent of particles in quiescent water in graduated cylinders to which Stoke's law for particle size determination can be applied. A standard test for determining micron-sized whiting (Pulverized Limestone Institute 3-61) is a very detailed sedimentation test, and its use is specified for material with 0.5% (or less) residue retained on a No. 325 sieve.

The *Wagner Turbidimeter test* is a type of sedimentation test in which density or turbidity of a particulate substance in liquid suspension is measured with the aid of light passing through the suspension and impinging on a photoelectric cell. This test determines surface area in cm.^{2}/g. by calculation.

3. *Air Permeability*. The Blaine Air Permeability Test is an expeditious and accurate method of estimating surface area in cm.^{2}/g., although calibrating it initially is complicated. This apparatus draws air as the medium through a bed of particulate matter of known porosity.

Micromesh sieves with ultrasonic equipment for cleaning the fine screen mesh have been introduced but are not widely used. Other methods, such as X-ray diffraction, nitrogen absorption, porosimeter, etc., provide useful supplemental information but are mainly research instruments.

LIME MATERIALS SPECIFICATIONS

The following are current ASTM lime materials specifications for specific uses:

C 258- 52.   Quicklime for Calcium Carbide Manufacture.
C   5- 59.   Quicklime for Structural Purposes.
C 46- 62.   Quicklime and Limestone for Sulfite Pulp Manufacture.
C 259- 52.   Hydrated Lime for Grease Manufacture.

C 207- 49. Hydrated Lime for Masonry Purposes.
C 141- 61. Hydraulic Hydrated Lime for Structural Purposes.
C   6- 49. Normal Finishing Hydrated Lime.
C 206- 49. Special Finishing Hydrated Lime.
C 433- 63. Quicklime and Hydrated Lime for Hypochlorite Bleach Manufacture.
C 415- 63. Quicklime and Hydrated Lime for Sand-Lime Products.
C 49- 57. Quicklime and Hydrated Lime for Silica Brick Manufacture.
C 53- 63. Quicklime and Hydrated Lime for Water Treatment.
C 379-56T. Fly Ash for Use as a Pozzolanic Material with Lime.
C 432-59T. Pozzolans for Use with Lime.

No ASTM specifications exist on some of the most important end-uses, like steel fluxing, road stabilization, sulfate pulp manufacture, leather tanning, etc., largely because there is so much variance in individual consumer requirements. Comments on the principal tolerances contained in these and other use requirements may be found in Chapter 11 among the use descriptions.

Lime: Chemical Analysis

Many of the following methods of test are also equally applicable to limestone and are a continuation of the chemical tests for limestone enumerated on p. 471.

Literature is rife with many different methods of test for chemical analysis of lime and limestone. But many of these tests are actually modifications and short-cuts to basic tests. Certainly the ASTM specification C25-58 contains as complete and authoritative group of routine tests as exists, with emphasis more on accuracy than expediency. These tests include separation of various chemical compounds and impurities of which lime and limestone are composed. Most of these tests are largely based on the reference text *Applied Inorganic Analysis* by Hillebrand, Lundell, Bright, and Hoffman, 2nd Ed. (1963), which provides an even more detailed explanation of various of these test procedures.

Descriptions of most of these ASTM tests follow:

1. *Insoluble matter (including SiO_2)*. This determination consists entirely or largely of the combined amount of SiO_2, Al_2O_3, and Fe_2O_3. It may also contain small to trace amounts of FeO, P_2O_5, Mn_3O_4, and TiO_2.

Place 0.5 g. of sample in an evaporating dish, preferably platinum type, mix with distilled water to a thin slurry, add 5 to 10 ml. of HCl (sp. gr. 1.18), and digest with aid of heat and agitation until sample is entirely dissolved. (In the case of limestone or hydrated lime, the volatiles should first be completely removed by igniting the sample in a covered platinum crucible in an electric muffle before digestion in acid.) Then evaporate solution to dryness and place dish and contents (covered) on a hot plate and heat one hour at $200°C$ for high calcium limestone and $120°C$ for dolomitic limestone. HCl and water are again added, and the sample is placed on water bath for ten minutes. SiO_2 is then separated by filtration and washed consecutively with HCl and then twice with cold water. Filtrate is again evaporated; further residue is extracted with HCl, and the resulting solution is again filtered through a second, smaller filter paper. Wet filter papers, containing the separated residues, are then transferred to a platinum crucible and ignited to constant weight in an electric muffle; on cooling, the increase in weight is reported as insoluble matter, including SiO_2.

2. *Silicon dioxide.* Total insoluble matter (item 1 above) is treated with 5 ml. of water, 5 ml. of HF, and one or two drops of H_2SO_4 and evaporated to dryness. Residue is then heated for two to three minutes in electric muffle and weighed. The difference between this weight and insoluble matter (above) gives weight of SiO_2 that was separated by volatilization with HF.

3. *Iron and aluminum oxides.* The residue, remaining from SiO_2 determination in item 2 above, is fused with a little Na_2CO_3 in a crucible and is then cooled and dissolved in dilute HCl. To this solution a few drops of bromine water or HNO_3 are added, and the solution is boiled until all Cl_2 and Br_2 are evolved. Then add sufficient HCl to assure the equivalent of 10 to 15 ml. of HCl(sp. gr. 1.18). Add a few drops of methyl red, dilute this filtrate to 200 to 250 ml., and heat to the boiling point. Neutralize the solution with NH_4OH until color of liquid changes to distinct yellow. Boil for one to two minutes, settle, filter, and immediately wash the precipitate several times with hot 2% NH_4Cl solution and then dry by suction. Precipitate collected on filter paper is dissolved in hot, dilute HCl and returned to original beaker from which precipitation was made. The solution is boiled again and reprecipitated with NH_4OH. The resulting moist precipitate is dried to constant weight in a platinum crucible and weighed as combined Al_2O_3 and Fe_2O_3.

4. *Total iron (standard procedure).* The combined Al_2O_3 and Fe_2O_3 precipitate (item 3) is fused at low temperature with 3 to 4 g. of

$Na_2S_2O_7$ or $K_2S_2O_7$. This melt is then treated with sufficient dilute H_2SO_4 to insure presence of not less than 5 g. of absolute acid and ample free water. Evaporate solution until acid fumes are vented. After cooling and redissolving in water, filter out any trace of SiO_2, wash, ignite, weigh, and correct for impurities with HF and H_2SO_4. Any weight of SiO_2 obtained here should be added to SiO_2 (item 2) and reduced from combined iron and alumina (item 3). Reduce filtrate with zinc and titrate with a 0.05 N $KMnO_4$ solution to determine iron.

5. *Total iron (alternative procedure)*. Dissolve 2 to 5 g. of combined Al_2O_3 and Fe_2O_3 (item 3) in HCl and evaporate to dryness. The residue is then treated with dilute HCl; any SiO_2 remaining is filtered off and previous results are corrected. The remaining residue is washed thoroughly with hot water. Iron is precipitated in a boiling solution with alkali, settled, filtered, and washed free of chlorides with hot water. Then after dissolving in H_2SO_4, calculate weight of Fe_2O_3.

6. *Aluminum oxide*. Subtract the calculated weight of Fe_2O_3 (items 4 or 5) from weight of $Al_2O_3 + Fe_2O_3$ (item 3) for weight of Al_2O_3. (If phosphorus is to be calculated, then deduct from Al_2O_3 the weight of P_2O_5 determined.)

7. *CaO by gravimetric method*. Add a few drops of NH_4OH to the combined filtrate from the $Al_2O_3 + Fe_2O_3$ precipitate (item 3) and boil. Thirty-five ml. of a saturated solution of $(NH_4)_2C_2O_4$ is added to the boiling solution, and boiling is continued until the precipitated CaC_2O_4 assumes a granular form. After settling, filter the precipitate and wash thoroughly with boiling water. Then ignite resulting solution with water to a volume of 100 ml. Add NH_4OH and boil liquid. (If a trace of $Al_2(OH)_6$ then precipitates, remove it by filtering, wash with 2% NH_4Cl, ignite, weigh, and add to weight of Al_2O_3 in item 6. Then repeat reprecipitation procedure with $(NH_4)_2C_2O_4$ and after ignition to constant weight in a covered platinum crucible, weigh as CaO.

8. *CaO by volumetric method*. Alkalize the filtrate from item 3 with NH_4OH, boil, and then add 35 ml. of a saturated solution of $(NH_4)_2C_2O_4$. After settling, filter off resultant CaC_2O_4 precipitate on 11 cm. paper and wash ten times in hot water. Transfer filter paper with precipitate back to same beaker in which precipitation was made and spread paper out against upper portion of beaker. Wash precipitate from paper with jet of hot water; fold paper, leaving it adhering to upper portion of beaker. Add to contents of beaker 50 ml. of 10% H_2SO_4; dilute with hot water to a volume of 250 ml.; and heat to 90°C. Titrate with specified strength $KMnO_4$ solution

until pink color end point is obtained. Then drop folded filter paper into solution which eliminates color and retitrate with more of same $KMnO_4$ solution until pink color again appears. By measuring amount of $KMnO_4$ solution consumed, calculate percentage of CaO. (One ml. of standard $KMnO_4$ solution is equivalent to 0.005 g. CaO.)

9. *Magnesium oxide*. Acidify the combined filtrates from the precipitation of CaO (items 7 and 8) with HCl, add water to 150 ml. volume, and heat to boiling. Add 10 ml. of a saturated solution of $Na(NH_4)HPO_4$ and continue boiling. Cool to room temperature and add NH_4OH drop by drop amid agitation, creating excess alkalinity until a crystalline ammonium-magnesium orthophosphate is precipitated. Allow liquid to cool and stand for twelve to forty-eight hours and filter. Dissolve this precipitate in hot dilute HCl; add water to 100 ml. volume; and reprecipitate by adding 1 ml. of the saturated solution of $Na(NH_4)HPO_4$ and NH_4OH drop by drop as before. Filter through paper or Gooch crucible, wash with dilute NH_4OH containing some ammonium nitrate, ignite, cool, and weigh as $Mg_2P_2O_7$. Recalculate to MgO.

(Loss on ignition, mechanical moisture in limestone, organic matter, and CO_2 determinations are included earlier on p. 471 in limestone section.)

10. *Mechanical moisture in hydrated lime.* 2.5 to 3.0 g. of hydrated lime are placed into a bottle and immediately stoppered. Since the test involves aspirating over the enclosed sample at 120°C, a slow stream of dry, CO_2-free air, sample bottle must be equipped with another interchangeable stopper with two entry tubes for conducting the hot airstream over the sample and for connection to a preceding apparatus train in a drying oven. In the order named, this train consists of a soda-lime absorption tower, a lime-water bottle, H_2SO_4 bottle, and two consecutive bulbs of P_2O_5—and then the hydrated lime bottle. The sample bottle is quickly placed in train by exchanging stoppers, and a slow current of Co_2-free air is drawn through the whole apparatus for two hours. The lime sample bottle is then removed, cooled, and placed in balance case for weighing. Just before weighing, the stopper is removed for an instant to relieve any vacuum that may exist in the bottle. The resulting loss in weight of the sample represents mechanical moisture or hygroscopic moisture as opposed to chemically combined moisture.

11. *Sulfur trioxide*. Two g. of sample are thoroughly mixed in 10 ml. of cold water until lightest particles are in suspension; 15 ml. of dilute HCl (1:1) are added, and the mixture is heated until solution

is complete. Filter off residue and wash it with hot water. The filtrate is diluted with water to 250 ml. volume, boiled, and 10 ml. of 10% boiling solution of $BaCl_2$ is added drop by drop amid stirring. After the solution stands overnight, filter, wash with boiling water, ignite, and weigh as $BaSO_4$. Then recalculate to SO_3.

12. *Total sulfur*. Mix 1 g. of lime sample with 0.5 g. of Na_2CO_3 in porcelain crucible and heat until sintered, continuing for fifteen minutes at 1000°C. Cool, place crucible in a 250 ml. beaker, and cover with hot water. Then add 10 ml. of bromine water and 30 ml. of HCl (1:1) and boil until solution is complete and all bromine is expelled. Remove crucible, add a few drops of methyl red to beaker, and alkalize with NH_4OH (1:1). Boil solution, filter, and wash with hot water. Add 5 ml. of HCl (1:1) to filtrate, adjust volume to 200 ml. with water, and bring to boil. Add 10 ml. of 10% hot $BaCl_2$ solution to boiling solution. After standing overnight, filter, wash with hot water, ignite, weigh as $BaSO_4$, and recalculate to S.

13. *Manganese*. Ten g. of sample are dissolved in 100 ml. of HNO_3 (1:1); filter and the residue is washed in hot water. The residue is fused with Na_2CO_3 in a platinum crucible, and the melt is added to the filtrate, along with 0.5 g. of sodium bismuthate or sodium persulfate. Heating precipitates MnO_2. More bismuthate is added for reprecipitation of remaining MnO_2. The total is calculated as an oxide.

If manganese separation is not made before the MgO determination, frequently, considerable MnO_2 will separate with the $Mg_2P_2O_7$. In this case, the weighed pyrophosphate filtrate (item 9) should be redissolved in HNO_3 and precipitated with bismuthate (as above), but this time in the form of $Mn_2P_2O_7$. A corresponding correction should be made: reduction of $Mg_2P_2O_7$ content and adding the equivalent of Mn to MnO_2 above.

14. *Phosphorus*. Ten g. of sample are dissolved in 80 ml. HNO_3 (1:1) and residue is filtered. After fusing with Na_2CO_3, residue is returned with more HNO_3 to main filtrate. The HNO_3 is neutralized with NH_4OH, and the solution then is reacidified with 1 ml. of HNO_3 (sp. gr. 1.42) for every 100 ml. of filtrate. At 40 C ammonium molybdate solution is added amid 10 min. of agitation. After standing for one to twelve hours at 4 C or less, filter and wash ten times with 1% KNO_3 solution. Precipitate is again returned to precipitating vessel and is redissolved in excess 0.1 N NaOH. After dilution to 100 to 200 ml. with water, three drops of phenolphthalein is added, and the pink color is discharged with standard acid. Titration is finished

6. Subtract this MgO equivalent from total MgO to obtain the percentage of unhydrated MgO on the "as-received" basis.

This procedure is primarily designed for analysis of Type S highly hydrated lime, and it is acknowledged that this method does not provide results of meticulous accuracy, but they are sufficiently accurate for the purpose intended. Values are to be reported to the nearest 0.5%. The following table illustrates this calculation:

Compound	Values from chemical analysis, %	Residual values	Factors	Calculated values, %
CO_2	0.40		$\times 1.275$	= equiv. CaO = 0.51
SO_3	0.48		$\times 0.700$	= equiv. CaO = 0.34
CaO	$42.79 - (0.51 + 0.34$ $= 0.85)$	41.94	$\times 0.3213$	= equiv. H_2O = 13.48
H_2O	$25.09 - 13.48$	11.61	$\times 2.238$	= equiv. MgO = 25.98
MgO	$30.68 - 25.98$			= equiv. unhydrated MgO = 4.70

The above calculation is based on the reasonable assumption that in hydrating quicklime all of the CaO is completely hydrated and that in dolomitic quicklime in particular only a portion of the MgO will hydrate under normal conditions. Therefore, all of the chemically combined water is initially apportioned to the CaO component; then, secondly to its equivalent MgO, and the remainder of MgO is assumed to be unhydrated. If the chemical analysis is readily available, this is a simple calculation (above). But if the chemical analysis is not needed and only the *free* MgO content is required, it is a tedious process to make all of the above necessary analytical separations. To preclude this a short direct chemical test for *free MgO* has been developed as follows:

A 2 to 3 g. sample of hydrated lime is placed in a tared 25 ml. platinum crucible and dried for two hours in a CO_2-free absorption train. After cooling in a desiccator, it is weighed. A standard autoclave, conforming to ASTM C 151, containing the customary amount of water, was heated and the water was brought to a boil. It is then boiled for a few minutes to dispel dissolved CO_2 in the water and displace CO_2 in the air with steam. The crucible with the dried sample is then autoclaved for one hour at 295 psi pressure and then removed and immediately placed in the same CO_2-free drying apparatus. After drying for three hours, it is cooled and weighed. The increase in weight between the two weighings is assumed to be a gain in H_2O assimilated by the free MgO. This weight is multiplied by the factor, 2.238, to obtain the weight of the resulting $Mg(OH)_2$, and it is then simple to calculate the percent of unhydrated (free MgO) in the first weighing.

This test assumes complete hydration in the autoclave, but certainly it is at least as accurate as the other procedure and much less time-consuming. Further details of this test are contained in the October 1957 ASTM *Bulletin*, p. 53, by E. Trattner, who developed this test.

Acid Neutralization. A tentative test, ASTM C400-57T, for measuring neutralization efficiency of quicklime and hydrated lime, expressed as neutralization coefficient, is outlined as follows:

Quicklime should be initially slaked in accordance with the manufacturer's instructions. In lieu of this, it should be ground to $-\frac{1}{4}$ in. and about 300 g. are mixed with either 500 ml. or 275 ml. of water for high calcium and dolomitic quicklime, respectively, to produce a putty consistency of 20 ± 10 mm. After remixing in mortar with pestle, transfer sufficient putty to provide 2.5 g. of lime solids to a 500 ml. elongated beaker. Weight of putty required to provide the 2.5 g. of quicklime equivalent can be calculated with the following formula:

$$W = 2.50 \frac{(U + V)}{U}$$

where:

W = weight of putty required in g.
U = g. of quicklime used to prepare putty.
V = g. of water used to prepare putty.

Add water (the amount: 50 ml. less $W - 2.50$ g.) to beaker and stir for ten minutes. Start electric stirrer in beaker and add 250 g. of 1.5% H_2SO_4 at one time. After thirty minutes record pH of mixture. Repeat above test three more times except vary the amount of acid each time, both less and more than 250 g., so that the resulting four pH recordings will provide a family of curves, preferably in the pH region of 4.4, as follows:

1. Plot pH versus time in minutes for each lime-acid ratio, which provides basis for calculating the lime requirement.

2. Plot pH in thirty minutes, as determined in item 1 above, against g. of lime used per 1000 g. of 1.5% H_2SO_4.

3. Interpolate the quantity of lime required to obtain a pH of 4.4 in thirty minutes from item 2 above, correct for actual acid concentration, and calculate *neutralization coefficient* as parts of lime required per million parts of 1.5% H_2SO_4 as follows:

$$N = \frac{1.5}{Z} \times X \times 1000$$

where:

N = neutralization coefficient from item 2 above.
X = g. of lime per l. of 1.5% H_2SO_4.
Z = concentration of H_2SO_4 to nearest 0.005%, as determined by analysis.

For *hydrated lime* the same procedure as above is employed except that slaking and preparation of putty consistency are eliminated; 2.5 g. of hydrate are mixed with 50 ml. of water in the same 500 ml. beaker.

To be practically applicable for *limestone* the conditions of the above test should be modified, such as adding less acid and/or extending reaction time limit so that neutralization occurs at a higher pH.

A standard *titration test* is generally employed for limestone in determining its *basicity factor* and *rate of neutralization* as follows:

2 g. of pulverized limestone, at least as fine as No. 30 mesh, is added to an Erlenmeyer flask containing 10 ml. of distilled water; 0.5N H_2SO_4 is then added in excess of 30 to 35 ml. of the amount necessary to decompose the limestone. Boil sample until neutralization is complete, replace water lost through evaporation, cool to room temperature, add two drops of phenolphthalein, and titrate to end point with 0.5N NaOH.

The basicity factor is then calculated as grams equivalent CaO per gram of limestone sample, a convenient common denominator to employ in the evaluation of all types of liming materials (p. 368). The equivalent CaO of high calcium limestone is 0.50 to 0.55 g., depending upon the purity of the stone or lime, but it can be more closely approximated by measuring the loss on ignition and exactly determined from procedure stipulated on p. 489.

The rate of neutralization can be measured in the above test by timing the reaction to completion. When employed comparatively with different limestones and limes, all conditions of the test must be identical to procure valid results.

The above titration procedure is equally adaptable for all limes.

Pozzolans. The reactivity of pozzolans with lime along with a definition of pozzolans is embraced in ASTM specification, C 432-59T, and is described as follows:

The term "pozzolan" is defined as a siliceous or alumino-siliceous material that in itself possesses little or no cementitious value, but that in finely divided form and in the presence of moisture will chemically react with alkali or alkaline earth hydroxides at ordinary temperatures to form compounds possessing cementitious properties.

Limes for use with these pozzolans may be any class of quick or hydrated lime, both high calcium and dolomitic.

Pozzolans must conform to the chemical and physical requirements stipulated in Table 13-7 when tested in accordance with the following methods:

For *moisture content* a weighed sample of pozzolan is dried at 105 to 110°C to constant weight, and the percent loss after drying is measured.

For the *water-soluble* determination place 10 g. of the dried sample in a 200 ml. Erlenmeyer flask and add 100 ml. of distilled water at 73 ± 3°F. Shake well until all lumps disappear and then agitate with mechanical stirrer for one hour. Pour material into a Gooch

Table 13-7. Chemical and physical requirements of pozzolans for use with lime

Moisture content, max., %	10
Water soluble fraction, max., %	10
Loss on ignition, max., %	10
Fineness, amount retained when wet-sieved on:	
No. 30 sieve, max., %	2
No. 200 sieve, max., %	10
Lime-pozzolanic compressive strength, min., psi:	
after seven days at 130 ± 3°F	600
after additional twenty-eight days at 73 ± 3°F	600
Hydraulic compressive strength, max., psi:	
after seven days at 130 ± 3°F	100
after additional 28 days at 73 ± 3°F	150
Soundness	after seven days specimens must remain hard and firm and show no signs of distortion

crucible, washing all residue from flask. Wash the residue several times to free it from adherent solution, and dry in oven at 105°C to constant weight. Calculate percent loss in weight as water soluble component.

For *loss on ignition* determination complies with Section 20 of ASTM C 114-61 for chemical analysis of portland cement, except that a 2 g. sample of the dried material shall be ignited to constant weight in an uncovered porcelain crucible at 900 to 1000°C.

For *fineness* measurement should be made in accordance with ASTM C 110 for physical testing of lime (p. 484) except that sample shall consist of 100 g. of the dried pozzolan.

For *lime-pozzolan strength* requirement ASTM C 109-63 procedure for compressive strength of hydraulic cement mortars and C 305-59T for mixing of cement mortars shall be pursued. Lime shall comply with hydrated lime for masonry purposes (ASTM C 207) and standard graded sand specified in ASTM C 109 (above). Batches shall be made of sufficient size to make six specimens for triplicate testing and shall consist of the following proportion of dry materials:

> Hydrated lime 180 g.
> Pozzolan (dry basis) 360 g.
> Graded standard sand 1480 g.

Sufficient water measured in ml. shall be added to yield a flow of 65 to 75 and shall be expressed as percentage by weight of the combined lime-pozzolan increment. Mixing procedure of mortar shall comply with C 305. Determination of mortar flow is in accordance with C 109, except that number of drops of flow table shall be ten drops in six seconds instead of twenty-five drops in fifteen seconds. Mortar is then molded and placed above water in a closed vapor oven at 130°F, protected from condensation dripping, for forty-eight hours. Then mortar bars are removed from molds and immersed in the 130°F water until they are seven days old. After cooling to 73°F in water or saturated air, the compressive strength is determined. Other samples for later testing, such as at twenty-eight days, are stored at 95 to 100 R.H. at 73°F until time for test.

For *hydraulic strength development* the same procedure should be followed (above), except that the mix is composed of:

> Pozzolan (dry basis) 540 g.
>
> Graded standard sand 1480 g.

If the specimens are too weak for removal from molds in forty-eight hours, strength shall be reported as "no strength."

A systematic *form* recommended by ASTM for reporting the analysis of limestone or lime products, adaptable for both producers and consumers is contained in Table 13-8. Other physical tests can be added at the bottom, and the form can be easily modified to a particular company's needs.

Table 13-8. Form for reporting analysis of limestone or lime products

Analysis of limestone or lime products
(Name of company)

Date   Lab. no.
Name   Date rec'd.
Material
Sample marked

| Constituents determined | | Per | Constituents calculated[a] | | Per |
Name	Formula	cent	Name	Formula	cent
Silicon dioxide	SiO_2		Calcium carbonate	$CaCO_3$	
Iron oxide	Fe_2O_3 ⎱		Calcium hydroxide	$Ca(OH)_2$	
Aluminum oxide	Al_2O_3 ⎰		Magnesium carbonate	$MgCO_3$	
Calcium oxide	CaO		Magnesium hydroxide	$Mg(OH)_2$	
Magnesium oxide	MgO		Calcium sulfate	$CaSO_4$	
Total sulfur	S		Unhydrated oxides		
Sulfur trioxide	SO_3				
Phosphorus pentoxide	P_2O_5				
Carbon dioxide	CO_2				
Water {at 120°C	H_2O				
{total	H_2O				
Insoluble matter					
Loss on ignition					
Available lime index					

Constituents calculated[a]

[a] Method for calculation *must* be noted for each constituent.

Sieve analysis

Sieve no.
Opening in microns
Percentage passing

REMARKS

Total neutralizing
  value in terms of   {Calculated
  $CaCO_3$ {Determined
Plasticity
Soundness

NOTE.—Unless otherwise noted all determinations have been made according to methods prescribed by the American Society for Testing Materials.

Signed

APPENDIX A

Chemical-physical characteristics of Ca and Mg compounds

Chemical formula	Chemical name	Common name	Mol. weight
Ca	Calcium	Calcium metal	40.08
$CaO \cdot Al_2O_3$	Monocalcium aluminate	—	158.02
$3CaO \cdot Al_2O_3$	Tricalcium aluminate	—	270.18
$CaO \cdot Al_2O_3 \cdot 2SiO_2$	Calcium aluminate di-silicate	—	278.14
$2CaO \cdot Al_2O_3 \cdot SiO_2$	Dicalcium aluminate silicate	—	274.16
$4CaO \cdot Al_2O_3 \cdot Fe_2O_3$	Tetracalcium aluminate ferrite	—	485.94
CaC_2	Calcium carbide	Carbide	64.08
$CaCO_3$	Calcium carbonate	High calcium lime-stone, precip. cal. carbonate	100.08
$Ca(HCO_3)_2$	Calcium bicarbonate	—	162.10
$CaCl_2 \cdot 6H_2O$	Calcium chloride	—	219.09
$Ca_3(C_6H_5O_7)_2 \cdot 4H_2O$	Calcium citrate	—	570.55
$CaCN_2$	Calcium cyanamide	Cyanimid	80.01
CaF_2	Calcium fluoride	—	78.07
$Ca(OH)_2$	Calcium hydroxide	Hydrated lime, slaked lime, hydrate	74.10
$Ca(OCl)_2$	Calcium hypochlorite	Bleach	143
$Ca(NO_3)_2 \cdot 4H_2O$	Calcium nitrate	Lime saltpeter	236.16
CaC_2O_4	Calcium oxalate	—	128.10
CaO	Calcium oxide	High calcium quick-lime, unslaked lime, hot lime	56.08
$Ca(H_2PO_4)_2 \cdot H_2O$	Monocalcium phosphate	—	252.14
$CaHPO_4 \cdot 2H_2O$	Dicalcium phosphate	Dibasic calcium phosphate	172.13
$Ca_3(PO_4)_2$	Tricalcium phosphate	—	310.29
$CaO \cdot SiO_2$	Monocalcium silicate	—	116.14
α-$2CaO \cdot SiO_2$	α-Dicalcium silicate	—	172.20
β-$2CaO \cdot SiO_2$	β-Dicalcium silicate	—	172.20

[a] in CO_2 saturated water at 20°C
[b] at 20°C
[c] at 10°C
[d] at 20°C
[e] at 15°C
d. = decomposes

Spec. grav.	Melting Pt. °C	max. solubility g./l.	% CaO	% MgO	% CO_2	% H_2O	% Other
					Chemical compounds		
1.54	851	d. in H_2O	100 Ca	—	—	—	—
2.98	1600	—	35	—	—	—	65 Al_2O_3
3.04	1535	—	62	—	—	—	38 Al_2O_3
2.76	1550	—	20	—	—	—	37 Al_2O_3 43 SiO_2
3.04	1590	—	41	—	—	—	37 Al_2O_3 22 SiO_2
3.77	1415	—	46	—	—	—	21 Al_2O_3 33 Fe_2O_3
2.22	2300	d. in H_2O	63 Ca	—	—	—	37 C
2.71–.94	d. CaO	0.013	56	—	44	—	—
—	—	1100[a]	31	—	49	20	—
1.65	—	394[c]	18 Ca	—	—	50	32 Cl_2
—	185, $4H_2O$	173	29	—	—	13	68 bal.
—	—	—	50 Ca	—	—	—	35 N_2, 15 C
3.16	1403	0.037	51 Ca	—	—	—	49 F_2
2.30	d. CaO	1.76[c]	76	—	—	24	—
—	—	—	39	—	—	—	61 OCl_2
1.82	42.5	502	24	—	—	31	45 N_2O_5
2.20	d.	0.014	44	—	—	—	56 C_2O_3
3.34–3.40	2580	1.31[c]	100	—	—	—	—
2.22	—	40[e]	22	—	—	—	78 H_3PO_4
2.31	—	0.28	33	—	—	10	57 H_3PO_4
3.14	1670	0.08	54	—	—	—	46 P_2O_5
2.92	1540	insol.	48	—	—	—	52 SiO_2
3.27	2130	insol.	65	—	—	—	35 SiO_2
3.28	—	insol.	65	—	—	—	35 SiO_2

(Continued)

Chemical-physical characteristics of Ca and Mg compounds (*Continued*)

Chemical formula	Chemical name	Common name	Mol. weight
$3CaO \cdot SiO_2$	Tricalcium silicate	—	228.27
$CaSi_2$	Calcium silicide	—	96.20
$CaOCl_2$	Calcium oxychloride	Chloride of lime, bleach	127
CaS	Calcium sulfide	—	72.14
$CaSO_4$	Calcium sulfate	Gypsum anhydride	136.14
$CaSO_4 \cdot \frac{1}{2}H_2O$	Calcium sulfate, hemi-hydrate	Gypsum, plaster	145.15
$CaSO_4 \cdot 2H_2O$	Calcium sulfate, di-hydrate	Gypsum rock	172.17
$Ca(HSO_3)_2$	Calcium bisulfite	Calcium acid sulfite	202.22
$CaCO_3 \cdot MgCO_3$	Calcium magnesium carbonate	Dolomitic limestone, dolomite	184.40
$CaCO_3 + MgO$	Calcium carbonate-magnesium oxide	Selectively calcined dolomite	140.40
$CaO \cdot MgO$	Calcium magnesium oxide	Dolomitic quicklime	96.40
$Ca(OH)_2 \cdot MgO$	Calcium hydroxide-magnesium oxide	Normal hydrated dolomitic lime (type N)	114.42
$Ca(OH)_2 \cdot Mg(OH)_2$	Calcium magnesium hydroxide	Highly hydrated dolomitic lime (type S)	132.44
$2CaO \cdot MgO \cdot 2SiO_2$	Dicalcium-magnesium disilicate	—	272.60
Mg	Magnesium	Magnesium metal	24.32
$Mg(HSO_3)_2$	Magnesium bisulfite	Magnesium acid sulfite	186.46
$MgCO_3$	Magnesium carbonate	Magnesite	84.32
$Mg(C_2H_3O_2)_2 \cdot 4H_2O$	Magnesium acetate	—	214.47
$MgBr_2$	Magnesium bromide	—	184.15
$MgCl_2 \cdot 6H_2O$	Magnesium chloride, six hydrate	Magnesium chloride	203.33
MgF_2	Magnesium fluoride	—	62.32
$Mg(OH)_2$	Magnesium hydroxide	Hydrate of magnesia	58.34
$Mg(NO_3)_2 \cdot 6H_2O$	Magnesium nitrate hydrate	—	256.43
MgO	Magnesium oxide	Magnesia, periclase	40.32
$MgO \cdot SiO_2$	Monomagnesium silicate	—	100.38
$2MgO \cdot SiO_2$	Dimagnesium silicate	—	140.70
$MgSO_4$	Magnesium sulfate	—	120.38
$MgSO_4 \cdot H_2O$	Magnesium sulfate, mono hydrate	—	138.40
$MgSO_4 \cdot 7H_2O$	Magnesium sulfate, seven hydrate	Epsom salts	246.49

Spec. grav.	Melting Pt. °C	max. solubility g./l.	% CaO	% MgO	% CO_2	% H_2O	% Other
							Chemical compounds
3.25	—	insol.	73	—	—	—	27 SiO_2
—	—	—	41.7	—	—	—	58.3 Si
—	—	—	44	—	—	—	56 Cl_2
2.25	—	0.150[c]	56 Ca	—	—	—	44 S
2.97	1450	1.76 (0°)	41	—	—	—	59 SO_3
2.45	—	0.885[d]	39	—	—	6	55 SO_3
2.32	1450	2.0[c]	32	—	—	21	47 SO_3
—	d.	0.07	28	—	—	19	63 SO_2
2.80 to 2.99	d.	—	30.4	21.8	47.8	—	—
—	—	—	40	29	31	—	—
—	—	—	58	42	—	—	—
—	—	—	49	35.3	—	15.7	—
—	—	—	42	31	—	27	—
2.94	—	—	41	15	—	—	44 SiO_2
1.74	650	—	—	100 Mg	—	—	—
—	—	—	—	22	—	9	69 SO_2
3.04	d., MgO	0.2[e]	—	48	52	—	—
1.454	d.	664 (68°)	—	19	—	25	56 CH_3COOH
3.72	700	532 (80°)	—	13 Mg	—	—	87 Br_2
1.56	—	366	—	12 Mg	—	53	35 Cl_2
2.9–3.2	1396	0.076	—	46 Mg	—	—	54 F_2
2.36	d.	—	—	69	—	31	—
1.46	95	578	—	16	—	42	42 N_2O_5
3.65	2800	0.0084[b]	—	100	—	—	—
3.28	1524	—	—	40	—	—	60 SiO_2
3.21	1890	—	—	57	—	—	43 SiO_2
2.66	1120	sol. as $7H_2O$	—	34	—	—	66 SO_3
2.57	—	sol. as $7H_2O$	—	29	—	13	58 SO_3
1.69	—	236[b]	—	16	—	31	33 SO_3

APPENDIX B

Lime conversion factors and equations in water treatment

1 ppm $Ca(OH)_2$ = 1.35 ppm alkalinity increase
1 ppm $Ca(OH)_2$ = 1.19 ppm free CO_2 reduction
1 ppm aluminum sulfate = 0.45 ppm alkalinity reduction
1 ppm aluminum sulfate = 0.40 ppm CO_2 increase
$\therefore$ 8 mg./l. of $Ca(OH)_2$ neutralizes acidifying effect of 1 grain each (or 17.1 mg./l.) of alum or iron salts
1.65 mg./l. of $Ca(OH)_2$ removes 1 mg./l. of CO_2 as $CaCO_3$ precipitate
3 mg./l. of $Ca(OH)_2$ removes 1 mg./l. of Mg hardness as $MgCO_3$
0.74 mg./l. of $Ca(OH)_2$ neutralizes 1 mg./l. of CO_2
For calculating the weight of lime slurry with varying percentages of water, the following formula may be used:

$$W = \frac{6237s}{100 - a + sa}$$

when:

W = weight in lb. of slurry per cu.ft.
s = specific gravity of dry lime solids
a = % water in slurry

Coagulation reactions:

$$Al_2(SO_4)_3 \cdot 18\ H_2O + 3\ Ca(OH)_2 = 3\ CaSO_4 + 2\ Al(OH)_3 + 18\ H_2O$$
$$4\ FeSO_4 \cdot 7\ H_2O + 4\ Ca(OH)_2 + 2\ O_2 = 4\ CaSO_4 + 4\ Fe(OH)_3 + 5\ H_2O$$
$$Fe_2(SO_4)_3 + 3\ Ca(OH)_2 = 3\ CaSO_4 + 2\ Fe(OH)_3$$
$$2\ FeCl_3 + 3\ Ca(OH)_2 = 3\ CaCl_2 + 2\ Fe(OH)_3$$

APPENDIX C

Alkalinity equivalents of limestone, lime, and other alkalis

Compound	Formula	$Al_2(OH)_6$	BaO	$Ba(OH)_2$	CaO	$Ca(OH)_2$	$CaCO_3$	$CaH_2(CO_3)_2$	MgO	$Mg(OH)_2$
Aluminum hydrate	$Al_2(OH)_6$	1.0000	0.3390	0.3034	0.9271	0.7017	0.5195	0.3207	1.2895	0.8913
Barium oxide	BaO	2.9494	1.0000	0.8948	2.7734	2.0697	1.5323	0.9461	3.8035	2.6289
Barium hydrate	$Ba(OH)_2$	3.2959	1.1174	1.0000	3.0559	2.3128	1.7123	1.0572	4.2503	2.9377
Calcium oxide	CaO	1.0785	0.3656	0.3272	1.0000	0.7568	0.5603	0.3459	1.3908	0.9613
Calcium hydrate	$Ca(OH)_2$	1.4250	0.4831	0.4323	1.3212	1.0000	0.7403	0.4571	1.8376	1.2701
Calcium carbonate	$CaCO_3$	1.9247	0.6525	0.5839	1.7845	1.3506	1.0000	0.6174	2.4821	1.7155
Calcium bicarbonate	$CaH_2(CO_3)_2$	3.1174	1.0569	0.9458	2.8904	2.1876	1.6196	1.0000	4.0202	2.7786
Magnesium oxide	MgO	0.7754	0.2629	0.2352	0.7189	0.5441	0.4028	0.2487	1.0000	0.6911
Magnesium hydrate	$Mg(OH)_2$	1.1219	0.3803	0.3403	1.0402	0.7873	0.5828	0.3598	1.4468	1.0000
Magnesium carbonate	$MgCO_3$	1.6216	0.5498	0.4920	1.5035	1.1379	0.8425	0.5201	2.0912	1.4454
Magnesium bicarbonate	$MgH_2(CO_3)_2$	2.8143	0.9541	0.8538	2.6094	1.9749	1.4621	0.9027	3.6293	2.5085
Sodium oxide	Na_2O	1.1922	0.4042	0.3617	1.1054	0.8366	0.6194	0.3824	1.5375	1.0627
Sodium hydrate	NaOH	1.5387	0.5217	0.4668	1.4267	1.0798	0.7994	0.4935	1.9843	1.3715
Sodium carbonate	Na_2CO_3	2.0385	0.6911	0.6184	1.8900	1.4305	1.0590	0.6538	2.6288	1.8169
Sodium carbonate, hydrated	$Na_2CO_3, 10H_2O$	5.5033	1.8658	1.6697	5.1025	3.8619	2.8592	1.7653	7.0969	4.9052
Sodium bicarbonate	$NaHCO_3$	3.2312	1.0955	0.9803	2.9958	2.2674	1.6787	1.0364	4.1669	2.8800
*Trisodium phosphate	Na_3PO_4	2.1028	0.7129	0.6380	1.9497	1.4756	1.0925	0.6745	2.7118	1.8743
Trisodium phosphate, hydrated	$Na_3PO_4, 12H_2O$	4.8747	1.6527	1.4790	4.5197	3.4207	2.5326	1.5636	6.2863	4.3449

(Continued)

505

APPENDIX C (Continued)

Alkalinity equivalents of limestone, lime, and other alkalis

Compound	Formula	MgCO₃	MgH₂(CO₃)₂	Na₂O	NaOH	Na₂CO₃	Na₂CO₃, 10H₂O	NaHCO₃	Na₃PO₄	Na₃PO₄ 12H₂O
Aluminum hydrate	$Al_2(OH)_6$	0.6166	0.3553	0.8387	0.6498	0.4905	0.1817	0.3094	0.4755	0.2151
Barium oxide	BaO	1.8187	1.0480	2.4737	1.9167	1.4468	0.5359	0.9128	1.4025	0.6050
Barium hydrate	$Ba(OH)_2$	2.0324	1.1711	2.7643	2.1419	1.6168	0.5989	1.0200	1.5672	0.6761
Calcium oxide	CaO	0.6650	0.3832	0.9046	0.7009	0.5290	0.1959	0.3337	0.5128	0.2212
Calcium hydrate	$Ca(OH)_2$	0.8787	0.5063	1.1952	0.9260	0.6990	0.2589	0.4410	0.6776	0.2923
Calcium carbonate	$CaCO_3$	1.1869	0.6839	1.6143	1.2508	0.9442	0.3497	0.5956	0.9152	0.3948
Calcium bicarbonate	$CaH_2(CO_3)_2$	1.9223	1.1076	2.6146	2.0259	1.5292	0.5664	0.9647	1.4824	0.6395
Magnesium oxide	MgO	0.4781	0.2755	0.6503	0.5039	0.3803	0.1409	0.2399	0.3687	0.1590
Magnesium hydrate	$Mg(OH)_2$	0.6918	0.3986	0.9409	0.7291	0.5503	0.2038	0.3472	0.5334	0.2301
Magnesium carbonate	$MgCO_3$	*1.0000*	0.5762	1.3601	1.0538	0.7955	0.2946	0.5018	0.7711	0.3326
Magnesium bicarbonate	$MgH_2(CO_3)_2$	1.7354	*1.0000*	2.3604	1.8289	1.3806	0.5113	0.8709	1.3382	0.5773
Sodium oxide	Na_2O	0.7352	0.4236	*1.0000*	0.7748	0.5848	0.2166	0.3689	0.5669	0.2445
Sodium hydrate	$NaOH$	0.9488	0.5467	1.2906	*1.0000*	0.7548	0.2796	0.4762	0.7317	0.3156
Sodium carbonate	Na_2CO_3	1.2570	0.7243	1.7097	1.3247	*1.0000*	0.3704	0.6308	0.9693	0.4181
Sodium carbonate, hydrated	$Na_2CO_3, 10H_2O$	3.3936	1.9554	4.6157	3.5764	2.6996	*1.0000*	1.7031	2.6169	1.1289
Sodium bicarbonate	$NaHCO_3$	1.9925	1.1481	2.7100	2.0998	1.3850	0.5871	*1.0000*	1.5364	0.6628
Trisodium phosphate	Na_3PO_4	1.2967	0.7471	1.7637	1.3665	1.0315	0.3821	0.6504	*1.0000*	0.4313
Trisodium phosphate, hydrated	$Na_3PO_4, 12H_2O$	3.0059	1.7320	4.0885	3.1679	2.3913	0.8857	1.5078	2.3181	*1.0000*

APPENDIX D

Alkali conversion tables for limes and other alkalis, fractional %'s

CaO	Ca(OH)$_2$	CaO·MgO	NaOH	Na$_2$CO$_3$	CaO·MgO	CaO	Ca(OH)$_2$	NaOH	Na$_2$CO$_3$
1	1.32	0.86	1.43	1.89	1	1.16	1.54	1.66	2.20
2	2.64	1.72	2.85	3.78	2	2.33	3.07	3.32	4.40
3	3.96	2.58	4.28	5.67	3	3.49	4.61	4.98	6.60
4	5.29	3.44	5.71	7.56	4	4.65	6.15	6.64	8.80
5	6.61	4.30	7.13	9.45	5	5.82	7.69	8.30	11.00
6	7.93	5.16	8.56	11.34	6	6.98	9.22	9.96	13.20
7	9.25	6.02	9.99	13.23	7	8.14	10.76	11.62	15.39
8	10.57	6.88	11.41	15.12	8	9.31	12.30	13.28	17.59
9	11.89	7.74	12.84	17.01	9	10.47	13.84	14.94	19.79
10	13.21	8.59	14.27	18.90	10	11.63	15.37	16.60	21.99
15	19.82	12.89	21.40	28.35	15	17.45	23.06	24.90	32.99
20	26.43	17.19	28.53	37.80	20	23.27	30.75	33.20	43.99
25	33.03	21.49	35.67	47.26	25	29.09	38.43	41.50	54.98
30	39.64	25.78	42.80	56.71	30	34.90	46.12	49.80	65.98
35	46.24	30.08	49.93	66.16	35	40.72	53.80	58.10	76.97
40	52.85	34.38	57.07	75.61	40	46.54	61.49	66.40	87.97
45	59.46	38.68	64.20	85.06	45	52.36	69.18	74.70	98.97
50	66.06	42.97	71.34	94.51	50	58.17	76.86	83.00	109.96
55	72.67	47.27	78.47	103.96	55	63.99	84.55	91.30	120.96
60	79.28	51.57	85.60	113.41	60	69.81	92.24	99.60	131.96
65	85.88	55.87	92.74	122.86	65	75.63	99.92	107.90	142.95
70	92.49	60.16	99.87	132.32	70	81.44	107.61	116.20	153.95
75	99.09	64.46	107.00	141.77	75	87.26	115.29	124.50	164.94
80	105.70	68.76	114.14	151.22	80	93.08	122.98	132.80	175.94
85	112.31	73.06	121.27	160.67	85	98.90	130.67	141.10	186.94
90	118.91	77.35	128.40	170.12	90	104.71	138.35	149.40	197.93
95	125.52	81.65	136.54	179.57	95	110.53	146.04	157.70	208.93
100	132.13	85.95	142.67	189.02	100	116.35	153.73	166.00	219.93

Alkali conversion tables for limes and other alkalis, fractional %'s *(Continued)*

NaOH	CaO	Ca(OH)$_2$	CaO·MgO	Na$_2$CO$_3$	Na$_2$CO$_3$	CaO	Ca(OH)$_2$	CaO·MgO	NaOH
1	.70	.93	.60	1.32	1	.53	.70	.45	.75
2	1.40	1.85	1.20	2.65	2	1.06	1.40	.91	1.51
3	2.10	2.78	1.81	3.97	3	1.59	2.10	1.36	2.26
4	2.80	3.70	2.41	5.30	4	2.12	2.80	1.82	3.02
5	3.50	4.63	3.01	6.62	5	2.65	3.49	2.27	3.77
6	4.21	5.56	3.61	7.95	6	3.17	4.19	2.73	4.53
7	4.91	6.48	4.22	9.27	7	3.70	4.89	3.18	5.28
8	5.61	7.41	4.82	10.60	8	4.23	5.59	3.64	6.04
9	6.31	8.33	5.42	11.92	9	4.76	6.29	4.09	6.79
10	7.01	9.26	6.02	13.25	10	5.29	6.99	4.55	7.55
15	10.51	13.89	9.04	19.87	15	7.94	10.48	6.82	11.32
20	14.02	18.52	12.05	26.50	20	10.58	13.98	9.09	15.10
25	17.52	23.15	15.06	33.12	25	13.23	17.47	11.37	18.87
30	21.03	27.78	18.07	39.75	30	15.87	20.97	13.64	22.64
35	24.53	32.41	21.08	46.37	35	18.52	24.46	15.91	26.42
40	28.04	37.04	24.10	53.00	40	21.16	27.96	18.19	30.19
45	31.54	41.67	27.11	59.62	45	23.81	31.45	20.46	33.97
50	35.05	46.30	30.12	66.24	50	26.45	34.95	22.73	37.74
55	38.55	50.93	33.13	72.87	55	29.10	38.44	25.01	41.51
60	42.05	55.57	36.15	79.49	60	31.74	41.94	27.28	45.29
65	44.56	60.20	39.16	86.12	65	34.39	45.43	29.56	49.06
70	49.06	64.83	42.17	92.74	70	37.03	48.93	31.83	52.83
75	52.57	69.46	45.18	99.37	75	39.68	52.42	34.10	56.61
80	56.07	74.09	48.19	105.99	80	42.32	55.92	36.38	60.38
85	59.58	78.72	51.21	112.62	85	44.97	59.41	38.65	64.16
90	63.08	83.35	54.22	119.24	90	47.61	62.91	40.92	67.93
95	66.59	87.98	57.23	125.86	95	50.26	66.40	43.20	71.70
100	70.09	92.61	60.24	132.49	100	52.90	69.90	45.47	75.48

Index

AASHO, 120, 469
Abandoned quarries, 36
Abundance of limestone, 1, 13
  calcium, 1
  magnesium, 1
Acid gases (reaction), 31, 193
Acid neutralization, see Neutralization, acids
Adams, F., 187, 300
Aggregate tests, 121, 462–470
  deleterious substances, 468, 470
Agricultural hydrates, 165, 382–384
Agricultural liming, 103–113, 381–384
  acid-loving plants, 109
  application, 107–108
  attributes, 105–107
  burned lime products, 381–384
  fineness, 111
  history, 381–382
  home use, 384
  materials, 112
  overliming, 110
  testing soils, 110
  total neutralizing power, 110
  trees, 115
Air entraining, 392
Air Reduction Co., 319, 356
Air separation, 60, 331–332
Air slaked lime, 165, 191, 291, 294
Alabama, 12, 402, 439, 442
Alaska, 439
Alkali, for neutralization, 368–369
  reactivity test, 467
  use of lime, 355, 445
Alumina, determination of, 488–489
  impurity, 17, 154–156
  manufacture, 85–86, 353–354
  reaction with lime, 193–194, 269, 274–275

American Assn. of State Highway Officials, see AASHO
American Public Works Association, 415
American Society of Testing Materials, 95, 120, 336, 460–462, 486–487
American Standards Association, 405, 407
American Water Works Association, 484, 493
Ammonium nitrate blasting, 47
Angle of repose (limes), 172, 174
Angstrom units, 146
Anthracite coal, 251
Apple storage, 378
Aragonite, 8, 29
Argentina, 418
Argillaceous limestone, 9, 17, 90
Arkansas, 446
Arsenic (impurity), 17
Ash Grove Lime & Portland Cement Co., 281
Asphalt filler, see Filler
Asphalt hot mix, 423–426
Austin chalk formation, 13
Australia, 418
Austria, 243, 443
Autoclaved lime, 165, 330, 333–334, 397
Available basicity, 367–368
Available lime, 81, 155, 165, 492
Azbe, V. J., 133–134, 140, 150, 152, 217, 220–221, 237, 243, 258, 449
Azbe award, 449
Azores, 401

Bäckström, H., 26
Bag packing, 332, 434
Baking Industry, 378
Barberton, Ohio, 45

Basalt, 117–118, 127
Base exchange, 105–106, 116, 195, 410
Base stabilization, *see* Lime stablization
Basic brick, 374
Basic oxygen furnace, 345, 348
Bassett, H., 177
Bastard limestone, 9
Bayer process, 353
Belgium, 226, 436, 443
Bellefonte, Pa., 45
Belt conveyor, 61–62
Benches (quarry), 42
Bishop, D., 299, 306
Bituminous coal, 247, 249–250, 253
Bituminous limestone, 9
Bituminous road surfaces, 124–125, 423–426, 467–468
  lime additive, 423–426
Black, Joseph, 3
Blaine air permeability test, 304–305, 485–486
Blast furnace slag, 79–82, 91, 99, 117–118, 401–402
Blasting, ammonium nitrate, 47
  delayed action, 49
  dynamite, 50
  other detonation methods, 49
Bleaches, 357
Bleaching, 355
Bleininger, A. V., 143
Blocked Iron Corp., 351
Boiling point, 172
Bordeaux mixture, 357
Brazil, 248
Brecciated limestone, 9
Bucket elevator, 61–62, 331
Building lime, definition, 165, 388–410
  mortar, 388–402
  plaster and stucco, 403–410
Building materials, 374–377
Bulk density, limestone, 21
  of limes, 169, 174
Bureau of Public Roads, 423, 468
By-product stone, 452

Calcareous marl, 77
Calcimatic kiln, 239
Calcination, theory of, 132–163
  atmosphere, 145
  calorific requirements, 136, 254–255

Calcination, theory of, dead-burning, 148, 268–270
  effect on ionic spacing, 153
  effect of limestone size, 150, 152
  effect of salts, 153
  effect of steam, 157, 248, 278
  fuels for, 246–252
  influence on porosity, 143, 146–147
  influence on shrinkage, 143–144, 147, 150–151
  influence on stone impurities, 154–156
  influence on surface area, 149
  loss of weight, 140, 147
  rate of heating, 138, 147
  reaction, 29, 133
  temperatures, 142–143, 147
Calcining effort, 152
Calcining zone, 212–215
Calcite, 8
Calcitic limestone, 9
Calcium (element), 1, 5, 105, 114, 115, 359
  aluminate, 97–98, 193–194
  arsenate, 356
  bicarbonate, 5, 30, 361
  carbide, 356, 446
  dioxide, 197
  ferrites, 197
  hemihydrate, 198
  nitrate, 114
  oxide (free), 81, 155, 165, 492
    determination of, 489–490
  peroxide, 193
  salts, 357
  silicates, 97–98, 193–197, 273–277, 412
California, 271, 406, 439
Cal-Nitro, 114
Cambrian era, 13
Canada, 352, 376, 379, 436, 443, 454
Canary Islands, 401
Captive production, 355, 445–446, 458
Carbide lime, 165
Carbonaceous impurity, 18, 140, 472
  carbon, reaction with, 197
Carbonaceous limestone, 9
Carbon dioxide, atmosphere, 145
  calcination measurement, 258–260
  determination of, 472
  effect of stone size, 150, 152
  loss of weight, 140, 147

Carbon dioxide, pressure, 134–135
  rate of carbonation, 191, 298
  reaction with lime, 191, 351, 355, 377–379
  reaction on mortar, 191, 389
Carbon monoxide, 192, 222–223
Cato, 3
Causticization, 193, 272, 354–355
Cellular concrete products, 376
Cellular structure, lime, 148–149, 168
  of limestone, 20–21
Cementation index, 274–275
Cement stone, 9, 19, 99
Central America, 436
Ceramic pottery (use), 90
Chalk, 9, 93–94, 99
Chemical analysis, dead-burned dolomite, 270
  hydrated lime, 338
  lime, 243–244, 277, 487–499
  limestone, 19, 155–157, 276, 471–472, 487–499
  oystershell, 271
  waste lime-limestone dust, 268
Chemical grade limestone, 10
Chemical properties, limes, 182–198
  of limestones, 29–32
Chemical stability, limes, 183
  of limestones, 29
Chemical tests, lime, 487–499
  of limestone, 471–472, 487–499
Chemical uses, lime, 340–381
  of limestone, 78–103
Chert, 17
Cherty limestone, 10
Chicken litter, 385
China, 388, 443
CLAIRA, 448
Clam shells, 10, 271
Classification (limestone), 8
Clay minerals, 195–196
Coal mine dusting, 101–102
Coarse aggregate, sizes, 68–69, 119
  uses of, 116–129
Coefficient of expansion, lime, 169, 175
  of limestone, 22
Coke, 250–251
Color, lime, 167–168, 172
  of limestone, 20
Colorado, 423

Commercial hydration, *see* Dry hydrate
Compact limestone, 10
Compost, 385
Compressive strength, tests, 464, 482
Concrete (lime additive), 402
Concrete products, 128, 376
Cone crusher, 60
Construction aggregates, 116–129
Construction projects, 402–403, 420
Conveying, 61–62, 66–68, 331
Coolers (rotary kiln), 228–229, 231
Cooling zone, 210–211
Copper, impurity, 17, 156
  use of lime, 85, 352
Coquina, 10
Coral limestone, 10
Coral reefs, 5
Core (lime), 141
Coring, 37
Corson Co., G. & W. H., 336
Costs, limestone production, 70–71, 73
  lime manufacture, 279–281, 338
CP-Grade of lime, 319
Cretaceous period, 13
Crete, 388
Crushing, 58–61
Crystal structure, hydrated lime, 173–174
  limestone, 8, 20, 21, 137, 143–144, 159–163
    coarse-grained, 161
    fine-grained, 159
    medium-grained, 160
    oölitic, 162
  quicklime, 148–149, 168
    effect of heat, 137, 139
    growth (crystallites), 146, 148–149
Czechoslavakia, 443

Dairy industry, 378
Davis, G., 307
Dead-burned dolomite, 157, 166, 268–270, 350
  lime, 157, 354
Decrepitation of stone, 137–138, 140
Definitions, aggregate, 120
  lime, 2, 165–167
  limestone, 2, 9
  minerals (limestone), 8

Dehydration of hydrate, 288, 318
  use of quicklime, 381
Delayed action blasting, 49
Delayed hydration, 295–297, 335, 407–
    408, 480–481
Denmark, 226
Density of limes, 169, 174
Deposits (limestone), 14–16
Dessicant, 183
Desulfurization, 349
Deval test, 463
Dielectric constant, 24
Diffuse reflecting power, 23
Diffusion constant (hydrates), 176
Dioscorides, 3
Disposal of animal bodies, 386
Disposal of overburden, 41
Dissociation temperatures, 133
  pressures, 134
Dolomite, 8
Dolomitic limestone, 9, 10, 15, 19
Dorry hardness test, 464
Dow processes, 352
Draft, 249
Drilling (limestone), rotary and per-
    cussion drills, 46
  secondary drilling, 47
  well drills and pneumatic hammers,
    46
Drill-lime stabilization, 419–422
Drop-ball (quarry), 50
Dry grinding, 61
Dry hydrate, 290, 326–338
Dust collection, 267, 331
Dyes (use), 359
Dynamite, 50

Eades, J. L., 195–197
Eckel, E. C., 217, 281
Economic factors, 430–458
Effect on metals, 197
Egypt, 388
Electrical resistivity, 24, 169
Electrolytic conductivity, 180–181
Electromotive force potential, 31
Emley, W., 143
Emley plasticity values, 301, 306, 308–
    311, 475–476
Essentiality (lime, limestone), 1
Ethylene glycol, 359

Exhaust gases, 223, 258–260
Exothermic reaction, 287
Expansion, *see* Delayed hydration
Explosives, 47
Exploration (limestone deposits),
    criteria, 34
  core drilling, 37
  how to proceed, 35
  object of search, 35
Extraction (limestone), costs, 52
  disposal of overburden, 41
  haulage, 51
  loading, 50
  mining, 43
  pumping, 52
  quarrying, 41
  stripping, 38
    dragline scraper, 39
    hydraulic method, 39
    power shovel, 40
    tractor-propelled bulldozer, 40

Fat lime, 166
Federal Specification Board, 120
Feeding equipment, 363, 366
Fertilizers, 106–107, 113–114, 386
  lime additive, 386
Ferrosilicon process, 352
Ferrous metals (use), lime, 344–351
  limestone, 79–85
Ferruginous limestone, 10
Fill, 127
Fillers, asphalt, 125, 424–425, 470
  fertilizer, 113
  industrial, 92–97, 100
  pesticide, 357
  tests, 466
Filter beds, 102
Fine aggregate, 125–126
Fineness modulus, 121–122, 466, 470
Finish coat (plaster), 404–405
Finishing lime, 166, 404, 407–409
Fink, G. J., 313–314
Finland, 226
Fischer, H. C., 146
Fish and Wildlife Service, 387
Floor covering (use), 96
Florida, 270, 362, 386, 406, 419, 455
Flotation uses, 352

Flow diagrams, 56–57, 202–203, 235, 328–330
Flow values, 477–478
Fluorine (impurity), 17
Fluo-solids kiln, 237–238
Fluxing, lime, 166, 344–351, 353
  limestone, 79–86, 345–347
Fluxstone, 10, 79–86
Flyash, 402
Food and food by-products, 377–379
Formation (limestone), 6
Fossils, 5
Fossiliferous limestone, 10
Foster, R. S., 138
France, 103, 275–277, 400–401, 436, 443, 447
Franklin, Ben, 382
Fruit industry, 378
Fuels, 246–252
Fuel statistics, 247–248
Fuel oil, 250, 252
Fuel requirements, 254–258
Fungicide, 357

Gas analysis, 253
General Motors Co., 319
Geologic, origin, 5
  nomenclature, 12
Germany (West), 103, 115, 223, 226, 241, 243, 248, 255–257, 263, 267, 276–277, 309, 349–350, 376, 384, 386, 398, 400, 405–406, 414, 418, 423, 430, 436, 443, 445, 447–448
German Lime Association, 443, 448
Glass stone, 10, 89
Glass manufacture, 86–89, 374
Glauconic limestone, 10
Gradations (limestone), 68–69, 81, 83, 85, 89, 102, 118–120, 152, 224
  determination of, 465–466, 484–485
Granite, 115, 117–118, 127
Grappier's cement, 275
Grate-kiln system, 234–235
Great Britain, 93, 103, 115, 190, 226, 248, 276, 360, 379, 382, 384, 398, 401, 405–406, 423, 448, 476, 481
Greece, 388, 401
Greenfeld, S., 146
Grim, Ralph, 195–197
Grinding, 60–61, 327

Grizzlies, 59
Ground burnt lime, 166, 382–384
Gypsum, 404
Gyratory crusher, 59–60

Halliburton Oil Well Cementing Co., 380
Halogens, reaction with, 193
Hammer mills, 58, 60
Hard-burned lime, 138, 143, 146, 166
Hardness, limestone, 22, 462–463
  of limes, 169, 175
Haslam, R., 176, 180, 198, 317
Haulage of stone, 51
Hawaii, 439
Heat balance, 253–260
Heat capacity, 24
Heat exchangers, 229–230
Heat of formation, limestone, 24
  of limes, 172, 175
Heat of fusion, 170
Heat of hydration, 175, 288
Heat losses, 255–258
Heat of solution in acid, limestone, 31
  of limes, 172, 175, 185–186
Heavy-media separation, 65
Hedin, Rune, 137, 139, 143, 146, 173, 176, 306–307, 310, 448–449
Herold, J., 177
Heterogeneous impurities (limestone), 17
High calcium limestone, 9, 10, 14, 16, 19
History, limestone, 3
  of lime, 3, 382, 388–389
Holmes, M., 307, 313–314
Homogeneous impurities (limestone), 17
Hungary, 243, 443
Hydrated lime, 166
Hydrates of calcium carbonate, 29
Hydration, chemical reaction, 288–289
  effect on particle size, 298–302
  effect on plasticity, 306–311, 336–337, 475–476
  effect on putty volume, 313
  effect on reactivity, 317
  effect on sedimentation, 314–317
  effect on surface area, 303–305
  effect on water retentivity, 311–312

Hydration, hydrated forms of lime, 290–291, 293
  hydrator machines, 331
  methods of, 322–338
  plant equipment, 327–337
  rates of, 294–298, 483–484
  theory, 287–320
  under pressure, 330
  water content, 289–290
Hydraulic hydrated lime, 166, 273–277, 398, 400
Hydraulic limestone, 11

Iceland spar, 11, 146, 150, 177
Illinois, 13, 19, 101, 104, 403, 455
Impact crusher, 58, 60
Impurities, from fuels, 154–155, 250–251
  in limes, 180
  in limestones, 6, 9, 17, 81, 121, 143, 145, 154–156
India, 402
Indiana, 13, 19, 104, 439, 455
Indiana limestone, 11
Industrial uses, see Chemical uses
Insecticide use, 356–357
Instrumentation, 260–263
Internal Revenue Code, 457
Ionization, 185–186
Iowa, 104, 455
Iron, determination of, 488–489
  impurity, 17, 89, 154
  reaction with lime, 197, 269
  with dead-burned dolomite, 269
Italy, 375, 388, 401, 443

Japan, 153, 403
Jaw crusher, 58–59
Johnston, J., 133, 198

Kalousek, G., 194
Kansas, 13, 19
Kennedy, Van Saun, 337
Kentucky, 104, 455
Kilns, 204–243, calcimatic, 239–240
  classification of, 206
  corson vibratory, 240
  ellernan, 240
  field or pot types, 204
  fluosolids, 237–238
  Hoffman (ring), 240–241

Kilns, miscellaneous, 243
  old mixed feed, 205
  rotary, 226–237, 253, 255–256
    grate-kiln system, 234–235
    heat recuperative equipment, 227–231
  vertical, 206–226, 253
    center burners, 216–218, 220–221
    indirect fired, 209
    modern mixed feed, 219–224, 255–257
    standard—direct fired, 206–208
Kiln capacities, 218, 223, 236–237, 240, 243, 253
Kiln charging, 216–217
Kiln dimensions, 218–219, 223, 226, 236–237, 240
Kiln discharge, 211–212
Kiln feed size, 224–225, 231–232, 235, 237, 239, 240, 269
Kiln rings, 233–234
Kiln zones, 206–208
Kline, 27
Knibbs, N. V. S., 140, 157, 298, 333
Kritzer hydrator, 333

Lacey, G., 281
Lavoisier, 3
Leather tanning, 380
Le Chatelier mold test, 481
Lias limestone, 11
Lime industry, captive production, 445–446
  competition, 430–431
  consumption, 441–442
  location, 440
  number and size of plants, 437–438, 445
  packaging, 434
  prices, 432–433
  profits, 442
  research, 447–449
  transportation, 434
  wartime, 449–451
  world production, 443
  world trade, 435–436
Lime manufacture, 100–101, 201–286
Lime solids, 289–290
Lime stabilization, 410–423
  construction procedures, 416–417, 420

Lime stabilization, drill-lime, 419–420
  effect on soil, 412–413
  growth, 417–419
  recommended use, 414–416
  scope of use, 417
Lime suspensions, 291–293
Lime water, 291
Lime-cement mortars, 390–395
  stuccos, 406–407
Lime-pozzolan reaction, *see* Lime-
    silica reaction
Lime-silica reaction, 81, 84, 86, 87, 90,
    97–99, 193–197, 273–277, 346, 353,
    374, 380, 401, 412
Lime-soil mixtures, 195–196, 412–413
Lime-sulfur sprays, 357
Limestone industry, captive produc-
    tion, 458
  competition, 451–452
  location, 455
  number and size of plants, 455
  packaging, 454
  prices, 453
  profits, 457
  taxes, 455
  transportation, 454
  wartime, 451
  world trade, 454
Limestone sand, 126
Limestone uses, 75–131
Liming ponds and lakes, 386
Liming soils, 103–113, 381–384
Limonite, 17
Linzell, H. K., 134, 146
Lithographic limestone, 11
Loading stone, 50
Los Angeles abrasion test, 462, 469
Loss on ignition, 140, 147, 471
Louisiana, 271, 446, 455
Lower Devonian age, 12
Luminescence, 23, 169
Luxembourg, 436

Macadam type road bases, 123–124
Magnesia hydrates, 198
Magnesian limestone, 11
Magnesite caustic calcined magnesite,
    358, 368
  dead-burned magnesite, 357–358
  mineral, 8–9

Magnesium (element), 1, 6, 106
  bicarbonate, 30
  insulation, 91
  metal (use of lime), 352
  oxide (limit of), 98, 407, 490
  salts, 357
  silicate, 197
  sulfate test, 466
Maine, 439
Manganese (impurity), 17, 156
  determination of, 491
Maps (limestone deposits), 14–16
  lime plants, 440
Marble, 6, 8, 11, 20, 22, 23, 77
Marcasite, 20
Marl, 11, 99
Masonry cement, 100, 391–392
Masons lime, 167, 388–402
Massachusetts, 13, 439
Mather, F., 142, 307, 314
Mayer, R., 148, 449
Melting points, 170
Mesazoic era, 13
Metallurgical grade limestone, 11
Metallurgical uses, limestone, 78–86
  lime, 344–353
Metallic calcium, 32, 359
Metallic magnesium, 32, 352
Mexico, 436
Michigan, 12, 101, 150, 386, 450, 455
Microtests, 486
Micron sized limestone, 60–61, 92–97
Miller, T. C., 305, 449
Milk-of-lime, 167, 291
Milling, *see* Air separation
Millisecond delays, 49
Mineral feed, 114
Mineral filler, 92–97
Mineral wool, 90–91
Minerals (limestone), 8
Mining layout (limestone), 43
  room and pillar, 44
  stope, 44
Minnesota, 101, 402
Mississippian age, 13
Mississippi, 415
Missouri, 13, 18, 104, 439, 442, 450, 455
Mitchell, J., 133
Modulus of elasticity, 22
Modulus of rigidity, 22

Moh's scale, 22
Mortars, ancient, 4, 388–389
  attributes, 396
  carbon dioxide absorption, 191, 389
  forms of lime, 396–398
  hydraulic lime, 398, 400
  lime-pozzolans, 401–402
  mix adjustments, 393–395
  properties, 399
  proportions, 390–392, 400
  specifications, 393, 397, 401
Murray, J. A., 138, 143, 145–146, 150,
    153, 158, 294, 301, 303, 317, 484

National Bureau of Standards, 407, 447,
    481
National Lime Association, 281, 283,
    417, 419, 447, 449
National Safety Council, 283
Natural gas, 250, 252–253
Netherlands, 436
Neutralization, acids, 102, 185–187, 365,
    367–373, 495–496
  descaling, 372–373
  rate of, 369
  sludge formation, 372
  sulfuric acid, 185, 372–373
  total neutralizing power, 110
  values (factors), 111, 368
New Jersey, 382, 386
New York, 13, 19, 104, 436, 450, 455
New Zealand, 278–279
Ney, P., 309
Nevada, 398, 439
Niagaran dolomite, 12
Nitrogen (plant food), 106–107, 113–
    114
Nitrogen atmosphere, 145
Nogareda, C., 197
Nonferrous metals, 85–86, 352–353, 445

Occurrence, calcium, 1, 5
  magnesium, 1, 6
  limestone (geographic), 13
Odor (of limes), 168
Off-highway trucks (quarry), 51
Office of Price Administration, 451
Ohio, 12, 19, 42, 101, 104, 112, 190,
    386, 398, 408, 439, 442, 455
Oklahoma, 419

Oölitic limestone, 6, 11, 143, 162
Open-shelf quarry, 41
Open-pit quarry, 42
Optical calcite, see Iceland spar
Ordovician age, 7, 12
Ore beneficiation, 351–352
Organic chemicals (use), 359
Organic content, 472
Origin (geologic), limestone, 5
  magnesium carbonate or dolomite, 6
Orthorhombic crystal, 8, 21
Overburden, 38, 56, 99
Oystershell, 11, 115, 270–271
  for lime, 270–271

Packaging, 434, 454
Packinghouse by-products, 378
Page impact test, 463
Paint (use), 93–94, 380
Paleozoic era, 13
Paper, see Pulp and paper
Particle size, 298–302
  determination of, 465–466, 474, 484–
    486
Pebble lime, 167
Pelletizing, 264, 266
Pelletized calcination, 278
Pennsylvania, 13, 19, 104, 204, 382–
    383, 439, 442, 455
Percentage depletion, 457
Periclase, 269, 358
Petroleum industry, 379
pH values, limestones, 30
  lime, 184–185
  soil liming, 105, 109–110, 383
  potable water, 190, 361
  lime-clay reaction, 195
  neutralization, 370, 371–372
Phosphatic limestone, 12
Phosphorus, determination of, 491–492
  element, 114
  impurity, 156
  in limestone, 17
  plant nutrient, 106
  reaction with lime, 197, 346
Photomicrographs, 158–163, 309–312, 486
Physical properties, limestone, 20–29
  lime, 167–182
Physical testing, limestone, 460–471,
    484–486

Physical testing, lime, 472–486
Pickle liquor treatment, 364, 368
Pig iron (use), 79–83, 349
Pigments, 380
Pisolitic limestone, 12
Pit & Quarry (publ.), 55, 74, 201, 284, 327, 339
Pitting and popping, 408, 480
Plant layouts, crushed stone, 56–57
  quicklime, 202–203
  hydrated lime, 328–330
Plant nutrients, 105–106
Plaster, coats, 404
  exterior plaster, 406
  foreign plaster, 405
  lime requirements, 407–409
  proprietary compounds, 405
  proportions of, 404, 406–407
  stucco mixes, 406
  trends, 405
Plasticity, *see* Hydration, effect on
Plasticity index, 412
Pliny, 388
Poisson's ratio, 22
Poland, 436, 443
Pore space, 21
Porosity, limestone, 21
  of quicklime, 147, 168
Portable plants, 68–69
Portland cement, composition, 97
  concrete, 121–122
  manufacture, 97–100, 267, 403
  origin, 388–389
  quicklime use, 267, 403
  raw materials, 98, 267
Potassium (impurity), 17, 156
  oxide, 155, 177, 180
  plant nutrient, 106
Poultry grits, 115
Power shovels (quarry), 50
Poz-O-Pac Co. of America, 416
Pozzolans, 196, 376, 380, 388, 401, 497–498
Pozzolanic reactivity, 197, 497–498
Pre-Cambrian, 13
Precast concrete, 128
Precipitated calcium carbonate, 272–273, 360
Preheaters (rotary kiln), 231
Preheating, 137

Preheating zone, 215–216
Pressure hydrate, 167, 330, 333–334
Prices, 117, 432–433, 453
Primary crushing, 58–59
Producer gas, 243, 252
Pulp and paper (use), 90, 95–96, 272, 354
Pulverized Limestone Institute, 485–486
Pulverized quicklime, 264, 349, 396
Pumping, 52
Putty, composition, 290
  standard consistency, 474
  volume, 313
  whiting use, 95
Pyramids (Egyptian), 3
Pyrite, 17, 20

Quarry layout, 41
Quartz, 17
Quicklime manufacture, costs, 279–281
  classification (sizes), 263–266
  definition, 167
  flexibility, 246
  flow sheet, 202–203
  heat balance, 253–260
  instrumentation, 260–263
  theory, *see* Calcination

Radioactive fallout, 116
Rail cars on incline (quarry), 51
Railroad ballast, 126
Ratio of solubility, 27
Ray, K., 307
Reaction of limestone with acids, 30
  acid gases, 30
  ammonia, 32
  chlorine, 32
  hydrochloric acid, 30
  phosphorus compounds, 32
Reactivity, 143, 146, 147, 154, 317–319, 369–371, 471
Ready-mixed lime mortar, 396
Recarbonation, 135, 140, 191
Refractive index, limestone, 23
  of lime, 169, 175
Refractory, dead-burned dolomite, 166, 268–270, 350
  dead-burned magnesite, 357–358
  dolomite (use), 84–85

Refractory, lime, quality of, 167, 359
  use of lime, 374
    used in lime kilns, 213, 232, 244–245,
      253
Residue of lime, 474
Retention silo, 333
Rhombohedral crystal, 8, 21, 173
Rinquist, G., 180
Rip rap, 127
Road construction, 122–126, 410–426
Road profiles, 123, 411
Road stone, 122–125
Rock Products (publ.), 55, 74, 201, 284,
      327, 339
Rod mill, 60
Roll crusher, 60
Roman lime, 167
Romans use, 3, 388
Roofing granules, 127
Rotary kiln, 226–237
Rubber (use), 94–95, 381
Rudolfs, W., 325
Russia, 443

Safety (plant), 73–74, 281–283
Sampling, 462, 473–474
Sand-carrying capacity, 481
Sand and gravel, 116–118
Sand-lime brick, 375–376
Scandinavia, 406, 436
Schaffer hydrator, 333
Schmatolla, 217
Screening, 62–63
Searles, A. B., 204
Secondary breaking, 50
Secondary crushing, 59
Sedimentation, *see* Settling
Seismic vibrations, 49
Selective quarrying, 43
Selectively calcined dolomite, 91,
      277–278
Self-fluxing sinters, 82–83, 351
Serpentine, 7
Settling, 314–316, 326
  tests, 482–483, 486
Sewage treatment, 363
Shell, 5, 77, 99, 115, 118, 270–271
Shell limestone, 12
Shrinkage of lime, 147, 150

Siderite, 20
Sieving tests, 465–466, 484–485
Silica, brick, 374
  determination of, 487–488
  as impurity, 17, 65, 80–81, 85, 101,
      154–155
  reaction, *see* Lime-silica reaction
Siliceous limestone, 12
Siliceous materials, 196, 376
Sintered lime, 157–158
Size, *see* Gradation
Slaking, equipment, 323–324
  gradients, 295–296
  methods, 322–326
  tests, 294–296, 303–304, 483–484
  theory of, *see* Hydration
Slow-firing, 245
Slurry, 291
Smelting uses, 85, 353
Snake-hole blasting, 49
Sodium as impurity, 17, 156
  oxide, 143, 154, 177, 180
Sodium chloride addition, 153–154
Sodium carbonate addition, 153–154
  use of lime, 359
Sodium sulfate test, 460
Soft-burned lime, 138, 143, 146, 167
Soil conservation, 103
Soil stabilization, *see* Lime stabilization
Solubility of lime, effect of inorganic
      salts, 177–178, 182
  effect of organic salts, 179
  effect of sugar, 179–180, 189
  magnesium oxide, 181
  rate of solution, 186–189
Solubility of limestone, CO_2-free water,
      25
  CO_2 pressure, 26
  effect of alkalis, 29
  effect of organics, 29
  effect of salts, 28
  magnesium carbonate, 27
Solubility products, 31
Soluble alkali salts, 17
Solval process, 355–356
Soundness, limestone, 466–467, 469
  of lime, 479
South Africa (Union of), 226, 352, 418,
      443, 481

Southwest Lime Co., 44
Spalls, 56, 225, 452
Spray lime, 167
Specific gravity, hydrated lime, 174
    limestone, 22, 464–465
    quicklime, 147, 150, 168–169
Specific surface area, hydrate, 301,
    303–305
Specific heat, limestone, 24
    of limes, 171–172, 175
Specifications, 120, 393, 397, 404, 409,
    460–462, 486–487, 497
Stability (lime), 183
Stalactite, 12
Stalagmite, 12
Staley, H., 146, 190, 298, 303, 317
Standards, *see* Specifications
Starfish control, 387
Statistics (limestone), 76–77, 79, 104,
    115, 117, 453, 456
    of lime, 341–344, 433, 435, 437, 438,
    441–442, 444–445
Steam, effect of, 157, 248, 278
Steel, use of limestone, 83–84, 345–347
    lime, 344–350, 446
Sterilization, 190
Storage, 65–68, 329, 332
Storage zone (kilns), 215–216
Stowe, R., 148, 449
Strength, limestone, 22, 464
Stripping, 38
    tests for, 467–468
Strontium, 17, 90, 116, 156
Structural uses, 388–410
Stucco, 128, 406
Subsidy (liming), 383
Sugar, beet sugar, 377, 445
    effect on solubility, 179–180, 189
    refining, 377
Sulfate pulp manufacture, 272, 354
Sulfite pulp manufacture, 90, 354
Sulfur, determination of, 490–491
    impurities, 17, 155, 192, 348
    in limestone, 17, 81
    reactions with lime, 192–193, 349, 379
Surface area, determination of, 485
    lime, 146, 149, 190
    limestone, 92–93
Surge piles, 66
Sweden, 226, 376, 418, 443, 447

Tailings, 65, 331–332, 338
Tariffs, 436, 454
Television (industrial), 263
Tennessee, 13, 104, 439, 450, 455
Tertiary grinding, 60
Texas, 13, 118, 270–271, 415, 439, 442,
    446, 450, 455
Texas Highway Department, 415, 417
Texture, limestone, 20
    of lime, 168
Thermal conductivity, 23, 169
Thermal expansion, 24
Thermal dissociation, 32, 133–134
Tonnage, limestone, 2, 75–77
    of lime, 341–343
Trace elements, limestone, 17, 107, 156
Trade waste treatment, cellulosic, 366
    chemical and explosive, 364–365
    food plants, 367
    miscellaneous, 367
    steel and metals, 364
Transportation, 434, 454
Trap rock, 117–118, 127
Trattner, E., 495
Travertine, 12, 20, 22
Tube mills, 60, 336
Tufa, 12
Tunnel blasting, 49
Type N hydrated lime, 397–398, 407–408
Type S hydrated lime, 397–398, 407–408

Unhydrated oxide tests, 481, 493–495
Upper (blanket) shooting, 49
Upper Silurian age, 12
Uranium (use), 352
U.S. Bureau of Mines, 76–78, 101, 233,
    281, 344, 432, 443, 446
U.S. Dept. of Agriculture, 103, 383
U.S. Dept. of Commerce, 68, 118–119
U.S. Engineers, 120, 415
U.S. exports, 435–436, 454
U.S. Geological Survey, 35
U.S. imports, 435–436, 454
Uses, limestone, 75–131
    lime, 340–429

Vapor pressure (hydrates), 175
Vaterite, 8
Velocity of sound, 23
Vermont, 439

Vertical kiln, 206–226
Vibrating screens, 62–63
Vienna lime, 167
Virginia, 13, 19, 270, 273, 382, 439, 442
Vitruvius, 3, 388

Wagner turbidimeter test, 486
War Manpower Commission, 451
War Production Board, 450
Wartime, 449–451
Washing (stone), 64–65
Water retentivity, 312, 477–478
Water transportation, 99, 101
Water treatment, coagulation, 362
  miscellaneous, 362
  neutralization of acid water, 362
  purification, 190–191, 361
  silica removal, 362

Water treatment, softening, 360–361
Weisz, W., 259
Wells, L., 312
Wet grinding, 61
Whitewash, 167, 381
Whiting, 12, 92–97
Whitman, W., 301–302, 307
Wisconsin, 13, 101, 104, 386, 439, 450
Wood (fuel), 248

Xenophon, 3

Young's modulus, 22
Yugoslavia, 443

Zalmanoff, N., 191
Zement Kalk Gips (publ.), 201